Achieving sustainable production of pig meat

Volume 1: Safety, quality and sustainability

It is widely recognised that agriculture is a significant contributor to global warming and climate change. Agriculture needs to reduce its environmental impact and adapt to current climate change whilst still feeding a growing population, i.e. become more 'climate-smart'. Burleigh Dodds Science Publishing is playing its part in achieving this by bringing together key research on making the production of the world's most important crops and livestock products more sustainable.

Based on extensive research, our publications specifically target the challenge of climate-smart agriculture. In this way we are using 'smart publishing' to help achieve climate-smart agriculture.

Burleigh Dodds Science Publishing is an independent and innovative publisher delivering high quality customer-focused agricultural science content in both print and online formats for the academic and research communities. Our aim is to build a foundation of knowledge on which researchers can build to meet the challenge of climate-smart agriculture.

For more information about Burleigh Dodds Science Publishing simply call us on +44 (0) 1223 839365, email info@bdspublishing.com or alternatively please visit our website at www.bdspublishing.com.

Related titles:

Achieving sustainable production of pig meat Volume 2: Animal breeding and nutrition
Print (ISBN 978-1-78676-092-0); Online (ISBN 978-1-78676-094-4, 978-1-78676-095-1)

Achieving sustainable production of pig meat Volume 3: Animal health and welfare
Print (ISBN 978-1-78676-096-8); Online (ISBN 978-1-78676-099-9, 978-1-78676-098-2)

Improving organic animal farming
Print (ISBN 978-1-78676-180-4); Online (ISBN 978-1-78676-182-8, 978-1-78676-183-5)

Chapters are available individually from our online bookshop: https://shop.bdspublishing.com

climate•SMART•publishing
IN AGRICULTURAL SCIENCE

BURLEIGH DODDS SERIES IN AGRICULTURAL SCIENCE

NUMBER 23

Achieving sustainable production of pig meat

Volume 1: Safety, quality and sustainability

Edited by Professor Alan Mathew
Purdue University, USA

burleigh dodds
SCIENCE PUBLISHING

Published by Burleigh Dodds Science Publishing Limited
82 High Street, Sawston, Cambridge CB22 3HJ, UK
www.bdspublishing.com

Burleigh Dodds Science Publishing, 1518 Walnut Street, Suite 900, Philadelphia, PA 19102-3406, USA

First published 2018 by Burleigh Dodds Science Publishing Limited

Library of Congress Control Number: 2017938496

British Library Cataloguing in Publication Data
A catalogue record for this book is available from the British Library

ISBN 978-1-78676-088-3 (print)
ISBN 978-1-78676-090-6 (online)
ISBN 978-1-78676-091-3 (online)
ISSN 2059-6936 (print)
ISSN 2059-6944 (online)

Typeset by Deanta Global Publishing Services, Chennai, India
Printed by Lightning Source

Contents

Series list

Title | Series number

Achieving sustainable cultivation of maize - Vol 1 001
From improved varieties to local applications
Edited by: Dr Dave Watson, CGIAR Maize Research Program Manager, CIMMYT, Mexico

Achieving sustainable cultivation of maize - Vol 2 002
Cultivation techniques, pest and disease control
Edited by: Dr Dave Watson, CGIAR Maize Research Program Manager, CIMMYT, Mexico

Achieving sustainable cultivation of rice - Vol 1 003
Breeding for higher yield and quality
Edited by: Prof. Takuji Sasaki, Tokyo University of Agriculture, Japan

Achieving sustainable cultivation of rice - Vol 2 004
Cultivation, pest and disease management
Edited by: Prof. Takuji Sasaki, Tokyo University of Agriculture, Japan

Achieving sustainable cultivation of wheat - Vol 1 005
Breeding, quality traits, pests and diseases
Edited by: Prof. Peter Langridge, The University of Adelaide, Australia

Achieving sustainable cultivation of wheat - Vol 2 006
Cultivation techniques
Edited by: Prof. Peter Langridge, The University of Adelaide, Australia

Achieving sustainable cultivation of tomatoes 007
Edited by: Dr Autar Mattoo, USDA-ARS, USA & Prof. Avtar Handa, Purdue University, USA

Achieving sustainable production of milk - Vol 1 008
Milk composition, genetics and breeding
Edited by: Dr Nico van Belzen, International Dairy Federation (IDF), Belgium

Achieving sustainable production of milk - Vol 2 009
Safety, quality and sustainability
Edited by: Dr Nico van Belzen, International Dairy Federation (IDF), Belgium

Achieving sustainable production of milk - Vol 3 010
Dairy herd management and welfare
Edited by: Prof. John Webster, University of Bristol, UK

Ensuring safety and quality in the production of beef - Vol 1 011
Safety
Edited by: Prof. Gary Acuff, Texas A&M University, USA & Prof.James Dickson, Iowa State University, USA

Ensuring safety and quality in the production of beef - Vol 2 012
Quality
Edited by: Prof. Michael Dikeman, Kansas State University, USA

Achieving sustainable production of poultry meat - Vol 1 013
Safety, quality and sustainability
Edited by: Prof. Steven C. Ricke, University of Arkansas, USA

Achieving sustainable production of poultry meat - Vol 2 014
Breeding and nutrition
Edited by: Prof. Todd Applegate, University of Georgia, USA

Achieving sustainable production of poultry meat - Vol 3 015
Health and welfare
Edited by: Prof. Todd Applegate, University of Georgia, USA

Introduction

Pig meat is the most widely-consumed meat in the world. Previous growth in production has relied, in part, on more intensive systems, but in meeting rising demand, these systems face challenges such as the ongoing threat of zoonotic diseases, the need to improve feed efficiency in the face of rising costs, the need to reduce the environmental impact of pig production and increasing concerns about animal welfare. These challenges are addressed by the three volumes of *Achieving sustainable production of pig meat*:

- Volume 1: Safety, quality and sustainability
- Volume 2: Animal breeding and nutrition
- Volume 3: Animal health and welfare

This volume, Volume 1, reviews the latest research on controlling pathogenic and non-pathogenic safety risks associated with pig meat. It then surveys the latest research on aspects of meat quality such as flavour, colour, texture and nutritional quality. Finally, it assesses ways of monitoring and reducing the environmental impact of pig production.

Part 1 Safety

The first part of the volume deals with safety issues associated with the production of pig meat. The focus of Chapter 1 is on zoonoses affecting pigs. Zoonoses are defined as diseases and infections that are transmitted between vertebrate animals and humans. Major food animal species occupy a special position within the framework of zoonotic disease. This is particularly the case in developed societies, where direct livestock contact has become relatively rare, and where for much of the population the food supply has become the predominant route of exposure to livestock-associated pathogens. The chapter provides a broad overview of the zoonotic hazards associated with pigs, including leptospirosis, tuberculosis, brucellosis, influenza A viruses (IAV) and *Streptococcus suis*. The chapter examines emerging zoonoses such as livestock-associated *Staphylococcus aureus*, hepatitis E and novel paramyxoviruses, as well as foodborne pathogens. The chapter concentrates on the agents and pathways which most contribute to zoonotic risk, and looks at how these risks may be modulated by changing conditions at the human-swine interface.

Complementing the preceding chapter's focus on the main zoonoses affecting pigs, Chapter 2 examines the challenge of effective control of zoonoses in pig production. *Salmonella* infections of zoonotic origin are one of the most frequent causes of zoonotic infections worldwide, and there are clear indications that, at least in some countries, pork is an important source of human salmonellosis. The chapter describes the epidemiology of *Salmonella* and possible interventions in the pig and pork production chain. The chapter addresses animal surveillance and monitoring, control of feed and environment, and the role of replacement animals and finisher herds. The chapter also considers vaccination as a reduction strategy and measures that can be adopted at the slaughterhouse stage. The chapter also includes a detailed case study on *Salmonella* reduction in Danish pig and pork production.

Moving from the challenge of controlling disease to a challenge resulting from disease control itself, Chapter 3 addresses the issue of antibiotic resistance in pig production. The chapter describes the practice of antibiotic use in pig production. The chapter begins with a brief history of antibiotic use in livestock production, before examining the current state of research focused on developing non-antibiotic means of controlling bacterial infections in livestock. Finally, the chapter provides a case study of research in phage therapy as a case study of a (re)emerging technology that could be utilized in biocontrol of bacterial pathogens in agriculture.

The final chapter of the section, Chapter 4, deals with detecting veterinary drug residues in pork. The United States Food and Drug Administration (FDA) new animal drug approval process evaluates veterinary drugs intended for use in food-producing animals for safety and effectiveness. As part of the human food safety evaluation process, new animal drugs are evaluated for microbial food safety, toxicology and residue chemistry. The chapter summarizes the evaluation process that includes the assignment of tolerance(s) and withdrawal periods. It also establishes the criteria for violative residues i.e. residues above the established tolerance limits which may have potential adverse health effects in humans.

Part 2 Quality

The focus of the second part of the book is on ensuring the quality of pig meat. Chapter 5 provides an overview of the process of producing meat of consistent quality from the modern pig. Pork producers have focused, over many years, on producing pork more efficiently in order to remain competitive and to satisfy consumer demand for lean pork. Increases in efficiency and leanness have been achieved through genetics, targeted nutrition, the use of entire males and metabolic modifiers. The chapter first discusses the importance of visual appearance, sensory quality and protein functionality in assessing pork products. It then discusses recent developments in genetics such as the identification of gene markers for tenderness as well as breeding to achieve a desirable muscle pH. The chapter also discusses the role of nutrition in such areas as minerals, vitamins, fishmeal and other supplements high in polyunsaturated fatty acids (PUFAs), as well as slaughter and post-mortem operations. The chapter concludes with a detailed case study on optimizing pork quality in Australia.

The focus of Chapter 6 is on the factors affecting pork flavour. Pork-related products such as bacon, sausage, and ham comprise a large portion of the meat products sold in today's market due to their desirable flavor. The application of various processing methods such as cooking, curing, deboning, grinding, canning, etc., as well as additives or spices applied during processing, greatly contribute to the characteristic aromas of specific pork-related products. The chapter provides an integrated overview of current research on essential flavor constituents in pork products and the factors affecting pork flavor. Despite large differences among animals, genetics, and methods used in processing and cooking, the chapter highlights common odorants that underpin pork flavour, emphasizing the heat-induced pathways for formation of pork odorants.

Moving from flavour to other features of pig meat, Chapter 7 considers the factors affecting the colour and texture of pig meat. In pork, colour and texture are the two most significant factors influencing consumer perceptions of quality. It is therefore important to understand the many factors that can affect pork colour and texture. The chapter explores the biological and environmental factors that affect colour and texture in pig meat, including both antemortem and postmortem factors. The chapter then discusses existing and potential methods for the measurement and assessment of the colour and texture attributes of pig meat, including imaging technology.

Concluding Part 2, Chapter 8 concentrates on the nutritional composition and value of pig meat. The chapter provides an overview of the nutritional content of pork, examining pork consumption and dietary guidance in the USA. The chapter looks at the challenge of studying the influence of nutrition on human health, concentrating on the effects of pork consumption on weight control and body composition as well as the effects of consumption on cardiometabolic health and other health indices.

Part 3 Sustainability

The focus of the third part of the book is on the sustainability of pig meat production. The subject of Chapter 9 is assessing the environmental impact of swine production. The swine production industry has reduced its environmental impacts during the past 50 years due to productivity gains. However, there are increasing demands on the industry's resource base, making clear the need for robust tools to continue to support the best decisions in the face of environmental challenges. The chapter presents a review of the environmental sustainability impacts of swine production, focused at the farm level because the majority of environmental impacts occur by this stage of the supply chain. Two case studies comparing European and US swine production and the adoption of gestation pens to replace gestation stalls demonstrate the value of using life cycle assessments (LCA).

Complementing the preceding chapter's focus on sustainability, Chapter 10 examines nutritional strategies to reduce emissions from waste in pig production. Gaseous emissions of ammonia, odour, and greenhouse gases from livestock housing and storage and application of manure are major concerns in the environmental sustainability of pig production. The chapter addresses dietary strategies to reduce these emissions. The chapter examines the relationship between nutrition and ammonia emissions and between nutrition and odour emissions. It then considers the impact of nutrition on greenhouse gas emissions before evaluating the effectiveness of dietary solutions for reducing gaseous emissions and improving animal performance. The volume's final chapter, Chapter 11, addresses organic pig production systems, welfare and sustainability. The chapter describes standards for organic pig production, as well as the current state of organic pig production. The chapter then moves on to consider the issue of animal welfare in organic pig production and its environmental impact, providing an authoritative overview of the contribution of organic farming to pig production.

Part 1

Safety

Chapter 1

Zoonoses associated with pigs

Peter R. Davies, University of Minnesota, USA

1 Introduction

In the course of more than 300 million years of evolution, diverse and complex relationships have arisen among vertebrate species and their associated macro- and microbiota. Inevitably, there is considerable overlap among the families, genera, species and subtypes of organisms that have evolved to have commensal, parasitic, pathogenic and/or symbiotic relationships with vertebrate hosts. Absolute host specificity is the exception rather than the rule, and many organisms have some capacity to be transmitted among vertebrate hosts and to act, more or less frequently, as causal agents of disease (Viana et al., 2014). Some multi-host organisms, such as cestode parasites, have life cycles that are fully dependent on transmission among two or more host species. Others, such as Japanese B encephalitis virus, have harnessed the mobility of arthropods to expedite intra- and inter-species transmission. Bacteria such as *Salmonella* spp., *Brucella* spp. and *Leptospira* spp. have diverse host relationships, with individual serovars or biovars exhibiting varying degrees of host adaptation and pathogenic potential across multiple host species. At an ecological level, individual vertebrate species may serve a range of roles in the maintenance, transmission and emergence of multi-host pathogens in different ecosystems and socio-cultural contexts (Caron et al., 2014; Viana et al., 2014).

http://dx.doi.org/10.19103/AS.2017.0030.01

The concept of zoonotic disease places an anthropocentric lens upon this multi-dimensional matrix of host, agent and environmental relationships. It is estimated that the majority of known human pathogens are multi-host organisms (Taylor et al., 2001). The World Health Organization defines zoonoses as diseases and infections that are transmitted between vertebrate animals and humans. With few exceptions, animal to human transmission of a pathogen constitutes only one dimension of human infection risk, and the proportion of human illness that is attributable to zoonotic transmission is highly variable. Although hundreds of zoonotic hazards (i.e. infectious agents capable of causing or contributing to human disease) have been identified, the actual burden of zoonotic disease (defined as the product of the frequency and severity of human infections) is mostly aggregated within a relatively small subset of hazards. For some (e.g. tuberculosis, brucellosis, rabies and hydatidosis), impact on public health has spurred substantial societal investments to reduce or eliminate the agents from domestic and/or wildlife animal reservoirs. Simplistically, for any hazard, the likelihood of human infection is a function of the prevalence of the agent in the contact animal populations, the sum of probabilities of transmission across the set of possible transmission routes and the frequency and nature of human–animal interactions. Consequently, efforts to reduce zoonotic disease risk can be targeted at the vertebrate host populations (e.g. pathogen elimination or vaccination programmes); at interventions affecting transmission routes (e.g. food hygiene and cooking for foodborne zoonoses) or at a societal level in relation to interactions with animals (e.g. rabies education; hygiene measures at petting zoos).

Major food animal species such as pigs, poultry and cattle occupy a special position within the framework of zoonotic disease. This is particularly the case in developed societies, where direct livestock contact has become relatively rare, and for much of the population, the food supply has become the predominant route of exposure to livestock-associated pathogens. Furthermore, the pig has singular significance as the preferred species for xenotransplantation to humans (Halperin, 2001). The following discussion provides a broad overview of zoonotic hazards associated with pigs. Emphasis is given to agents and pathways which most contribute to zoonotic risk, and how these risks may be modulated by changing conditions at the human–swine interface.

2 Overview of zoonotic hazards associated with pigs

Numerous infectious agents associated with swine have some capacity to infect humans (Uddin Khan et al., 2013; Smith et al., 2011; Pappas, 2013). The impact of these organisms on human health ranges from negligible to extensive, and their relative importance is greatly influenced by environmental, cultural and societal factors, including methods of animal rearing (Davies, 2011, 2012; Pappas, 2013). As with all sectors of animal and public health, appropriate allocation of resources to address zoonotic pathogens requires prioritization to identify the hazards that present the greatest risk (Brookes et al., 2015). Both the criteria influencing prioritization and the ranking of pathogens are influenced by the social context and perceptions of stakeholder groups (Ng and Sargeant, 2016). For example, a Canadian study ranked cysticercosis 57th in importance among 62 zoonotic agents considered (Ng and Sargeant, 2016), whereas cysticercosis (*Taenia solium/Cysticercus cellulosae*) has been estimated to cause 50 000 annual deaths globally and is recognized as a major threat to public health in low-income countries in Latin America, Asia and sub-Saharan

Table 1 Major zoonotic agents associated with swine: Impact on swine health and modes of transmission to humans

Agent	Impact on swine health[a]	Modes of transmission to humans				
		Direct[a]	Foodborne[a]	Environmental[a]	Arthropods[a]	Xenotransplantation[a]
Leptospira interrogans	+++	+++	–	+++	–	+/–
Brucella suis	+++	+++	+/–	–	–	+/–
Streptococcus suis	+++	++	+	–	–	+/–
Erysipelas rhusopathiae	+++	+	–	++	–	+/–
Salmonella choleraesuis	+++	+/–	?	+	–	+/–
Salmonella (non *choleraesuis*)	++	+	++++	+	–	+/–
Clostridium difficile	+	?	–	+/–	–	+/–
Listeria monocytogenes	+/–	–	+++	+/–	–	+/–
Staphylococcus aureus	+/–	+++	+/–	+	–	+/–
Campylobacter spp.	–	–	+?	+/–	–	+/–
Yersinia enterocolitica	–	–	++++	+/–	–	+/–
Influenza A	+++	++++	–	–	–	+/–
Nipah	+++	+++	–	–	–	+/–
Japanese B encephalitis	+++	+	–	–	++++	+/–
Hepatitis E	–	+/–	?	?	–	+/–
Endogenous retroviruses	–	–	–	–	–	++?
Porcine cytomegalovirus	+	–	–	–	–	++?
Lymphotrophic herpesvirus type 1	–	–	–	–	–	++?
Trichinella spp.	+/–	–	+++	–	–	+/–
Toxoplasma gondii	+	–	+++	–	–	+/–
Taenia solium	+/–	–	+++	–	–	+/–

[a] Qualitative assessment of impact from – (no impact) to ++++ (highly important).

Africa (Eddi et al., 2003; Phiri et al., 2003; Trevisan et al., 2016). *Streptococcus suis* type 2 infection was ranked above foodborne bacteria to be the third-most important zoonosis in Vietnam (Trang do et al., 2015), but to date has been an insignificant zoonosis in the Americas. Prioritization is implicit in the content of this chapter, and the emphasis given to respective agents reflects the perceptions and biases of the author. The aim is to provide a global context for swine-associated zoonoses, and it is to be expected that others would perceive the relative importance of these agents differently. Table 1 lists the agents considered to have had the most impact on human health, together with an indication of the relative importance of routes of transmission categorized into foodborne, occupational, environmental and xenotransplantation pathways, as well as an indication of the relative importance of the agents to swine health.

For other zoonotic hazards associated with swine (Table 2), the prevalence and severity of human disease are of relatively less consequence generally, although higher incidence may occur in specific settings, such as 'Pig-bel' in New Guinea (Duke et al., 2013). For some agents, such as Hendra virus and transmissible spongiform encephalopathies, the ability to infect swine has been demonstrated only under artificial conditions, and no role for swine has been established in the ecology and epidemiology of the agent/disease under natural conditions (Li et al., 2010; Hedman et al., 2016). Similarly, neonatal pigs are susceptible to experimental infection with Zika virus, but the epidemiological significance is unknown (Darbellay et al., 2017). For other agents, the zoonotic risk is minuscule in relation to importance to animal health. For example, food and mouth disease (FMD) virus is unequivocally zoonotic (Bauer, 1997), but human infections are exceedingly rare and usually mild and self-limiting. Therefore, FMD has negligible medical significance as a zoonosis. The prevalence of zoonotic hazards in swine can vary greatly among ecological settings, and particularly between wild or feral populations compared to domesticated pigs (Ruiz-Fons, 2017). The housing and management of domesticated food animals is a substantial anthropogenic influence upon the profile of infectious diseases of swine including zoonotic and foodborne pathogens (Davies, 2011, 2012). Given the diversity of host, agent and environmental interactions involving pigs globally, it is to be expected that different approaches to animal rearing (e.g. intensive confinement production versus outdoor production) would involve trade-offs regarding risks of zoonotic infections. Clearly, the nature and frequency of human interactions with pigs vary enormously among societies. At one extreme, religious authority forbids contact with pigs and their products for much of the Middle East populace, while pigs play a central societal role in New Guinea and some Pacific Island communities (Dwyer, 2006). Therefore, this general overview of swine-associated zoonoses needs to be interpreted in the context of local environments where the relative importance of pathogens may differ substantially.

3 Classical zoonoses: leptospirosis, brucellosis and tuberculosis

Several globally significant zoonotic diseases have been recognized since antiquity and in many societies have motivated interventions to protect public health. Among these classical zoonoses, the incidence and routes of transmission vary geographically and over time, with zoonotic risk representing a greater or lesser component of overall risk in different settings.

Table 2 Other zoonotic agents linked to swine

	Impact on swine health[a]	Comments
Bacillus anthracis	+	Risk for outdoor pigs in endemic regions
Burkholderia pseudomallei	+	Risk for outdoor pigs in endemic regions
Pasteurella multocida	+++	Commensal and opportunistic pathogen of pigs
Coxiella burnetti	–?	Feral pigs a potential reservoir in some regions
Enterococcus spp.	–	Primary concern related to transfer of antibiotic resistance genes
Enterobacteriaceae	Variable	Primary concern related to transfer of antibiotic resistance genes
Mycobacterium bovis	–	Now rare due to eradication programmes for cattle
Mycobacterium avium spp.	+/–	Ubiquitous – lesions may lead to carcass condemnation
Clostridium perfringens type C	–	Important cause of necrotic enteritis (Pig-bel) in New Guinea
Foot-and mouth-disease virus	++++	Human infections rare
Vesicular stomatitis virus	+	Human infections rare
Rotaviruses	++	Zoonotic risk not established
Menangle virus	+	Only one reported outbreak in Australia
Noroviruses	–	Zoonotic risk not established
Sapoviruses	–	Zoonotic risk not established
Pseudorabies virus	++++	Zoonotic risk not established
Rabies virus	+	Rare outbreaks in swine
Hendra virus	–	Experimental infection of pigs only
Ebola Reston virus	–	Human clinical disease not reported
Dermatophytes	+/–	Uncommon in modern industry
Sarcocystis spp.	+/–	
Balantidium coli	–	Zoonotic risk not established
Cryptosporidia spp.	+	
Giardia spp.	+/–	

[a] Qualitative assessment of impact from – (no impact) to ++++ (highly important).

3.1 Leptospirosis

Leptospirosis has been nominated as the world's most widespread zoonosis. It is an occupational disease recognized since antiquity, but is also considered to be an emerging disease (Plank and Dean, 2000; Wasinski and Dutkiewicz, 2013; Adler, 2015). Leptospires occur worldwide and possess an impressive arsenal of strategies to elude control, including antigenic variation; ability to infect a broad range of species; a sophisticated scale of host-adaptation characterized by long-term asymptomatic carriage by infected hosts; a substantial capacity to survive in moist environments, especially water surfaces, and multiple modes of transmission and entry into susceptible hosts. The central feature of leptospiral biology is persistent renal infection in mammals which shed organisms in the urine, although infection also occurs in birds and reptiles (Jobbins and Alexander, 2015).

The taxonomy of *Leptospira* is complex and is still not completely resolved (Levett, 2015). Traditional serology-based taxonomy has now been complemented by genomic classification, and currently 21 species are recognized. However, much of the historic body of leptospiral research has been based on serotyping, which is complicated by considerable cross-reactivity among serovars. Traditional serogroups and serovars of *Leptospira* do not enable prediction of genomospecies, and it is recommended that isolates be characterized by both genomospecies and serovar (Levett, 2015). For example, the swine-adapted serovar *L. Pomona* occurs within 2 genomospecies, *L. interrogans* and *L. noguchii* (Levett, 2015).

Conventionally, mammalian hosts are classified as maintenance hosts or accidental (spillover) hosts of leptospires (Hathaway, 1981). Hathaway et al. (1983) suggested that a maintenance host be defined by the ability to act as a natural source of leptospiral infection for its own species, and a maintenance population be defined as a species that acts as a continuous reservoir of a serovar in a specific ecosystem (Hathaway et al., 1983). Typically, a species that acts as a maintenance host for one (or few) serovar(s) will be an accidental host for other serovars. Other than reproductive disease, clinical disease in maintenance hosts tends to be inapparent or mild, and infection leads to chronic renal colonization and shedding of the leptospires in urine. In contrast, clinical disease in accidental hosts (such as man) tends to be more severe. Typically, renal infection and urinary shedding are of much shorter duration in accidental infections than for infections of maintenance hosts. Like other species, swine are maintenance hosts for several host-adapted serovars and can be accidental hosts for other serovars (Ellis, 2015).

Although the major routes of transmission of leptospirosis are well established, their relative importance is not, as it likely varies with epidemiological circumstances, and is strongly influenced by geo-climatic context (Mwachui et al., 2015). Amongst epidemiological scenarios in which human leptospirosis can arise, three principal patterns have been described (Levett, 2001):

1 In urban environments, most often when urban infrastructure is poor or disrupted, human risk is largely due to transmission from rodents.
2 In moist tropical regions, environmental contamination and exposure are of considerable importance and involve a diverse range of serovars and mammalian reservoirs.
3 In temperate regions, and particularly in developed countries, human leptospirosis is dominated by occupational exposure to domestic animals, and involves relatively few serovars.

Swine have been identified as a maintenance host of the serovars Pomona (and the closely related Kennewicki), Bratislava and Tarrasovi (Strutzberg-Minder and Kreienbrock, 2011). Most of the information about leptospiral infection of swine is from developed countries, and Pomona is both the most widespread serovar and is of greatest significance with respect to zoonotic risk. The epidemiology of Bratislava infection in pigs is not well understood, but urinary transmission is less effective than with Pomona and venereal transmission may be more important (Ellis, 2015). Tarrasovi appears to have had some historical significance in some European countries (Akkermans, 1991), but has been scarcely reported in recent decades. Vaccination of breeding animals against leptospirosis is routinely practised in modern swine production to prevent abortions in sows, but vaccination of growing pigs is not typically performed. The importance of leptospirosis in swine appears to have declined in some developed countries in recent decades, which may in part be attributed to confinement housing (Ellis, 2015). Other factors may include better rodent control, all-in/all-out management of growing pigs and housing on fully slatted floors.

Swine infected with leptospires are an occupational risk for farmers, veterinarians, abattoir workers, hunters and others having direct contact with pigs or swine tissues or urine. Human infection is most likely where animal infection is prevalent, and use of personal protective equipment such as gloves, masks and protective eyewear should be emphasized in known infected herds. Particular attention should be paid to handling of aborted foetuses and placentas in outbreaks of leptospiral reproductive disease.

3.2 Tuberculosis

Tuberculosis (TB), caused by the human-adapted *Mycobacterium tuberculosis*, is among the most pressing human health challenges globally. Zoonotic TB, caused by the related mycobacteria *M. bovis* or *M. caprae*, was recently estimated to account for <1.4% of global TB incidence (Muller et al., 2013). Swine are susceptible to infection with both *M. bovis* and *M. caprae* (Garcia-Jimenez et al., 2013), but in general probably play a relatively minor role in the epidemiology of zoonotic TB. The prevalence of infection in pigs is usually related to the prevalence in other species with which they are in contact, both in domesticated and wild settings (Nugent et al., 2014; Barandiaran et al., 2015). In the twentieth century, TB lesions due to *M. bovis* in slaughtered pigs in the USA declined from around 15% in 1924 to 1% in 1970, and generally are now rare in countries that have successfully controlled *M. bovis* in bovines (Pesciaroli et al., 2014). In developed countries, industry evolution away from mixed species farms (including dairy/swine operations where pigs were often fed unpasteurized milk products such as whey) and implementation of stricter biosecurity measures have greatly reduced the likelihood of inter-species transmission of many pathogens (Davies, 2012), including the risk of exposure to *M. bovis* and *M. caprae*. However, in developing countries where pigs are reared less intensively and often in contact with cattle, *M. bovis* can be the most common mycobacterial species found in tuberculous lesions of pigs (Barandiaran et al., 2015). Moreover, in some developed countries, the reversion to less biosecure outdoor production systems may lead to higher risk of TB infection in domestic swine (Bailey et al., 2013). Wild boar play an important role as a reservoir of zoonotic TB in some regions, such as the Iberian peninsula (Naranjo et al., 2008; Pesciaroli et al., 2014). In other regions, although prevalence of infection can be high, swine are typically considered incidental or dead-end hosts and thought to play a minimal role in TB epidemiology (Nugent et al., 2015). This serves as an example of the danger of generalizations regarding the epidemiology of inter-species, including zoonotic, transmission which can vary widely depending on local factors such as species demographics and contact patterns among species.

Swine are also commonly infected with mycobacteria of the *M. avium–intracellulare* complex, notably *M. avium* subsp. *avium* and *M. avium* subsp. *hominisuis*. The four subspecies of the *M. avium* complex are all closely related, but display varied host ranges and pathogenicity (Rindi and Garzelli, 2014). These organisms are generally considered ubiquitous in the environment, including water sources, and have the capacity to infect a broad range of host species (Biet et al., 2005). An ongoing challenge is determining whether, and how frequently, zoonotic transmission may occur due to the multiple potential sources of human exposure. It is estimated that all humans may experience low level exposure to *M. avium* species on a daily basis, but few clinical infections result, primarily in immunocompromised patients. Infections in immunocompetent people tend to be limited to granulomatous lymphadenitis, whereas severe and fatal systemic infections are common in severely immunocompromised people. Although genetically identical variants have been detected among isolates from humans and swine (Leao et al., 2014), this does not constitute proof of zoonotic transmission given the likelihood of frequent environmental exposures for both species. Studies have generally reported broad diversity in *M. avium* isolates and there remains little evidence that zoonotic transmission is epidemiologically significant relative to environmental exposures (Vluggen et al., 2016).

3.3 Classical zoonoses: brucellosis

Brucellosis, like leptospirosis, was recognized in antiquity, yet remains a zoonosis of global significance, with a conservative estimate of 500 000 human cases occurring annually (Godfroid et al., 2005; de Figueiredo et al., 2015). There is evidence for human cases of foodborne brucellosis, or 'Malta fever', dating back two thousand years, and even of infection in hominid ancestors over two million years ago (Capasso, 2002; D'Anastasio et al., 2011). The disease has probably been endemic in the Mediterranean region and parts of Asia since the dawn of civilization (Moreno, 2014). The history of zoonotic brucellosis is intimately linked to the domestication of livestock and the evolution of animal husbandry, and practices of production and consumption of foods of animal origin (Mantur and Amarnath, 2008). It has been observed that the most virulent *Brucella* species for humans are those associated with domesticated livestock species (with the exception of the non-zoonotic *B. ovis*), whereas *Brucella* species endemic in wild animals are less prone to cause human disease (Moreno, 2014). Dissemination of zoonotic *Brucella* to the New World and Australia was a consequence of colonization and introduction of domestic ungulates from Europe (Moreno, 2014).

The taxonomy of *Brucella* remains incompletely resolved, and there are currently at least ten species recognized in the genus. Collectively, these infect more than 80 mammalian species including humans, livestock, rodents and marine mammals (Godfroid et al., 2005; Olsen and Palmer, 2014). Although some events of transmission among people have been documented (including via sexual, trans-placental and trans-mammary routes, blood transfusion and due to occupational exposures of physicians), person-to-person transmission of *Brucella* is extremely rare (Mesner et al., 2007). Consequently, *Brucella* are not considered to be self-sustaining pathogens in human populations, and virtually all human infections are derived from animal sources via direct contact or through consumption of contaminated food, particularly unpasteurized milk products (Olsen and Palmer, 2014; de Figueiredo et al., 2015; Casalinuovo et al., 2016). Although *Brucella* can be recovered from environmental samples from locations frequented by infected hosts, neither environmental persistence (with the exception of the non-zoonotic *B. microti*) nor vectors are thought to play significant roles in *Brucella* epidemiology (Moreno, 2014; Olsen and Palmer, 2014). However, the organisms are capable of surviving intra-cellularly in the reticuloendothelial system, and persistent infection of hosts associated with granulomatous inflammation in diverse tissues is a hallmark of the disease (de Figueiredo et al., 2015).

Brucella species display variable affinity for, and pathogenicity in, different mammalian species. With the exceptions of *B. melitensis* (mostly associated with small ruminants) and *B. abortus* (bovine), *Brucella* species have generally been named according to their preferred (or originally recognized) host (e.g. *B. suis, B. ovis, B. canis, B. ceti, B. pinnipedialis, B. neotomae, B microti*). However, all are multi-host pathogens and there is considerable variability among and within these species, including with respect to their host range and capacity for human infection. Human cases of brucellosis are predominantly caused by *B. melitensis, B. abortus* and *B. suis*, and in most developed countries systematic programmes have been implemented to control or eradicate these three agents from livestock populations. Consequently, the impact of human brucellosis remains greatest in regions where livestock infection remains endemic, including parts of Africa, Latin America and the Middle East (Olsen and Palmer, 2014). Mammalian wildlife are also epidemiologically significant reservoirs, with major zoonotic transmission risks associated with hunting and slaughtering wild or feral pigs in some developed countries (Godfroid et al., 2013). Wildlife reservoirs are also an important obstacle to control programmes as sources of re-infection of domestic livestock (Szulowski et al., 2013; Pedersen et al., 2014, 2015; Fournier, et al., 2015). Notably, in some developed countries where brucellosis is well controlled in domestic livestock, the expansion of feral swine populations has emerged to be a pre-eminent challenge in prevention of human brucellosis (Eales et al., 2010). Apart from direct zoonotic infection of hunters and others contacting wild or feral swine, human cases may indirectly result from infection of other mammalian hosts with *B. suis*, particularly in dogs that are involved in pig hunting (Mor et al., 2016) and cattle (Mantur and Amarnath, 2008; Musser et al., 2013; Szulowski et al., 2013).

The pathogenesis of brucellosis has recently been reviewed in detail (de Figueiredo et al., 2015). With respect to zoonotic risk, key features are the low infectious dose (10–100 organisms) and ability to translocate rapidly across mucosal epithelia, creating a high likelihood of infection of people exposed to contaminated animal tissues or fluids, or food products. Portals of entry of the organisms include via ingestion, conjunctival and respiratory mucosae and abraded skin (Olsen and Palmer, 2014). Risk of direct transmission is implicit for all groups exposed to susceptible species or their tissues, including farmers, veterinarians, abattoir workers, butchers and laboratory staff. Inhalation can also be an important route of exposure in some circumstances, which has led to recognition of the potential use of *Brucella* as a bioterrorism agent (Doganay and Doganay, 2013).

Within each of the three major zoonotic species, multiple biovars have been identified, including 5 biovars of *B. suis*. Notably, of the 5 recognized *B. suis* biovars, only biovars 1 to 3 infect *Suidae*, with biovars 4 and 5 associated, respectively, with wild Cervidae in the Arctic, and wild rodents (Fretin et al., 2008). With respect to zoonotic risk associated with swine, biovars 1 and 3 predominate in human *B. suis* infections, while biovar 2 appears to have limited capacity to cause infections in healthy humans (Fretin et al., 2008). The three biovars infecting pigs vary in geographical distribution, host range and pathology, with biovars 1 and 3 predominant in North America, while biovar 2 is endemic in wild boar in Europe, where some specific variants also infect hares (Kreizinger et al., 2014). Detailed information of the pathophysiology and clinical signs of brucellosis in swine are reviewed in detail elsewhere (Olsen et al., 2012). Features relevant to transmission and zoonotic risk include prolonged bacteraemia and widespread distribution of organisms among tissues, including reproductive organs and semen.

Prevention of zoonotic brucellosis has several dimensions with respect to swine. Although swine can be infected with *B. melitensis* and *B. abortus* (Olsen et al., 2012), and may even act as reservoirs of *B. abortus* (Stoffregen et al., 2007), the predominant concern is *B. suis*, specifically biovars 1 and 3. Globally, the consumption of unpasteurized milk and milk products remains a major risk that is readily mitigated via pasteurization. As swine (unlike cattle, sheep, goats and camels) are not dairy animals, their contribution to the burden of brucellosis is minor in many regions. In commercial swine production, establishing herds and industries free from *B. suis* is the preferred approach and has been reliably achieved in most developed countries. Maintenance of freedom has also not been problematic in confinement production, provided that basic biosecurity measures are in place to prevent introduction of infected animals or semen. If so, the predominant residual risk of introduction to herds is linked to exposure to wildlife reservoirs, notably in outdoor production in areas with wild or feral swine populations (Olsen et al., 2012). Environments most suitable for outdoor production (e.g. the south east of the USA) are also likely to be favourable for feral populations (Wyckoff et al., 2009; Engeman et al., 2011).

In developed countries where brucellosis has been eliminated or well controlled in commercial livestock, wildlife reservoirs are likely of utmost concern to specific segments of the population, particularly hunters. Although data on prevalence in wild or feral swine populations are sparse, all such animals should be assumed likely to be infected, although zoonotic risk will be influenced by the biovars present, with biovar 2 presenting minimal risk of human infection. Education of hunters to understand the risks and use appropriate personal protective equipment, including goggles and gloves, when handling and dressing carcasses is essential. Education should also extend to appropriate disposal of tissues that could serve as sources of infection for other animal species (Fretin et al., 2013).

4 Emerging zoonoses: influenza A viruses and *Streptococcus suis*

Emerging diseases are defined as diseases that have newly appeared in a population, or are rapidly increasing in incidence or geographical range (Morse, 2004). Awareness and concern about disease emergence in humans and animals have increased exponentially in the last 20 years. A PubMed search (5 December 2016) of reviews using 'emerging infectious disease' as a search term yielded 172 reviews, with the oldest being from 1994, and more than half (88) being published since 2010. The role of animal reservoirs of potential human pathogens is a focal point of these concerns.

4.1 Influenza A viruses in swine

The influenza A viruses (IAV) constitute a unique global health concern manifest by significant seasonal epidemics in most countries and sporadic pandemics. The complex epidemiology of IAV has been regularly and extensively reviewed (Peiris and Yen, 2014). Core elements are the pivotal role of aquatic birds as reservoir hosts, the genetic lability of the viruses due to both genetic recombination and reassortment and the capacity of some IAV to infect domesticated mammals with which humans commonly interact, including swine, dogs and horses. Avian species harbour diverse IAV that are broadly characterized into subtypes based on their haemagglutinin (16 variants) and neuraminidase (9 variants) antigens. Most IAV subtypes are host restricted. Human pandemics to date have involved H1, H2 or H3 subtypes, although sporadic zoonotic infections have been reported with H5, H6, H7, H9 and H10 variants (Peiris and Yen, 2014). The human pandemic potential of novel IAV variants is a pre-eminent concern in global health. Spillover events of IAV among reservoir species occur continually, but very rarely give rise to sustained transmission of IAV among humans (Lipsitch et al., 2016).

Swine have been at the hub of concerns about zoonotic influenza for 100 years, following recognition of clinical disease in pigs concurrent with the 1918 influenza pandemic, and even more so since the novel H1N1 pandemic of 2009 (Vincent et al., 2014). Although exchange of IAV between swine and humans has likely been occurring with some regularity over at least a century, the genomics era has provided powerful new tools to understand the epidemiology of the disease and underpins more sophisticated programmes for surveillance (Vincent et al., 2014). It is increasingly recognized that transmission of IAV from humans to swine plays an important role in IAV epidemiology in swine (Nelson et al., 2014; Nelson and Vincent, 2015). The IAV that are adapted to swine (swine influenza viruses, or SIV) are also among the most significant pathogens of swine in many countries (Torremorell et al., 2012), but the following discussion will be limited to zoonotic risks of transmission.

Frequent and direct contact with swine in any setting will increase the risk of bidirectional transmission of influenza viruses, and may occur at farms, abattoirs, in live markets or at exhibitions where pigs are shown (Myers et al., 2006; Bowman et al., 2014; Choi et al., 2015; Fragaszy et al., 2016). Accurate estimates of incidence of infection and clinical disease are not available, and many human infections may be asymptomatic or go unrecognized. However, it appears that cases of human disease typically have not generated secondary cases, and propagated outbreaks of SIV have not been described in humans following swine exposure. That is, occupational exposure of people to influenza viruses that have adapted to pigs has caused only sporadic infections that reflect limited ability of SIVs for person-to-person transmission. It is important to underline that although influenza viruses from humans and pigs may share a common origin, they are neither genetically nor antigenically identical. Furthermore, distinct SIV viruses circulate in swine in different regions of the world (Lewis et al., 2016). Brown (2000) explained that

> Following interspecies transmission to pigs, some influenza viruses may be extremely unstable genetically, giving rise to variants which could be conducive to the species barrier being breached a second time. Eventually, a stable lineage derived from the dominant variant may become established in pigs. Genetic drift occurs particularly in the genes encoding the external glycoproteins, but does not usually result in the same antigenic variability that occurs in the prevailing strains in the human population. Adaptation of a newly transmitted influenza virus to pigs can take many years. Both human H3N2 and avian H1N1 were detected in pigs many years before they acquired the ability to spread rapidly and become associated with disease epidemics in pigs. (p. 29)

More recently, it was described that viruses of different origin took different evolutionary pathways after introduction into pigs, without displaying either parallel or convergent evolutionary trends (Dunham et al., 2009).

A controversial aspect of many zoonotic risks in pigs relates to the influence that changing swine industry demographics and management may have had on emerging disease risks, and particularly increases in herd size. However, traditional pig rearing practices on small farms are also not without concern regarding influenza risk, particularly commingling of pigs and chickens and lack of biosecurity practices (Netrabukkana et al., 2016). A mathematical modelling study predicted that workers in 'concentrated animal feed operations' (CAFOs) in the USA may provide a bridge for influenza transmission between large swine farms and the general community and could amplify a pandemic (Saenz et al., 2006). However, the study did not consider the fact that the size of this 'pig to human bridge' is dramatically smaller than in previous times. A 2009 study estimated 54 000 CAFO workers in the USA (Schmidt, 2009), which represent less than 0.02% of the population and just under one person per 1000 pigs. In 1900, around 40% of the US population of 100 million was involved in agriculture and three-quarters of US farms reared hogs. By the middle of the twentieth century, when SIV had already been endemic in the swine industry for several decades, two million US farms sold hogs, meaning over 1% of the US population (conservatively assuming one person per farm) had direct contact with swine (one person per 33 pigs). If we accept that we are dealing with a two-way biological 'bridge', not only were many more people directly exposed to pigs, but roughly the same number of pigs (60–70 million) were exposed to considerably more people.

More comprehensive and balanced analyses, accompanied by more extensive surveillance of inter-species transmission events in different production conditions, will be necessary to advance understanding of the influence of production conditions on the risk of zoonotic IAV infections. Elevated risk of IAV should be assumed for all people having direct contact with swine, and strict policies should be enforced to exclude workers with influenza-like symptoms from swine facilities. Annual vaccination against seasonal influenza is recommended for swine farm workers, although the impact of this measure is unknown.

4.2 *Streptococcus suis*

Streptococcus suis is a commensal inhabitant of the oropharynx and nasal cavities of pigs, an important opportunistic pathogen of weaned pigs, and a rare, though sometimes severe, zoonotic pathogen (Feng et al., 2014). As is characteristic of commensal bacteria, there is enormous genetic diversity within the species, and within serotypes (Auger et al., 2016; Du et al., 2017), and there is ongoing uncertainty about the species classification of some putative serotypes (Goyette-Desjardins et al., 2014; Okura et al., 2016). There is strong evidence for geographic heterogeneity in the distribution of sero/MLST types of *S. suis* in pigs (Goyette-Desjardins et al., 2014; Huong et al., 2014a), and in the incidence of human clinical cases (Goyette-Desjardins et al., 2014; Huong et al., 2014a). A comprehensive retrospective study identified reports of 1584 human cases until 2012, of which 36% were in Thailand, 30% in Vietnam and 22% in China (22%), with only 10% in the European region and 0.5% in the Americas (Huong et al., 2014a).

Of 35 known serotypes of *S. suis*, serotype 2 is the most important for human health. However, many other serotypes have been isolated from cases of swine and human infections, and serotype 14 and serotype 9 are relatively common among cases of *S. suis* disease in humans and pigs, respectively. In addition to the predominant serotype 2, cases of human disease have been reported with serotypes 1, 4, 5, 9, 16, 21, 24 and 31 (Goyette-Desjardins et al., 2014; Willemse et al., 2016). *S. suis* infection is not notifiable in most countries, and it is widely considered to be an under-reported disease, in part due to misidentification of the bacteria (Gottschalk et al., 2010; Huong et al., 2014a). Since 2005, human *S. suis* infection has been notifiable in Hong Kong, where the annual incidence in adults was estimated to be 0.09/100 000 for the general population and 32/100 000 for people with occupational exposure to pigs and raw pork (Ma et al., 2008). In a 1988 publication, annual risk was estimated to be 3/100 000 for pig farmers and abattoir workers in the Netherlands, approximately 1500 times greater than for the general population (Arends and Zanen, 1988). More recent data from a prospective cohort study calculated similar risks of 0.007/100 000 for the overall Dutch population, and 3–5/100 000 for the occupationally exposed (Schultsz et al., 2012; van Samkar et al., 2015). Globally, the vast majority of reported cases have occurred in adults, and predominantly in males, presumably due to more frequent exposure to pigs (Arends and Zanen, 1988; Huong et al., 2014a).

Generally, reported human cases have been sporadic, with the exception of two epidemics in China where an outbreak in 2005 involved over 200 cases and 39 deaths (Yang et al., 2006). Although the USA has the world's second-largest swine industry, to date only five human cases of *S. suis* infection (four serotype 2, one serotype 5) have been documented in the USA. Three of the type 2 infections were likely to have been acquired internationally, and the type 5 infection occurred in a medically compromised pig farmer (Fowler et al., 2013; Gomez et al., 2014). In contrast, *S. suis* type 2 infection was reported to be the most common cause of bacterial meningitis of humans in Vietnam (Mai et al., 2008). The most common manifestations of *S. suis* type 2 in people are meningitis (68%), septicaemia (25%), arthritis (12%) and endocarditis (12%) (Huong et al., 2014a). The case-fatality rate calculated in a meta-analysis of published reports was 13%, but varied among countries, and hearing loss and vestibular dysfunction were common clinical sequelae (Huong et al., 2014a).

Human *S. suis* infections in most of the world are primarily of occupational origin linked to contact with live pigs or processing of pigs, whereas in Asian countries, foodborne transmission may play a greater role (Arends and Zanen, 1988; Huong et al., 2014a; Willemse et al., 2016). The apparent spatial heterogeneity in *S. suis* type 2 risk for humans may be attributable in part to both variation in the pathogen and to socio-cultural factors linked to pork production and consumption. A study of whole genomes of 98 *S. suis* type 2 isolates from human patients and pigs with invasive disease in the Netherlands found that zoonotic isolates had smaller genomes than non-zoonotic isolates, but contained more virulence factors. The authors identified a zoonotic clone and estimated that it diverged from a non-zoonotic clone in the late nineteenth century (Willemse et al., 2016). The large outbreak of *S. suis* type 2 disease in China 2005 was caused by a particular MLST type, sequence type (ST) 7 (Du et al., 2017). This variant, which is yet to be reported in other countries, appears to be of enhanced virulence with a high case fatality rate (>60%) occurring in patients with streptococcal toxic shock syndrome. In contrast, type 2 isolates from human cases in the Netherlands were of ST1 or ST20 (Schultsz et al., 2012). Therefore, genetic variation within *S. suis* appears to contribute to the variability in risk of human infection and clinical manifestations of the disease across different regions. Importantly, within a region, the diversity of isolates from human cases may be much less than among isolates from invasive swine infections, indicating that certain variants have greater propensity for zoonotic infection (Schultsz et al., 2012).

Given the minimal risk for the general population seen in the Netherlands, foodborne transmission of *S. suis* appears to play a negligible role, but evidence suggests it is more important in Asia (Huong et al., 2014a; Willemse et al., 2016). *S. suis* infections occur predominantly in young pigs, and marketing, slaughter and consumption of weanling pigs are likely to increase risk of exposure. Cultural practices, such as consumption of raw or undercooked pork or pig blood (Lee et al., 2008; Huong et al., 2014b; Hatrongjit et al., 2015), also influence infection risk and likely contribute to the heterogeneity of human infection risk in different societies.

Prevention efforts should involve raising awareness about the low, but real, risk of *S. suis* among people with frequent exposure to pigs and pork, including butchers. As skin wounds have been implicated as portals of entry, it is recommended to wear gloves when handling carcasses or raw pork, cover skin wounds and rapidly clean, disinfect and cover any skin wounds that occur in farm or processing environments.

5 Emerging zoonoses: livestock-associated *Staphylococcus aureus*

It has been more than a decade since the revelation that livestock can harbour methicillin-resistant *Staphylococcus aureus* (MRSA), which can be transmitted to humans (Voss et al., 2005). 'Livestock-associated MRSA' (LA-MRSA), particularly in pigs, has since been at 'the eye of the storm' regarding the human health impacts of antibiotic use in livestock. Although the overall burden of human illness to date appears to be modest (van Cleef et al., 2013; Nair et al., 2016; van Cleef et al., 2016; Becker et al., 2017), LA-MRSA has become a politically charged subject that has inspired calls for eradication efforts and stricter regulation of antibiotic use in swine production. The prevalence of infected herds varies widely among countries (Anon., 2009; Sun et al., 2015), and Norway is to date the only country to implement a 'search and destroy' approach to depopulate positive swine farms (Grontvedt et al., 2016). Concerns about MRSA in pigs in Denmark have led to complex legal disputes about revealing the identity of positive farms and consideration of the potential harmful effects of 'stigmatization' of farmers and other groups (Ploug et al., 2015).

S. aureus is part of the normal commensal flora of the skin and upper respiratory tract of humans, pigs and many other mammals and birds. However, *S. aureus* is a premier 'opportunistic' pathogen of people, causing diverse clinical syndromes that range from trivial to fatal and are influenced by host and bacterial factors (Messina et al., 2016). Approximately one-quarter to one-third of healthy people in developed countries harbour *S. aureus* in the nose, skin or other sites (Sivaraman et al., 2009), and the risk of clinical disease is higher in those who are colonized (Safdar and Bradley, 2008). Meta-analyses of human infections concluded that MRSA cases incurred higher medical costs and poorer outcomes than cases infected with methicillin-susceptible *S. aureus* (MSSA) (Whitby et al., 2001; Cosgrove et al., 2003), but invasive infections with MSSA are also of great medical significance (Ericson et al., 2015). Prior to the zoonotic concerns raised by the emergence of ST398 MRSA in Europe, animals were not thought to play any role in the epidemiology of *S. aureus* infections in people. Furthermore, the epidemiology of *S. aureus* in pigs had been largely ignored due to its insignificance as a swine pathogen.

Numerous studies confirm that people working with live pigs (farmers, veterinarians, abattoir workers and their families) are more likely to be culture-positive for *S. aureus* and MRSA than the general population, and the predominant genotypes are typically those found in the corresponding animal populations (Armand-Lefevre et al., 2005; Khanna et al., 2008; van Cleef et al., 2010; Frana et al., 2013; Verkade et al., 2013). The majority of research of LA-MRSA has addressed multilocus ST398 MRSA that predominates in pigs and other livestock in much of Europe. However, it is increasingly apparent that other genotypes adapted to pigs can acquire methicillin resistance and be transmitted to people, including ST9 (particularly in Asia), ST5 (particularly in North America), ST1 and ST97 (in some parts of Europe) (Gomez-Sanz et al., 2010; Franco et al., 2011; Chuang and Huang, 2015; Hau et al., 2015). Notably, ST398, ST9 and ST5 MSSA are all prevalent in the US swine industry, and most farms and many individual pigs harbour multiple STs (Linhares et al., 2015; Sun et al., 2015).

The realization that people exposed to livestock were at greatly elevated risk of colonization with ST398 MRSA led to changes to patient-screening practices in the Netherlands, a country with a low endemic MRSA incidence and where a 'search-and-destroy' policy is used to control MRSA in hospital settings (van Rijen et al., 2008). The decision to screen patients with occupational exposure to pigs or cattle in 2006 led to marked increases in positive screening results and subsequent costs in managing these patients (van Rijen et al., 2008; Wassenberg, 2010; Wulf et al., 2012). An extensive review of LA-MRSA in Europe concluded that in some countries with low prevalence of human MRSA infection, livestock-associated isolates make a major contribution to the overall MRSA burden (Scientific Opinion of the Panel on Biological Hazards, 2009). In 2007, 30% of all MRSA human isolates typed by a national institute in the Netherlands belonged to the ST398 lineage, although most isolates were from screening at-risk patients on admission to hospitals rather than from clinical infections (Huijsdens et al., 2009).

Although there has been a tendency to assume that ST398 infections in humans are of livestock origin (Rasigade et al., 2011), it is now well established that distinct human- (most notably of spa type t571) and swine-adapted variants of ST398 *S. aureus* occur, and in some settings the human-adapted variants are maintained in populations without a livestock reservoir and can be prevalent causes of infection (McCarthy et al., 2012; Uhlemann et al., 2012; Uhlemann et al., 2013). Similarly, swine-adapted ST1 and ST5 isolates appear to be genetically distinct from human clinical isolates (Franco et al., 2011; Hau et al., 2015), and attribution of host origin based on ST alone is ill advised (Davies et al., 2011).

A lingering question surrounding zoonotic transmission of *S. aureus* has been whether positive nasal swabs indicate true colonization versus repeated transient contamination occurring following pig exposure. However, longitudinal studies in the Netherlands, Germany and the USA have provided increasing evidence that long-term colonization with LA-MRSA or LA-MSSA for up to two years occurs in some veterinarians and farmers, while repeated transient contamination occurs in others (Verkade et al., 2013; Sun et al., 2017; van Cleef et al., 2015; Walter et al., 2017). Thus, host factors likely influence the outcome of exposure to animal populations. Furthermore, persistently colonized individuals harbour higher numbers of *S. aureus* (Sun et al., 2017) and can be a source of secondary transmission to their families and social networks (Verkade et al., 2014; van Cleef et al., 2015).

LA-MRSA/MSSA causes a similar spectrum of clinical syndromes in people as do isolates of human provenance, including severe and even fatal cases. However, all MRSAs are not created equal and LA-MRSA strains carry fewer virulence genes (Mutters et al., 2016), are less transmissible among humans (Wassenberg et al., 2011; Hetem et al., 2013) and typically cause less severe disease than human strains (Becker et al., 2017). The incidence of human ST398 MRSA clinical infections (from mild through severe) was recently estimated to be <3/100 000 inhabitants per year in Denmark and Germany (Goerge et al., 2017). For comparison, the annual incidence of invasive and fatal MRSA cases in the USA was estimated to be 31.8 and 6.1 per 100 000, respectively, (Klevens et al., 2007), and a study in the pig-dense state of Iowa found only 1% of *S. aureus* isolates, and 0.24% of MRSA isolates were of spa types linked to livestock (Nair et al., 2016). In Denmark, which has high population densities of both swine and humans, and the vast majority of herds are MRSA-positive, the overall incidence of all MRSA infections (~10/100 000) did not differ between pig-dense and other regions, although the likelihood of ST398 infection was elevated in pig-dense areas, including in people with no known pig contact (Larsen et al., 2015). Reported fatalities attributed to LA-MRSA globally have been of the order of one per year, versus an estimate over 18 000 deaths due to MRSA in the USA alone (Klevens et al., 2007), where to date no fatal case of LA-MRSA has been reported.

One anomaly of LA-MRSA epidemiology is that exposure risk is overwhelmingly concentrated in people working with animals, yet approximately a third of reported clinical cases have occurred in people without livestock contact (Becker et al., 2017). A recent review cited that 24–86% of nasal swabs of pig farmers were positive for ST398 MRSA across multiple studies, but found only 12 reports of clinical infections, mostly mild, in people with livestock contact (Goerge et al., 2017). The 3 fatal cases documented were all in medically compromised swine workers, a feature common to most of the severe human infections reported to date. It remains to be demonstrated whether livestock workers are at overall increased risk of clinical *S. aureus* infections compared to the general population. Regarding the occurrence of LA-MRSA cases in people not exposed to animals, human-to-human transmission is the most likely explanation (Walter et al., 2017). To date, there is little convincing evidence that foodborne or environmental sources contribute significantly to human exposure risk.

Zoonotic transmission of *S. aureus* is likely to remain a controversial issue despite a relatively modest incidence of related disease, and will remain more important in countries with a low prevalence of MRSA in human infections, such as the Netherlands. There is little to suggest that LA-MRSA poses an imminent threat to public health, even in countries where prevalence in livestock is high. However, exposure risk is constant for swine workers on positive farms, and farms are places where minor injuries are frequent. Common-sense procedures should be in place to mitigate potential risks. Workers should be encouraged to use personal protective equipment and practise good personal hygiene with regular hand washing with soap and water. It is important to pay appropriate attention to existing and new skin wounds (clean and cover with bandages until healed) and seek medical attention if infections develop. Similar to Hepatitis E virus (HEV), exposure of workers is likely to be very frequent in many countries, infection risk appears to be small and attention should be focused on medically compromised individuals.

6 Emerging zoonoses: hepatitis E and novel paramyxoviruses

6.1 Hepatitis E virus

The HEV was the fifth viral agent identified to cause hepatitis in humans, but is considered the most frequent cause of acute hepatitis globally (Purcell and Emerson, 2008; Hoofnagle et al., 2012). An estimated 20 million people acquire HEV infection annually, of which some three million suffer from acute hepatitis, and as much as a third of the human population may have experienced infection (Khuroo et al., 2016a; Thiry et al., 2017). HEV is a single-stranded RNA virus that is genetically diverse and infects many animal species (Khuroo et al., 2016a). The classification of Hepeviridae remains dynamic, and there is evidence for host adaptation within the family. Orthohepevirus A viruses have been isolated from humans, swine, rabbits, deer, mongoose and camels, and include several defined genotypes. Genotypes 1 and 2 appear to be restricted to humans and are responsible for the majority of illness in developing countries. Genotypes 3 and 4 have been found in pigs, rabbits, deer, mongoose and people; genotypes 5 and 6 in wild boar and genotype 7 in camelids. Other Orthohepeviruses (B, C, D) infect birds, rats, ferrets and bats, but have not been described in swine and are of unknown zoonotic significance (Khuroo et al., 2016a).

In humans, acute HEV hepatitis is usually self-limiting, although severe and fatal cases occur in pregnant women, the immunocompromised or people with cirrhosis. The overall case fatality rate is estimated to be 1%, but can exceed 20% in infections with genotype 1 HEV in the third trimester of pregnancy (De Schryver et al., 2015). There are two distinct epidemiologic patterns of HEV disease. In regions with poor sanitary infrastructure, genotypes 1 and 2 can cause explosive epidemics via faecal–oral transmission (mostly waterborne) among humans. Sporadic infections with genotypes 3 or 4 occur worldwide and can be of zoonotic origin. In the USA, HEV has been detected in various mammalian and avian species, but infection is most prevalent in pigs (Dong et al., 2011). Both wild and domestic pigs are commonly seropositive for HEV, and Cervidae may be important reservoirs in some countries. Comparison of sequences of HEV genotypes 3 and 4 from humans and pigs shows very close similarity, suggesting potential for inter-species transmission without genetic modification. The zoonotic potential of genotypes 5 and 6, which have been found in wild boar in Japan, remains to be confirmed (Doceul et al., 2016).

The incidence of HEV infection is greatly underestimated, as many cases are asymptomatic (Khuroo et al., 2016a). Prior to the recognition of animal reservoirs of HEV, cases in developed countries were typically attributed to travel to endemic regions and autochthonous cases mostly went unrecognized (Hartl et al., 2016). Although HEV is commonly described as an emerging disease (Arends et al., 2014), apparent increases in incidence may be a result of better awareness and case ascertainment. In the USA, Denmark and the Netherlands, all countries with large swine industries, data from blood donors suggest that the prevalence of infection has been decreasing rather than increasing in recent decades (Hogema et al., 2014; Holm et al., 2015; Teshale et al., 2015). A systematic review of European studies of autochthonous infections found cases were generally older than the overall population and predominantly male (Lewis et al., 2010). No major route of transmission was identified, but zoonotic transmission from pigs is implicated. In the USA, active laboratory-based surveillance identified only 227 incident human cases of HEV hepatitis over three years (~0.07/100 000 per year), of which 56% were genotype 3 and were deemed to be autochthonous (Drobeniuc et al., 2016). This low incidence suggests minimal risk of clinical infection for healthy individuals in the general population, but particular concerns relate to immunocompromised individuals (Meng, 2016; Yugo and Meng, 2013).

The HEV infection in swine was first reported from pigs in Nepal in 1995, and soon thereafter in the USA (Clayson et al., 1995; Meng et al., 1997). Subsequently, it has been confirmed that HEV is endemic in swine populations across five continents, with more diversity observed in wild versus commercial populations (Thiry et al., 2017). Although infection occurs at high incidence in swine, infections are asymptomatic and HEV is of negligible significance to swine health. However, infected pigs have a viraemic phase of 1–2 weeks and will shed the virus in faeces for several weeks. Incidence is generally highest in young growing pigs, but infection can also occur in market age animals entering the food supply. The essentially ubiquitous presence of HEV in pig populations raises the question of occupational risk, and a meta-analysis of 12 studies of IgG seroprevalence found that exposure risk was higher in people with swine contact (De Schryver et al., 2015), although increased risk of clinical disease has not yet been demonstrated (Lewis et al., 2010).

The HEV has been repeatedly detected in pork products (Meng et al., 1997), and some authors infer that HEV may be a frequent foodborne infection based on detection of the virus in pork products and close genetic similarity between viruses from pigs and people (Pavio et al., 2015). However, the fraction of HEV cases attributable to zoonotic and foodborne transmission remains uncertain and is likely to be highly variable among countries (Khuroo et al., 2016b). Culinary preferences contribute to higher risk in some countries, particularly where undercooked pork or raw pig livers are consumed (Jori et al., 2016). Other high-risk items include dishes incorporating pig intestines such as chitterlings, and consumption of undercooked pork was associated with seroprevalence in one study in the USA (Cossaboom et al., 2016). Overall, there is no question that the exposure to pigs and pig products contributes to the incidence of human HEV infections in many countries. The relative importance of zoonotic transmission is highly variable across societies and is greatest in developed countries (Khuroo et al., 2016b). Although the incidence of zoonotic infections appears to be very low in developed countries, it is important to raise awareness in occupationally exposed groups and for immunocompromised individuals to avoid handling of high-risk products (e.g. livers, chitterlings) and consumption of raw or undercooked pork products.

6.2 Novel paramyxoviruses: Nipah virus and Menangle virus

The paramyxovirus now known as Nipah virus is arguably the most alarming instance of the emergence of a zoonotic agent associated with swine, manifesting as a neurological syndrome with a high case fatality rate in humans. The initial epidemic in Malaysia in 1998 caused at least 265 human cases and 105 deaths (Tan and Chua, 2008), and Nipah virus is listed by the WHO among the eight pathogens likely to cause severe outbreaks in the future (Sweileh, 2017). Similar to two other zoonotic paramyxoviruses identified in Australia in the 1990s (Menangle virus in pigs; Hendra virus in horses), Nipah virus is an endemic and asymptomatic infection of Pteropid bats, and infections of livestock were spillover events from bats (Clayton, 2017).

Menangle virus was isolated in a single outbreak of deformed stillborn piglets, with two clinically mild infections in associated workers, but has not been further implicated in swine or human disease (Kirkland et al., 2001). In contrast, although Nipah virus was successfully eliminated from the Malaysian swine industry, it remains an important pathogen in south Asia, with outbreaks being reported in Malaysia, Singapore, India, Bangladesh and possibly the Philippines (Clayton, 2017). The Nipah epidemic in Malaysia manifested as an outbreak of encephalitis mostly affecting swine farm workers, and was initially suspected of being Japanese B encephalitis virus. Recognition of clinical disease in pigs was delayed due to modest impact on mortality and clinical similarities to other swine diseases, which enabled spread of the disease to other regions of the peninsula and to Singapore (Hayman and Johnson, 2014). Human-to-human transmission was rare, and the epidemic was controlled by aggressive depopulation of affected herds, with no subsequent outbreaks reported. However, Nipah virus remains a pathogen of concern in other areas of Asia, particularly Bangladesh and India, where human-to-human transmission appears to be more common, and bat-to-human transmission, without any livestock intermediary, appears to be predominant (Hayman and Johnson, 2014; Clayton, 2017). Consumption of date palm sap contaminated by bats has been implicated, and most outbreaks have been relatively small, but with a high case fatality rate (Hayman and Johnson, 2014).

Several hypotheses have been presented regarding the factors that precipitated the Malaysian outbreak in 1998. These include complex ecological events related to El Niño, deforestation and widespread fires that preceded the outbreak, and could have led to increased concentration of bats in the outbreak region (Chua et al., 2002). The index farm was a very large herd (of the order of 30 000 animals) where pigs in open air enclosures were adjacent to or underneath mango trees likely to harbour bats. This unusual opportunity for bat-to-pig transmission, coupled with the large herd size, may have created unique conditions to establish a propagated epidemic in the pigs (Pulliam et al., 2012). Intensification of swine production is ongoing in many developing countries, and increases in herd sizes in the absence or rigorous biosecurity may increase vulnerability to emerging diseases if contact with wildlife is not precluded (Davies, 2012).

7 Foodborne pathogens

The WHO estimated that in 2010 the collective burden of foodborne illness caused by 31 hazards globally was 33 million disability-adjusted-life-years, involving approximately 600 million illnesses and 420 000 deaths (Havelaar et al., 2015). Several of the most important foodborne hazards (e.g. *Salmonella Typhi*, noroviruses, Hepatitis A virus, *Vibrio cholerae)* have no significant livestock reservoirs, but livestock including swine are central to the epidemiology of others. Although foodborne transmission is a facet of some of the zoonotic pathogens discussed earlier, it is the predominant or sole route of human infection for several agents that contribute considerably to the overall burden of foodborne illness. These can be divided into two groups:

1. Enteric bacteria that are common commensals or normal flora of healthy livestock. Foodborne transmission is mostly a consequence of inadvertent contamination of edible products in the harvest and post-harvest sectors (although lymph node infection may be important in some settings). These include *Salmonella, Campylobacter, Listeria monocytogenes, Yersinia enterocolitica* and some variants of *E. coli*.
2. Foodborne parasites that occur naturally in edible tissues as an inherent feature of their biology. In swine, these include *T. solium, Trichinella spiralis* and *Toxoplasma gondii*.

A holistic 'farm to table' paradigm in which all participants in the food supply continuum, from producers to consumers, bear responsibilities for reducing the risk of foodborne disease is generally viewed as the preferred approach to address foodborne illnesses (Davies et al., 2004). The logic of controlling potential foodborne pathogens at the farm level has broad appeal, but the cost and efficacy of controlling pathogens in farm settings are highly variable and dictated by their respective epidemiology. In general, efforts at pre-harvest control in swine and other species have been disappointing for enteric bacteria, but highly effective for parasitic zoonoses (Davies, 2011).

7.1 *Salmonella, Campylobacter* and other enteric bacteria

Diarrheal syndromes dominate the profile of foodborne bacterial diseases, although sporadic invasive infections and secondary complications can occur with all enteric pathogens. Risk is higher with *Listeria,* which has been described as an enteroinvasive pathogen due to its propensity to cause systemic disease, particularly in immunocompromised individuals (Barbuddhe and Chakraborty, 2009). There are also disputed concerns about the potential zoonotic origin of specific strains of *E. coli* associated with poultry causing extra-intestinal infections (Ewers et al., 2012; Bonten and Mevius, 2015).

In terms of incidence of human infections, *Salmonella* and *Campylobacter* dwarf other zoonotic agents globally, including developed countries (Havelaar et al., 2015; Alban, 2016). However, active surveillance data in the USA (Table 3) indicate reductions in the incidence of most major enteric bacterial pathogens over the last 20 years (Henao et al., 2015). Severity of disease is also an important consideration in addition to incidence. While listeriosis has relatively low incidence in the USA (Table 3), it accounts for an estimated 19% of deaths attributed to foodborne infections (Scallan et al., 2011).

With the exception of *Y. enterocolitica*, which is linked to swine (Bancerz-Kisiel et al., 2016), these species are common among all livestock and many vertebrate wildlife species. As they are faecal borne, all share the potential for environmental dispersal, including to crops and produce. Furthermore, all exhibit broad genetic diversity within and among host species, yet the potential for zoonotic transmission can be restricted to particular subtypes or strains (Lianou and Koutsoumanis, 2013). This may depend on both the capability to persist or multiply on food products or the environment along the food supply chain, as well as the virulence potential of individual variants in humans. Marked differences have been observed in the ability of different foodborne bacteria to survive events in the processing of pigs (Nesbakken et al., 2008). Therefore, zoonotic risk cannot be directly inferred from the prevalence of an organism in a livestock reservoir. Notably, *Campylobacter* spp. (particularly *C. coli* more than *C. jejuni*) are ubiquitous as part of the normal intestinal flora of pigs. However, current evidence indicates that pigs and pork play a negligible role in the epidemiology of human campylobacteriosis (Tyson et al., 2016; Thepault et al., 2017).

Non-typhoidal *Salmonella* remains the pre-eminent foodborne pathogen globally and afflicts almost all sectors of the food industry. Due to the multiplicity of potential sources

Table 3 Annual incidence (cases per 100 000) of bacterial foodborne pathogens in the United States, and observed reduction since 1996–98

Pathogen	Rate per 100 000	% reduction from 1996–98
Salmonella	15.3	5
Campylobacter	13.3	24
Non-O157 Shiga-toxin producing *E. coli*	1.43	M
O157 Shiga-toxin producing *E. coli*	0.91	33
Yersinia	0.28	60
Listeria	0.24	45

Adapted from FoodNet 2014 Annual Foodborne Illness Surveillance Report: https://www.cdc.gov/foodnet/reports/annual-reports-2014.html.

of infection, the proportion of cases attributable to meat in general, or to pork specifically, is difficult to establish and likely varies widely among countries (Guo et al., 2011). The complexity of the epidemiology of *Salmonella* is difficult to overstate and most aspects of *Salmonella* in the swine industry have been regularly reviewed (Davies et al., 2004; Sanchez et al., 2007; Wales et al., 2011; De Busser et al., 2013; Wilhelm et al., 2016; Young et al., 2016; Wales and Davies, 2017). At an applied level, the enduring question is the balance of emphasis that should be given to pre-harvest versus post-harvest interventions to control *Salmonella* in the pork supply (Alban and Stark, 2005). The feasibility of pre-harvest control of *Salmonella* has been demonstrated in Sweden using a programme based on exclusion of *Salmonella* from animal populations. *Salmonella* is notifiable and controlled using intensive bacteriologic screening of imported feed, feed mills, breeding stock suppliers and slaughter plants, backed up by regulatory measures that include rejection of feed shipments, closure of feed mills and herd/flock depopulation (Hopp et al., 1999; Sundstrom et al., 2014). However, this resource-intensive approach has not been adopted in any major swine-producing country or outside of Scandinavia. Pre-harvest control efforts in other European countries based on serological testing and intervening in high seroprevalence herds have had discouraging outcomes and do not appear to be cost-effective (Davies, 2011; Dahl, 2013). The greatest hurdle to pre-harvest control is the lack of validated and reliable management interventions, despite considerable research effort.

The costs and benefits associated with different approaches may be dependent on the particular epidemiological circumstances in each country or region (Dahl, 2013). It was suggested that pre-harvest interventions should focus on preventing transmission among farms rather than reducing prevalence on farms (Dahl, 2013). Again, the feasibility of this approach will be strongly influenced by existing herd prevalence and industry structure, and it is unlikely to be cost-effective in regions of medium-to-high prevalence. In contrast, improvements in slaughter hygiene appear to be a much more reliable and cost-effective alternative in most settings, and likely have contributed to the progress seen in reducing foodborne bacteria in the USA over the last 20 years (Table 3) in the absence concomitant programmes for pre-harvest control (Davies, 2011). In less developed settings, emphasis on slaughter hygiene and avoidance of time–temperature abuse in the distribution, retail and consumer segments should be the over-riding priority to address foodborne bacteria.

The rationale for pre-harvest control advanced for *Salmonella* equally applies to other zoonotic enteric bacteria, but no such attempts have been undertaken and there has been insufficient research to assess the influence of swine management on risk (Fosse et al., 2009; Davies, 2011). Given the questionable outcomes of substantial efforts in on-farm control of *Salmonella*, current knowledge indicates control efforts should be directed primarily at hygiene in slaughter and processing and at consumer education. For *Listeria*, attention to contamination in food processing facilities is of central importance, particularly in ready-to-eat products (Sofos and Geornaras, 2010). In the absence of verified pre-harvest interventions for enteric bacteria, educational efforts that target high-risk groups represent another strategic opportunity to reduce disease (Ong et al., 2012).

As stated elsewhere, using farm-management interventions to reduce the occurrence of organisms that are normal flora or common commensals of pigs is a challenging undertaking (Davies, 2010).

Development of valid and reliable pre-harvest interventions to control bacterial foodborne pathogens remains the pre-eminent challenge for researchers in pork safety and may require different research approaches. Given the complexity and variability of production settings, epidemiological studies face great limitations beyond hypothesis generation, and more fundamental studies to understand the biological phenomena of intestinal colonization may give better opportunities to deliver the 'holy grail' of cost-effective pre-harvest control of foodborne bacteria.

7.2 Foodborne parasites

The foodborne parasites *T. solium, T. spiralis* and *T. gondii* have collectively had enormous impact on human health over millennia, and remain major health problems in some parts of the world. The three agents share key attributes in that carnivorism plays a central role in their life cycles, and direct pig-to-pig transmission (other than vertical transmission of *T. gondii* to piglets) is negligible, if cannibalism is prevented. The relatively simple epidemiology (apart from *T. gondii*), of these parasites relative to bacteria such as *Salmonella*, renders them much more amenable to control by modifications in production. Management changes related to intensification of swine production have greatly reduced their impact on human health in developed countries (Davies, 2011).

The cestode parasite *T. solium* is arguably the most important zoonotic pathogen of pigs globally, although it is essentially eliminated from swine in developed countries. It has a typical cestode life cycle in which one species (humans) is the sole definitive host, in which the tapeworm completes its life cycle to become adult and lay eggs that are shed in faeces (Gripper and Welburn, 2017). Another species (pigs) is the intermediate host and becomes infested by ingesting the tapeworm eggs shed in human faeces. The metacestode forms spread systemically in intermediate hosts to become encysted in multiple tissues including muscle and the brain (cysticercosis). To complete the cycle, people acquire intestinal tapeworm infections (taeniasis) by eating undercooked infested pork. However, people can also acquire the more clinically significant cysticercosis form of infection via faecal–oral transmission of eggs from infested people, or by autoinfection (Gripper and Welburn, 2017). This is particularly important due the frequent involvement of neural tissues with associated clinical signs including seizures (Gripper and Welburn, 2017). The Food and Agricultural Organization cites estimates that 50 million people worldwide harbour the adult tapeworm, and up to 50 000 deaths per year are attributable to cysticercosis (Eddi et al., 2003). *T. solium* disease is primarily a concern in regions with poor sanitary infrastructure and thrives where traditional free range and scavenging pig production occurs (Phiri et al., 2003; Gabriel et al., 2017). Although long neglected, more systematic and integrated approaches to control the disease in endemic settings are now being implemented (Garcia et al., 2016; Thys et al., 2016; Gabriel et al., 2017; Okello et al., 2017; Winskill et al., 2017).

There are at least nine recognized species within the *Trichinella* genus of nematodes (Pozio and Zarlenga, 2013) that collectively infest at least 150 mammalian species. *T. spiralis* is of primary importance in domesticated swine, but other species occur in wild animals including bush pigs and warthogs that may enter the 'bushmeat' supply (Mukaratirwa et al., 2013). All genotypes appear to reproduce comparably in carnivore hosts, and variably in other species, and many genotypes found in wildlife have low infectivity and persistence in pigs (Kapel, 2000). Human infections occur only via ingestion of raw or undercooked meats. Many infections are asymptomatic and undiagnosed, but multiple clinical manifestations are possible, and severe cases can be fatal. *Trichinella* occur in all continents other than Antarctica, and cultural practices related to meat consumption greatly influence disease risk (Pozio, 2007). The incidence of human cases of trichinosis has dropped dramatically over the last 50 years in the USA and Europe due to virtual elimination from the commercial swine industry (Kennedy et al., 2009; Davies, 2011).

This has been a consequence of modernization of production practices, and trichinosis in Western Europe similarly is now limited to settings where traditional pig-rearing practices are maintained, but not in industrialized pig farms (Pozio et al., 1996). The decline in swine trichiniasis has been attributed to several factors including restrictions on feeding uncooked garbage to pigs, improved farm-management practices such as rapid removal of dead pigs and rodent control and reduced contact with wildlife. Outdoor pigs will be at higher risk of exposure through access to wildlife reservoirs (Davies, 2011).

T. gondii is a protozoan parasite that typically results in asymptomatic infection of mammals and birds. It is estimated that a third of the human population becomes infected, yet disease is rare. Human infection of humans with *T. gondii* can occur by ingestion or inhalation of sporulated oocysts that are dispersed in the faeces of cats, ingestion of undercooked meat containing cysts of bradyzoites or by trans-placental transmission of tachyzoites to the foetus (Jones and Dubey, 2012). The societal cost of toxoplasmosis is largely due to the high medical costs of congenital cases that can require lifelong institutional care (Hampton, 2015).

Because infection is mostly asymptomatic, or clinical signs may appear long after infection (congenital infections, recrudescence of latent infection in immunocompromised people), the relative contributions of different routes of transmission remain speculative. An expert panel estimated that 20–60% of human infections may be foodborne across different regions (Hald et al., 2016). This was consistent with a multi-centre case–control study in Europe, which estimated that 30–60% of infections were meatborne. In Chile, use of a serological test with an oocyte-specific antigen led to an estimate of 43% of infection due to oocysts, implying that foodborne infection was the more common route (Munoz-Zanzi et al., 2010).

Swine can be infected by ingesting oocysts that originate in cat faeces and may contaminate feed, water or soil, or by consuming tissues of animals (rodents, birds, other pigs) that contain cysts. The common occurrence of *T. gondii* infection of swine has been reported widely and seroprevalence increases with age (Davies, 2011; Guo et al., 2015). Therefore, food products derived from breeding stock (e.g. fresh sausages) may present elevated risk compared to products from market age animals. In both Europe and North America, prevalence has declined markedly over recent decades in association with intensification of swine production (Davies, 2011). Prevalence of *T. gondii* is higher in conventionally reared pigs, sheep and poultry than in cattle and is greater in meat products from organic than from conventionally reared meat animals because of outdoor access, which poses substantially greater opportunities for exposure to infected rodents, wildlife and oocyst-contaminated feed, water or environmental surfaces. Risk factors related to *T. gondii* exposure for livestock include farm type, feed source, presence of cats, methods of rodent and bird control, methods of carcass handling and water quality. A meta-analysis of studies in the USA showed that outdoor-raised pigs and goats had the highest prevalence (>30%) (Guo et al., 2016). This review serves as a useful resource and information repository for informing quantitative risk assessment studies for *T. gondii* infection in humans through meat consumption.

Similar to *Trichinella*, potential routes of infection are limited and implementing management changes to reduce the risk of infection should be relatively straightforward. These include not keeping cats on finishing sites, total confinement in bird-proofed buildings (likely to be cat proof), feed storage in enclosed silos, use of pelleted feed (including heat treated) and effective rodent control. Outdoor production of both pigs and poultry inherently involves higher risks of *T. gondii* exposure, and management of toxoplasma risk should be an explicit goal of such systems.

8 Summary

The challenge of meeting the future food and fibre demands of the human race is daunting, and livestock production contributes to pressing problems such as global warming, land degradation, air and water pollution and loss of biodiversity (Gerber and Steinfeld, 2008). Developing the most appropriate systems for future livestock production must also consider potential threats to human health from zoonotic diseases. The most vulnerable situation for disease emergence in swine may be rapid expansion and intensification of industries in developing countries without concomitant investment in biosecurity to prevent inter-species transmission of pathogens (Davies, 2012). The relative importance of swine-associated zoonoses varies enormously among societies; therefore, the needs for investment in research and other strategies to mitigate zoonotic risks are similarly diverse. Prioritization of needs is therefore necessary, and should be mostly driven by the actual rather than perceived potential risks. In some cases, where substantial risk reduction has been achieved in developed countries (e.g. foodborne parasites and some classical zoonoses such as brucellosis), barriers to progress may be more sociological than technological, or else limited by available resources. In others, most notably, control of enteric foodborne pathogens in farm settings, novel and more mechanistic approaches to understanding host–agent interactions and host adaptation are indicated.

9 Where to look for further information

Adler, B., 2014. *Leptospira and leptospirosis*. Springer, ISBN:3662450593, 9783662450598.

Bari, L. and Ukuku, D. O., 2015. *Foodborne pathogens and food safety*. CRC Press, ISBN:1498724108, 9781498724104.

Bueno-Marí, R., Gouveia Almeida, A. P., and Navarro, J. C., 2015. *Emerging zoonoses: eco-epidemiology, involved mechanisms and public health implications*. Frontiers Media SA, ISBN: 978-2-88919-618-0.

Fong, I. W., 2017. *Emerging zoonoses: A worldwide perspective-emerging infectious diseases of the 21st century*, Springer, 2017. ISBN:3319508903, 9783319508900.

Pappas, G., 2013. Socio-economic, industrial and cultural parameters of pig-borne infections. *Clin. Microbiol. Infect.* 19, 605–10.

Sing, A., 2014. *Zoonoses - infections affecting humans and animals: focus on public health aspects*. Springer, ISBN:940179457X, 9789401794572.

Webster, R. G., Monto, A. S., Braciale, T. J. and Lamb, R. A., 2014. *Textbook of Influenza*, John Wiley & Sons, ISBN:111863683X, 9781118636831.

10 References

Adler, B., 2015. History of leptospirosis and leptospira. *Curr. Top. Microbiol. Immunol.* 387, 1–9.

Akkermans, J. P., 1991. Changed health problems in a changing pig-farming concern in The Netherlands until 1980. *Tijdschr. Diergeneeskd.* 116, 1168–74.

Alban, L., 2016. Regulatory issues associated with preharvest food safety: European union perspective. *Microbiol. Spectr.* 4, (5): PFS-0003-2014. 10.1128/microbiolspec.PFS-0003-2014.

Alban, L. and Stark, K. D., 2005. Where should the effort be put to reduce the *Salmonella* prevalence in the slaughtered swine carcass effectively? *Prev. Vet. Med.* 68, 63–79.
Anon., 2009. Analysis of the baseline survey on the prevalence of methicillin-resistant *Staphylococcus aureus* (MRSA) in holdings with breeding pigs, in the EU, 2008 [1] - part A: MRSA prevalence estimates. *EFSA Journal* 7, 1376–457.
Arends, J. E., Ghisetti, V., Irving, W., Dalton, H. R., Izopet, J., Hoepelman, A. I. and Salmon, D., 2014. Hepatitis E: An emerging infection in high income countries. *J. Clin. Virol.* 59, 81–8.
Arends, J. P. and Zanen, H. C., 1988. Meningitis caused by *Streptococcus suis* in humans. *Rev. Infect. Dis.* 10, 131–7.
Armand-Lefevre, L., Ruimy, R. and Andremont, A., 2005. Clonal comparison of *Staphylococcus aureus* isolates from healthy pig farmers, human controls, and pigs. *Emerg. Infect. Dis.* 11, 711–14.
Auger, J. P., Fittipaldi, N., Benoit-Biancamano, M. O., Segura, M. and Gottschalk, M., 2016. Virulence studies of different sequence types and geographical origins of *Streptococcus suis* serotype 2 in a mouse model of infection. *Pathogens* 5(3):48, 10.3390/pathogens5030048.
Bailey, S. S., Crawshaw, T. R., Smith, N. H. and Palgrave, C. J., 2013. *Mycobacterium bovis* infection in domestic pigs in Great Britain. *Vet. J.* 198, 391–7.
Bancerz-Kisiel, A., Socha, P. and Szweda, W., 2016. Detection and characterisation of *Yersinia enterocolitica* strains in cold-stored carcasses of large game animals in Poland. *Vet. J.* 208, 102–3.
Barandiaran, S., Martinez Vivot, M., Perez, A. M., Cataldi, A. A. and Zumarraga, M. J., 2015. Bovine tuberculosis in domestic pigs: Genotyping and distribution of isolates in Argentina. *Res. Vet. Sci.* 103, 44–50.
Barbuddhe, S. B. and Chakraborty, T., 2009. *Listeria* as an enteroinvasive gastrointestinal pathogen. *Curr. Top. Microbiol. Immunol.* 337, 173–95.
Bauer, K., 1997. Foot- and-mouth disease as zoonosis. *Arch. Virol. Suppl.* 13, 95–7.
Becker, K., Ballhausen, B., Kahl, B. C. and Kock, R., 2017. The clinical impact of livestock-associated methicillin-resistant *Staphylococcus aureus* of the clonal complex 398 for humans. *Vet. Microbiol.* 200, 33–8.
Biet, F., Boschiroli, M. L., Thorel, M. F. and Guilloteau, L. A., 2005. Zoonotic aspects of *Mycobacterium bovis* and *Mycobacterium avium-intracellulare* complex (MAC). *Vet. Res.* 36, 411–36.
Bonten, M. J. and Mevius, D., 2015. Less evidence for an important role of food-producing animals as source of antibiotic resistance in humans. *Clin. Infect. Dis.* 60, 1867.
Bowman, A. S., Nelson, S. W., Page, S. L., Nolting, J. M., Killian, M. L., Sreevatsan, S. and Slemons, R. D., 2014. Swine-to-human transmission of influenza A(H3N2) virus at agricultural fairs, Ohio, USA, 2012. *Emerg. Infect. Dis.* 20, 1472–80.
Brookes, V. J., Del Rio Vilas, V. J. and Ward, M. P., 2015. Disease prioritization: What is the state of the art? *Epidemiol. Infect.* 143, 2911–22.
Brown, I. H., 2000. The epidemiology and evolution of influenza viruses in pigs. *Vet. Microbiol.* 74, 29–46.
Capasso, L., 2002. Bacteria in two-millennia-old cheese, and related epizoonoses in roman populations. *J. Infect.* 45, 122–7.
Caron, A., Grosbois, V., Etter, E., Gaidet, N. and de Garine-Wichatitsky, M., 2014. Bridge hosts for avian influenza viruses at the wildlife/domestic interface: An eco-epidemiological framework implemented in southern Africa. *Prev. Vet. Med.* 117, 590–600.
Casalinuovo, F., Ciambrone, L., Cacia, A. and Rippa, P., 2016. Contamination of bovine, sheep and goat meat with *Brucella* spp. Ital. *J. Food Saf.* 5, 5913.
Choi, M. J., Torremorell, M., Bender, J. B., Smith, K., Boxrud, D., Ertl, J. R., Yang, M., Suwannakarn, K., Her, D., Nguyen, J., Uyeki, T. M., Levine, M., Lindstrom, S., Katz, J. M., Jhung, M., Vetter, S., Wong, K. K., Sreevatsan, S. and Lynfield, R., 2015. Live animal markets in Minnesota: A potential source for emergence of novel influenza A viruses and interspecies transmission. *Clin. Infect. Dis.* 61, 1355–62.

Chua, K. B., Chua, B. H. and Wang, C. W., 2002. Anthropogenic deforestation, el nino and the emergence of Nipah virus in Malaysia. *Malays. J. Pathol.* 24, 15–21.
Chuang, Y. Y. and Huang, Y. C., 2015. Livestock-associated meticillin-resistant *Staphylococcus aureus* in Asia: An emerging issue? *Int. J. Antimicrob. Agents* 45, 334–40.
Clayson, E. T., Innis, B. L., Myint, K. S., Narupiti, S., Vaughn, D. W., Giri, S., Ranabhat, P. and Shrestha, M. P., 1995. Detection of hepatitis E virus infections among domestic swine in the Kathmandu valley of Nepal. *Am. J. Trop. Med. Hyg.* 53, 228–32.
Clayton, B. A., 2017. Nipah virus: Transmission of a zoonotic paramyxovirus. *Curr. Opin. Virol.* 22, 97–104.
Cosgrove, S. E., Sakoulas, G., Perencevich, E. N., Schwaber, M. J., Karchmer, A. W. and Carmeli, Y., 2003. Comparison of mortality associated with methicillin-resistant and methicillin-susceptible *Staphylococcus aureus* bacteremia: A meta-analysis. *Clin. Infect. Dis.* 36, 53–9.
Cossaboom, C. M., Heffron, C. L., Cao, D., Yugo, D. M., Houk-Miles, A. E., Lindsay, D. S., Zajac, A. M., Bertke, A. S., Elvinger, F. and Meng, X. J., 2016. Risk factors and sources of foodborne hepatitis E virus infection in the united states. *J. Med. Virol.* 88, 1641–5.
Dahl, J., 2013. Controlling salmonella in live pigs and at slaughter: The Danish experience. *Vet. J.* 197, 529–30.
D'Anastasio, R., Staniscia, T., Milia, M. L., Manzoli, L. and Capasso, L., 2011. Origin, evolution and paleoepidemiology of brucellosis. *Epidemiol. Infect.* 139, 149–56.
Darbellay, J., Lai, K., Babiuk, S., Berhane, Y., Ambagala, A., Wheler, C., Wilson, D., Walker, S., Potter, A., Gilmour, M., Safronetz, D., Gerdts, V. and Karniychuk, U., 2017. Neonatal pigs are susceptible to experimental Zika virus infection. *Emerg. Microb. Infect.* 6, e6.
Davies, P. R., 2010. Pork safety: Achievements and challenges. *Zoonoses Public. Health* 57 Suppl 1, 1–5.
Davies, P. R., 2011. Intensive swine production and pork safety. *Foodborne Pathog. Dis.* 8, 189–201.
Davies, P. R., 2012. One world, one health: The threat of emerging swine diseases. A North American perspective. *Transbound Emerg. Dis.* 59, 18–26.
Davies, P. R., Scott Hurd, H., Funk, J. A., Fedorka-Cray, P. J. and Jones, F. T., 2004. The role of contaminated feed in the epidemiology and control of *Salmonella enterica* in pork production. *Foodborne Pathog. Dis.* 1, 202–15.
Davies, P. R., Wagstrom, E. A. and Bender, J. B., 2011. Lethal necrotizing pneumonia caused by an ST398 *Staphylococcus aureus* strain. *Emerg. Infect. Dis.* 17, 1152–3.
De Busser, E. V., De Zutter, L., Dewulf, J., Houf, K. and Maes, D., 2013. *Salmonella* control in live pigs and at slaughter. *Vet. J.* 196, 20–7.
de Figueiredo, P., Ficht, T. A., Rice-Ficht, A., Rossetti, C. A. and Adams, L. G., 2015. Pathogenesis and immunobiology of brucellosis: Review of *Brucella*-host interactions. *Am. J. Pathol.* 185, 1505–17.
De Schryver, A., De Schrijver, K., Francois, G., Hambach, R., van Sprundel, M., Tabibi, R. and Colosio, C., 2015. Hepatitis E virus infection: An emerging occupational risk? *Occup. Med. (Lond)* 65, 667–72.
Doceul, V., Bagdassarian, E., Demange, A. and Pavio, N., 2016. Zoonotic hepatitis E virus: Classification, animal reservoirs and transmission routes. *Viruses* 8, E270.
Doganay, G. D. and Doganay, M., 2013. *Brucella* as a potential agent of bioterrorism. *Recent. Pat. Antiinfect Drug Discov.* 8, 27–33.
Dong, C., Meng, J., Dai, X., Liang, J. H., Feagins, A. R., Meng, X. J., Belfiore, N. M., Bradford, C., Corn, J. L., Cray, C., Glass, G. E., Gordon, M. L., Hesse, R. A., Montgomery, D. L., Nicholson, W. L., Pilny, A. A., Ramamoorthy, S., Shaver, D. D., Drobeniuc, J., Purdy, M. A., Fields, H. A., Kamili, S. and Teo, C. G., 2011. Restricted enzooticity of hepatitis E virus genotypes 1 to 4 in the United States. *J. Clin. Microbiol.* 49, 4164–72.
Drobeniuc, J., Greene-Montfort, T., Tejada-Strop, A., Hayden, T. M. and Kamili, S., 2016. Laboratory based surveillance identifies 227 cases of hepatitis E in the United States over a period of three years. *Hepatology* 63, 228.

Du, P., Zheng, H., Zhou, J., Lan, R., Ye, C., Jing, H., Jin, D., Cui, Z., Bai, X., Liang, J., Liu, J., Xu, L., Zhang, W., Chen, C. and Xu, J., 2017. Detection of multiple parallel transmission outbreak of *Streptococcus suis* human infection by use of genome epidemiology, China, 2005. *Emerg. Infect. Dis.* 23, 204–11, 10.3201/eid2302.160297.

Duke, T., Poka, H., Myers, S., Radcliffe, J. and Pavlin, B. I., 2013. Pigbel in the 21st century: Still here, and still in need of an effective surveillance system. *P. N. G. Med. J.* 56, 136–40.

Dunham, E. J., Dugan, V. G., Kaser, E. K., Perkins, S. E., Brown, I. H., Holmes, E. C. and Taubenberger, J. K., 2009. Different evolutionary trajectories of European avian-like and classical swine H1N1 influenza A viruses. *J. Virol.* 83, 5485–94.

Dwyer, P. D., 2006. People, pigs and parasites in New Guinea: Relational contexts and epidemiological possibilities. *Parasitol. Int.* 55 Suppl, S167–73.

Eales, K. M., Norton, R. E. and Ketheesan, N., 2010. Brucellosis in northern Australia. *Am. J. Trop. Med. Hyg.* 83, 876–8.

Eddi, C., Nari, A. and Amanfu, W., 2003. *Taenia solium* cysticercosis/taeniosis: Potential linkage with FAO activities; FAO support possibilities. *Acta Trop.* 87, 145–8.

Ellis, W. A., 2015. Animal leptospirosis. *Curr. Top. Microbiol. Immunol.* 387:99–137.

Engeman, R., Betsill, C. and Ray, T., 2011. Making contact: Rooting out the potential for exposure of commercial production swine facilities to feral swine in North Carolina. *Ecohealth* 8, 76–81.

Ericson, J. E., Popoola, V. O., Smith, P. B., Benjamin, D. K., Fowler, V. G., Benjamin Jr., D. K., Clark, R. H. and Milstone, A. M., 2015. Burden of invasive *Staphylococcus aureus* infections in hospitalized infants. *JAMA Pediatr.* 169, 1105–11.

Ewers, C., Bethe, A., Semmler, T., Guenther, S. and Wieler, L. H., 2012. Extended-spectrum beta-lactamase-producing and AmpC-producing *Escherichia coli* from livestock and companion animals, and their putative impact on public health: A global perspective. *Clin. Microbiol. Infect.* 18, 646–55.

Feng, Y., Zhang, H., Wu, Z., Wang, S., Cao, M., Hu, D. and Wang, C., 2014. *Streptococcus suis* infection: An emerging/reemerging challenge of bacterial infectious diseases? *Virulence* 5, 477–97.

Fosse, J., Seegers, H. and Magras, C., 2009. Prevalence and risk factors for bacterial food-borne zoonotic hazards in slaughter pigs: A review. *Zoonoses Public Health* 56, 429–54.

Fournier, A., Young, I., Rajic, A., Greig, J. and LeJeune, J., 2015. Social and economic aspects of the transmission of pathogenic bacteria between wildlife and food animals: A thematic analysis of published research knowledge. *Zoonoses Public Health* 62, 417–28.

Fowler, H. N., Brown, P., Rovira, A., Shade, B., Klammer, K., Smith, K. and Scheftel, J., 2013. *Streptococcus suis* meningitis in swine worker, Minnesota, USA. *Emerg. Infect. Dis.* 19, 330–1.

Fragaszy, E., Ishola, D. A., Brown, I. H., Enstone, J., Nguyen-Van-Tam, J. S., Simons, R., Tucker, A. W., Wieland, B., Williamson, S. M., Hayward, A. C., Flu Watch Group, Wood, J. L. and Combating Swine Influenza (COSI) Consortium, 2016. Increased risk of A(H1N1)pdm09 influenza infection in UK pig industry workers compared to a general population cohort. *Influenza Other Respir. Viruses* 10, 291–300.

Frana, T. S., Beahm, A. R., Hanson, B. M., Kinyon, J. M., Layman, L. L., Karriker, L. A., Ramirez, A. and Smith, T. C., 2013. Isolation and characterization of methicillin-resistant *Staphylococcus aureus* from pork farms and visiting veterinary students. *PLoS One* 8, e53738.

Franco, A., Hasman, H., Iurescia, M., Lorenzetti, R., Stegger, M., Pantosti, A., Feltrin, F., Ianzano, A., Porrero, M. C., Liapi, M. and Battisti, A., 2011. Molecular characterization of spa type t127, sequence type 1 methicillin-resistant *Staphylococcus aureus* from pigs. *J. Antimicrob. Chemother.* 66, 1231–5.

Fretin, D., Mori, M., Czaplicki, G., Quinet, C., Maquet, B., Godfroid, J. and Saegerman, C., 2013. Unexpected *Brucella suis* biovar 2 infection in a dairy cow, belgium. *Emerg. Infect. Dis.* 19, 2053–4.

Fretin, D., Whatmore, A. M., Al Dahouk, S., Neubauer, H., Garin-Bastuji, B., Albert, D., Van Hessche, M., Menart, M., Godfroid, J., Walravens, K. and Wattiau, P., 2008. *Brucella suis* identification and biovar typing by real-time PCR. *Vet. Microbiol.* 131, 376–85.

Gabriel, S., Dorny, P., Mwape, K. E., Trevisan, C., Braae, U. C., Magnussen, P., Thys, S., Bulaya, C., Phiri, I. K., Sikasunge, C. S., Makungu, C., Afonso, S., Nicolau, Q. and Johansen, M. V., 2017. Control of *Taenia solium* taeniasis/cysticercosis: The best way forward for sub-Saharan Africa? *Acta Trop.* 165, 252–60.

Garcia, H. H., Gonzalez, A. E., Tsang, V. C., O'Neal, S. E., Llanos-Zavalaga, F., Gonzalvez, G., Romero, J., Rodriguez, S., Moyano, L. M., Ayvar, V., Diaz, A., Hightower, A., Craig, P. S., Lightowlers, M. W., Gauci, C. G., Leontsini, E., Gilman, R. H. and Cysticercosis Working Group in Peru, 2016. Elimination of *Taenia solium* transmission in northern Peru. *N. Engl. J. Med.* 374, 2335–44.

Garcia-Jimenez, W. L., Benitez-Medina, J. M., Fernandez-Llario, P., Abecia, J. A., Garcia-Sanchez, A., Martinez, R., Risco, D., Ortiz-Pelaez, A., Salguero, F. J., Smith, N. H., Gomez, L. and Hermoso de Mendoza, J., 2013. Comparative pathology of the natural infections by *Mycobacterium bovis* and by *Mycobacterium caprae* in wild boar (sus scrofa). *Transbound Emerg. Dis.* 60, 102–9.

Gerber, P. J. and Steinfeld, H., 2008. Global environmental consequences of the livestock sector's growth. *Outlook Agric.* 37, 7–13.

Godfroid, J., Cloeckaert, A., Liautard, J. P., Kohler, S., Fretin, D., Walravens, K., Garin-Bastuji, B. and Letesson, J. J., 2005. From the discovery of the Malta fever's agent to the discovery of a marine mammal reservoir, brucellosis has continuously been a re-emerging zoonosis. *Vet. Res.* 36, 313–26.

Godfroid, J., Garin-Bastuji, B., Saegerman, C. and Blasco, J. M., 2013. Brucellosis in terrestrial wildlife. *Rev. Sci. Tech.* 32, 27–42.

Goerge, T., Lorenz, M. B., van Alen, S., Hubner, N. O., Becker, K. and Kock, R., 2017. MRSA colonization and infection among persons with occupational livestock exposure in europe: Prevalence, preventive options and evidence. *Vet. Microbiol.* 200, 6–12.

Gomez, E., Kennedy, C. C., Gottschalk, M., Cunningham, S. A., Patel, R. and Virk, A., 2014. *Streptococcus suis*-related prosthetic joint infection and streptococcal toxic shock-like syndrome in a pig farmer in the United States. *J. Clin. Microbiol.* 52, 2254–8.

Gomez-Sanz, E., Torres, C., Lozano, C., Fernandez-Perez, R., Aspiroz, C., Ruiz-Larrea, F. and Zarazaga, M., 2010. Detection, molecular characterization, and clonal diversity of methicillin-resistant *Staphylococcus aureus* CC398 and CC97 in Spanish slaughter pigs of different age groups. *Foodborne Pathog. Dis.* 7, 1269–77.

Gottschalk, M., Xu, J., Calzas, C. and Segura, M., 2010. *Streptococcus suis*: A new emerging or an old neglected zoonotic pathogen? *Future Microbiol.* 5, 371–91.

Goyette-Desjardins, G., Auger, J. P., Xu, J., Segura, M. and Gottschalk, M., 2014. *Streptococcus suis*, an important pig pathogen and emerging zoonotic agent-an update on the worldwide distribution based on serotyping and sequence typing. *Emerg. Microbes Infect.* 3, e45.

Gripper, L. B. and Welburn, S. C., 2017. Neurocysticercosis infection and disease-A review. *Acta Trop.* 166, 218–24.

Grontvedt, C. A., Elstrom, P., Stegger, M., Skov, R. L., Skytt Andersen, P., Larssen, K. W., Urdahl, A. M., Angen, O., Larsen, J., Amdal, S., Lotvedt, S. M., Sunde, M. and Bjornholt, J. V., 2016. Methicillin-resistant *Staphylococcus aureus* CC398 in humans and pigs in Norway: A 'One Health' perspective on introduction and transmission. *Clin. Infect. Dis.* 63, 1431–8.

Guo, C., Hoekstra, R. M., Schroeder, C. M., Pires, S. M., Ong, K. L., Hartnett, E., Naugle, A., Harman, J., Bennett, P., Cieslak, P., Scallan, E., Rose, B., Holt, K. G., Kissler, B., Mbandi, E., Roodsari, R., Angulo, F. J. and Cole, D., 2011. Application of bayesian techniques to model the burden of human salmonellosis attributable to U.S. food commodities at the point of processing: Adaptation of a Danish model. *Foodborne Pathog. Dis.* 8, 509–16.

Guo, M., Dubey, J. P., Hill, D., Buchanan, R. L., Gamble, H. R., Jones, J. L. and Pradhan, A. K., 2015. Prevalence and risk factors for *Toxoplasma gondii* infection in meat animals and meat products destined for human consumption. *J. Food Prot.* 78, 457–76.

Guo, M., Mishra, A., Buchanan, R. L., Dubey, J. P., Hill, D. E., Gamble, H. R., Jones, J. L. and Pradhan, A. K., 2016. A systematic meta-analysis of *Toxoplasma gondii* prevalence in food animals in the United States. *Foodborne Pathog. Dis.* 13, 109–18.

Hald, T., Aspinall, W., Devleesschauwer, B., Cooke, R., Corrigan, T., Havelaar, A. H., Gibb, H. J., Torgerson, P. R., Kirk, M. D., Angulo, F. J., Lake, R. J., Speybroeck, N. and Hoffmann, S., 2016. World Health Organization estimates of the relative contributions of food to the burden of disease due to selected foodborne hazards: A structured expert elicitation. *PLoS One* 11, e0145839.

Halperin, E. C., 2001. Non-human to human organ transplantation: Its biologic basis and a potential role for radiation therapy. *Int. J. Cancer* 96, 76–89.

Hampton, M. M., 2015. Congenital toxoplasmosis: A review. *Neonatal Netw.* 34, 274–8.

Hartl, J., Wehmeyer, M. H. and Pischke, S., 2016. Acute hepatitis E: Two sides of the same coin. *Viruses* 8, E299.

Hathaway, S. C., 1981. Leptospirosis in New Zealand: An ecological view. *N. Z. Vet. J.* 29, 109–12.

Hathaway, S. C., Blackmore, D. K. and Marshall, R. B., 1983. Leptospirosis and the maintenance host: A laboratory mouse model. *Res. Vet. Sci.* 34, 82–9.

Hatrongjit, R., Kerdsin, A., Gottschalk, M., Takeuchi, D., Hamada, S., Oishi, K. and Akeda, Y., 2015. First human case report of sepsis due to infection with *Streptococcus suis* serotype 31 in Thailand. *BMC Infect. Dis.* 15, 392.

Hau, S. J., Sun, J., Davies, P. R., Frana, T. S. and Nicholson, T. L., 2015. Comparative prevalence of immune evasion complex genes associated with beta-hemolysin converting bacteriophages in MRSA ST5 isolates from swine, swine facilities, humans with swine contact, and humans with no swine contact. *PLoS One* 10, e0142832.

Havelaar, A. H., Kirk, M. D., Torgerson, P. R., Gibb, H. J., Hald, T., Lake, R. J., Praet, N., Bellinger, D. C., de Silva, N. R., Gargouri, N., Speybroeck, N., Cawthorne, A., Mathers, C., Stein, C., Angulo, F. J., Devleesschauwer, B. and World Health Organization Foodborne Disease Burden Epidemiology Reference Group, 2015. World health organization global estimates and regional comparisons of the burden of foodborne disease in 2010. *PLoS Med.* 12, e1001923.

Hayman, D. T. and Johnson, N., 2014. Nipah virus a virus with multiple pathways of emergence. In: Johnson, N. (Ed.), *Role of Animals in Emerging Infectious Diseases*. Elsevier Academic Press Inc., San Diego, USA, pp. 293–315.

Hedman, C., Bolea, R., Marin, B., Cobriere, F., Filali, H., Vazquez, F., Pitarch, J. L., Vargas, A., Acin, C., Moreno, B., Pumarola, M., Andreoletti, O. and Badiola, J. J., 2016. Transmission of sheep-bovine spongiform encephalopathy to pigs. *Vet. Res.* 47, 14.

Henao, O. L., Jones, T. F., Vugia, D. J., Griffin, P. M. and Foodborne Diseases Active Surveillance Network (FoodNet) Workgroup, 2015. Foodborne diseases active surveillance network-2 decades of achievements, 1996–2015. *Emerg. Infect. Dis.* 21, 1529–36.

Hetem, D. J., Bootsma, M. C., Troelstra, A. and Bonten, M. J., 2013. Transmissibility of livestock-associated methicillin-resistant *Staphylococcus aureus*. *Emerg. Infect. Dis.* 19, 1797–802.

Hogema, B. M., Molier, M., Slot, E. and Zaaijer, H. L., 2014. Past and present of hepatitis E in The Netherlands. *Transfusion* 54, 3092–6.

Holm, D. K., Moessner, B. K., Engle, R. E., Zaaijer, H. L., Georgsen, J., Purcell, R. H. and Christensen, P. B., 2015. Declining prevalence of hepatitis E antibodies among Danish blood donors. *Transfusion* 55, 1662–7.

Hoofnagle, J. H., Nelson, K. E. and Purcell, R. H., 2012. Hepatitis E. *N. Engl. J. Med.* 367, 1237–44.

Hopp, P., Wahlstrom, H. and Hirn, J., 1999. A common *Salmonella* control programme in Finland, Norway and Sweden. *Acta Vet. Scand. Suppl.* 91, 45–9.

Huijsdens, X. W., Bosch, T., van Santen-Verheuvel, M. G., Spalburg, E., Pluister, G. N., van Luit, M., Heck, M. E., Haenen, A. and de Neeling, A. J., 2009. Molecular characterisation of PFGE non-typable methicillin-resistant *Staphylococcus aureus* in The Netherlands, 2007. *Euro Surveill.* 14, 19335.

Huong, V. T., Ha, N., Huy, N. T., Horby, P., Nghia, H. D., Thiem, V. D., Zhu, X., Hoa, N. T., Hien, T. T., Zamora, J., Schultsz, C., Wertheim, H. F. and Hirayama, K., 2014a. Epidemiology, clinical manifestations, and outcomes of *Streptococcus suis* infection in humans. *Emerg. Infect. Dis.* 20, 1105–14.

Huong, V. T., Hoa, N. T., Horby, P., Bryant, J. E., Van Kinh, N., Toan, T. K. and Wertheim, H. F., 2014b. Raw pig blood consumption and potential risk for *Streptococcus suis* infection, Vietnam. *Emerg. Infect. Dis.* 20, 1895–8.

Jobbins, S. E. and Alexander, K. A., 2015. Evidence of *Leptospira* sp. infection among a diversity of African wildlife species: Beyond the usual suspects. *Trans. R. Soc. Trop. Med. Hyg.* 109, 349–51.

Jones, J. L. and Dubey, J. P., 2012. Foodborne toxoplasmosis. *Clin. Infect. Dis.* 55, 845–51.

Jori, F., Laval, M., Maestrini, O., Casabianca, F., Charrier, F. and Pavio, N., 2016. Assessment of domestic pigs, wild boars and feral hybrid pigs as reservoirs of hepatitis E virus in Corsica, France. *Viruses* 8(8):236. doi:10.3390/v8080236.

Kapel, C. M., 2000. Host diversity and biological characteristics of the *Trichinella* genotypes and their effect on transmission. *Vet. Parasitol.* 93, 263–78.

Kennedy, E. D., Hall, R. L., Montgomery, S. P., Pyburn, D. G., Jones, J. L. and Centers for Disease Control and Prevention (CDC), 2009. Trichinellosis surveillance – United States, 2002–2007. *MMWR Surveill. Summ.* 58, 1–7.

Khanna, T., Friendship, R., Dewey, C. and Weese, J. S., 2008. Methicillin resistant *Staphylococcus aureus* colonization in pigs and pig farmers. *Vet. Microbiol.* 128, 298–303.

Khuroo, M. S., Khuroo, M. S. and Khuroo, N. S., 2016a. Hepatitis E: Discovery, global impact, control and cure. *World J. Gastroenterol.* 22, 7030–45.

Khuroo, M. S., Khuroo, M. S. and Khuroo, N. S., 2016b. Transmission of hepatitis E virus in developing countries. *Viruses* 8, E253.

Kirkland, P. D., Love, R. J., Philbey, A. W., Ross, A. D., Davis, R. J. and Hart, K. G., 2001. Epidemiology and control of Menangle virus in pigs. *Aust. Vet. J.* 79, 199–206.

Klevens, R. M., Morrison, M. A., Nadle, J., Petit, S., Gershman, K., Ray, S., Harrison, L. H., Lynfield, R., Dumyati, G., Townes, J. M., Craig, A. S., Zell, E. R., Fosheim, G. E., McDougal, L. K., Carey, R. B., Fridkin, S. K. and Active Bacterial Core surveillance (ABCs) MRSA Investigators, 2007. Invasive methicillin-resistant *Staphylococcus aureus* infections in the united states. *JAMA* 298, 1763–71.

Kreizinger, Z., Foster, J. T., Ronai, Z., Sulyok, K. M., Wehmann, E., Janosi, S. and Gyuranecz, M., 2014. Genetic relatedness of brucella suis biovar 2 isolates from hares, wild boars and domestic pigs. *Vet. Microbiol.* 172, 492–8.

Larsen, J., Petersen, A., Sorum, M., Stegger, M., van Alphen, L., Valentiner-Branth, P., Knudsen, L. K., Larsen, L. S., Feingold, B., Price, L. B., Andersen, P. S., Larsen, A. R. and Skov, R. L., 2015. Meticillin-resistant *Staphylococcus aureus* CC398 is an increasing cause of disease in people with no livestock contact in Denmark, 1999 to 2011. *Euro Surveill.* 20(37): doi:10.2807/1560-7917.ES.2015.20.37.30021.

Leao, C., Canto, A., Machado, D., Sanches, I. S., Couto, I., Viveiros, M., Inacio, J. and Botelho, A., 2014. Relatedness of *Mycobacterium avium* subspecies hominissuis clinical isolates of human and porcine origins assessed by MLVA. *Vet. Microbiol.* 173, 92–100.

Lee, G. T., Chiu, C. Y., Haller, B. L., Denn, P. M., Hall, C. S. and Gerberding, J. L., 2008. *Streptococcus suis* meningitis, United States. *Emerg. Infect. Dis.* 14, 183–5.

Levett, P. N., 2001. Leptospirosis. *Clin. Microbiol. Rev.* 14, 296–326.

Levett, P. N., 2015. Systematics of leptospiraceae. *Curr. Top. Microbiol. Immunol.* 387, 11–20.

Lewis, H. C., Wichmann, O. and Duizer, E., 2010. Transmission routes and risk factors for autochthonous hepatitis E virus infection in Europe: A systematic review. *Epidemiol. Infect.* 138, 145–66.

Lewis, N. S., Russell, C. A., Langat, P., Anderson, T. K., Berger, K., Bielejec, F., Burke, D. F., Dudas, G., Fonville, J. M., Fouchier, R. A., Kellam, P., Koel, B. F., Lemey, P., Nguyen, T., Nuansrichy, B., Peiris, J. M., Saito, T., Simon, G., Skepner, E., Takemae, N., ESNIP3 consortium, Webby, R. J., Van Reeth, K., Brookes, S. M., Larsen, L., Watson, S. J., Brown, I. H. and Vincent, A. L., 2016. The global antigenic diversity of swine influenza A viruses. *eLife* 5:e12217.

Li, M., Embury-Hyatt, C. and Weingartl, H. M., 2010. Experimental inoculation study indicates swine as a potential host for Hendra virus. *Vet. Res.* 41, 33.

Lianou, A. and Koutsoumanis, K. P., 2013. Strain variability of the behavior of foodborne bacterial pathogens: A review. *Int. J. Food Microbiol.* 167, 310–21.

Linhares, L. L., Yang, M., Sreevatsan, S., Munoz-Zanzi, C. A., Torremorell, M. and Davies, P. R., 2015. The effect of anatomic site and age on detection of *Staphylococcus aureus* in pigs. *J. Vet. Diagn. Invest.* 27, 55–60.
Lipsitch, M., Barclay, W., Raman, R., Russell, C. J., Belser, J. A., Cobey, S., Kasson, P. M., Lloyd-Smith, J. O., Maurer-Stroh, S., Riley, S., Beauchemin, C. A., Bedford, T., Friedrich, T. C., Handel, A., Herfst, S., Murcia, P. R., Roche, B., Wilke, C. O. and Russell, C. A., 2016. Viral factors in influenza pandemic risk assessment. *Elife* 5, 10.7554/eLife.18491.
Ma, E., Chung, P. H., So, T., Wong, L., Choi, K. M., Cheung, D. T., Kam, K. M., Chuang, S. K. and Tsang, T., 2008. *Streptococcus suis* infection in Hong Kong: An emerging infectious disease? *Epidemiol. Infect.* 136, 1691–7.
Mai, N. T., Hoa, N. T., Nga, T. V., Linh le, D., Chau, T. T., Sinh, D. X., Phu, N. H., Chuong, L. V., Diep, T. S., Campbell, J., Nghia, H. D., Minh, T. N., Chau, N. V., de Jong, M. D., Chinh, N. T., Hien, T. T., Farrar, J. and Schultsz, C., 2008. *Streptococcus suis* meningitis in adults in Vietnam. *Clin. Infect. Dis.* 46, 659–67.
Mantur, B. G. and Amarnath, S. K., 2008. Brucellosis in india – a review. *J. Biosci.* 33, 539–47.
McCarthy, A. J., van Wamel, W., Vandendriessche, S., Larsen, J., Denis, O., Garcia-Graells, C., Uhlemann, A. C., Lowy, F. D., Skov, R. and Lindsay, J. A., 2012. *Staphylococcus aureus* CC398 clade associated with human-to-human transmission. *Appl. Environ. Microbiol.* 78, 8845–8.
Meng, X. J., 2016. Expanding host range and cross-species infection of hepatitis E virus. *PLoS Pathog.* 12, e1005695.
Meng, X. J., Purcell, R. H., Halbur, P. G., Lehman, J. R., Webb, D. M., Tsareva, T. S., Haynes, J. S., Thacker, B. J. and Emerson, S. U., 1997. A novel virus in swine is closely related to the human hepatitis E virus. *Proc. Natl. Acad. Sci. U. S. A.* 94, 9860–5.
Mesner, O., Riesenberg, K., Biliar, N., Borstein, E., Bouhnik, L., Peled, N. and Yagupsky, P., 2007. The many faces of human-to-human transmission of brucellosis: Congenital infection and outbreak of nosocomial disease related to an unrecognized clinical case. *Clin. Infect. Dis.* 45, e135–40.
Messina, J. A., Thaden, J. T., Sharma-Kuinkel, B. K. and Fowler Jr., V. G., 2016. Impact of bacterial and human genetic variation on *Staphylococcus aureus* infections. *PLoS Pathog.* 12, e1005330.
Mor, S. M., Wiethoelter, A. K., Lee, A., Moloney, B., James, D. R. and Malik, R., 2016. Emergence of *Brucella suis* in dogs in New South Wales, Australia: Clinical findings and implications for zoonotic transmission. *BMC Vet. Res.* 12, 199.
Moreno, E., 2014. Retrospective and prospective perspectives on zoonotic brucellosis. *Front. Microbiol.* 5, 213.
Morse, S. S., 2004. Factors and determinants of disease emergence. *Rev. Sci. Tech.* 23, 443–51.
Mukaratirwa, S., La Grange, L. and Pfukenyi, D. M., 2013. *Trichinella* infections in animals and humans in sub-Saharan Africa: A review. *Acta Trop.* 125, 82–9.
Muller, B., Durr, S., Alonso, S., Hattendorf, J., Laisse, C. J., Parsons, S. D., van Helden, P. D. and Zinsstag, J., 2013. Zoonotic *Mycobacterium bovis*-induced tuberculosis in humans. *Emerg. Infect. Dis.* 19, 899–908.
Munoz-Zanzi, C. A., Fry, P., Lesina, B. and Hill, D., 2010. *Toxoplasma gondii* oocyst-specific antibodies and source of infection. *Emerg. Infect. Dis.* 16, 1591–3.
Musser, J. M., Schwartz, A. L., Srinath, I. and Waldrup, K. A., 2013. Use of serology and bacterial culture to determine prevalence of *Brucella* spp. in feral swine (sus scrofa) in proximity to a beef cattle herd positive for *Brucella suis* and *Brucella abortus*. *J. Wildl. Dis.* 49, 215–20.
Mutters, N. T., Bieber, C. P., Hauck, C., Reiner, G., Malek, V. and Frank, U., 2016. Comparison of livestock-associated and health care-associated MRSA-genes, virulence, and resistance. *Diagn. Microbiol. Infect. Dis.* 86, 417–21.
Mwachui, M. A., Crump, L., Hartskeerl, R., Zinsstag, J. and Hattendorf, J., 2015. Environmental and behavioural determinants of leptospirosis transmission: A systematic review. *PLoS Negl Trop. Dis.* 9, e0003843.
Myers, K. P., Olsen, C. W., Setterquist, S. F., Capuano, A. W., Donham, K. J., Thacker, E. L., Merchant, J. A. and Gray, G. C., 2006. Are swine workers in the United States at increased risk of infection with zoonotic influenza virus? *Clin. Infect. Dis.* 42, 14–20.

Nair, R., Wu, J., Carrel, M., O'Brien, A., Quick, M., Farina, S., Wardyn, S., Thapaliya, D., Grenier, D. and Smith, T. C., 2016. Prospective multicenter surveillance identifies *Staphylococcus aureus* infections caused by livestock-associated strains in an agricultural state. *Diagn. Microbiol. Infect. Dis.* 85, 360–6.

Naranjo, V., Gortazar, C., Vicente, J. and de la Fuente, J., 2008. Evidence of the role of European wild boar as a reservoir of *Mycobacterium tuberculosis* complex. *Vet. Microbiol.* 127, 1–9.

Nelson, M. I. and Vincent, A. L., 2015. Reverse zoonosis of influenza to swine: New perspectives on the human-animal interface. *Trends Microbiol.* 23, 142–53.

Nelson, M. I., Wentworth, D. E., Culhane, M. R., Vincent, A. L., Viboud, C., LaPointe, M. P., Lin, X., Holmes, E. C., Detmer, S. E. and 2014. Introductions and evolution of human-origin seasonal influenza a viruses in multinational swine populations. *J. Virol.* 88, 10110–19.

Nesbakken, T., Eckner, K. and Rotterud, O. J., 2008. The effect of blast chilling on occurrence of human pathogenic *Yersinia enterocolitica* compared to campylobacter spp. and numbers of hygienic indicators on pig carcasses. *Int. J. Food Microbiol.* 123, 130–3.

Netrabukkana, P., Robertson, I. D., Kasemsuwan, S., Wongsathapornchai, K. and Fenwick, S., 2016. Assessing potential risks of influenza A virus transmission at the pig-human interface in thai small pig farms using a questionnaire survey. *Transbound Emerg. Dis.* 63, e135–9.

Ng, V. and Sargeant, J. M., 2016. Prioritizing zoonotic diseases: Differences in perspectives between human and animal health professionals in North America. *Zoonoses Public Health* 63, 196–211.

Nugent, G., Gortazar, C. and Knowles, G., 2015. The epidemiology of *Mycobacterium bovis* in wild deer and feral pigs and their roles in the establishment and spread of bovine tuberculosis in New Zealand wildlife. *N. Z. Vet. J.* 63 Suppl 1, 54–67.

Nugent, G., Yockney, I. J., Whitford, E. J. and Cross, M. L., 2014. Assessing the effectiveness of tuberculosis management in brushtail possums (*Trichosurus vulpecula)*, through indirect surveillance of *Mycobacterium bovis* infection using released sentinel pigs. *Vet. Med. Int.* 2014, 361634.

Okello, A. L., Thomas, L., Inthavong, P., Ash, A., Khamlome, B., Keokamphet, C., Newberry, K., Gauci, C. G., Gabriel, S., Dorny, P., Thompson, R. A., Lightowlers, M. W. and Allen, J., 2017. Reprint of 'assessing the impact of a joint human-porcine intervention package for *Taenia solium* control: Results of a pilot study from northern Lao PDR'. *Acta Trop.* 165, 261–7.

Okura, M., Osaki, M., Nomoto, R., Arai, S., Osawa, R., Sekizaki, T. and Takamatsu, D., 2016. Current taxonomical situation of *Streptococcus suis*. *Pathogens* 5(3):45. doi: 10.3390/pathogens5030045.

Olsen, S. C., Garin-Bastuji, B., Blasco, J. M., Nicola, A. M. and Samartino, L., 2012. Brucellosis. In: Zimmerman, J. J., Karriker, L. A., Ramirez, A., Schwartz, K. J. and Stevenson, G. W. (Eds.), *Diseases of Swine*. John Wiley & Sons, Inc., Chichester, UK, pp. 697–708.

Olsen, S. C. and Palmer, M. V., 2014. Advancement of knowledge of *Brucella* over the past 50 years. *Vet. Pathol.* 51, 1076–89.

Ong, K. L., Gould, L. H., Chen, D. L., Jones, T. F., Scheftel, J., Webb, T. H., Mody, R. K. and Mahon, B. E., 2012. Changing epidemiology of *Yersinia enterocolitica* infections: Markedly decreased rates in young black children, foodborne diseases active surveillance network (FoodNet), 1996–2009. *Clin. Infect. Dis.* 54 Suppl 5, S385–90.

Pappas, G., 2013. Socio-economic, industrial and cultural parameters of pig-borne infections. *Clin. Microbiol. Infect.* 19, 605–10.

Pavio, N., Meng, X. J. and Doceul, V., 2015. Zoonotic origin of hepatitis E. *Curr. Opin. Virol.* 10, 34–41.

Pedersen, K., Pabilonia, K. L., Anderson, T. D., Bevins, S. N., Hicks, C. R., Kloft, J. M. and Deliberto, T. J., 2015. Widespread detection of antibodies to *Leptospira* in feral swine in the united states. *Epidemiol. Infect.* 143, 2131–6.

Pedersen, K., Quance, C. R., Robbe-Austerman, S., Piaggio, A. J., Bevins, S. N., Goldstein, S. M., Gaston, W. D. and DeLiberto, T. J., 2014. Identification of brucella suis from feral swine in selected states in the USA. *J. Wildl. Dis.* 50, 171–9.

Peiris, M. and Yen, H. L., 2014. Animal and human influenzas. *Rev. Sci. Tech.* 33, 539–53.

Pesciaroli, M., Alvarez, J., Boniotti, M. B., Cagiola, M., Di Marco, V., Marianelli, C., Pacciarini, M. and Pasquali, P., 2014. Tuberculosis in domestic animal species. *Res. Vet. Sci.* 97 Suppl, S78–85.

Phiri, I. K., Ngowi, H., Afonso, S., Matenga, E., Boa, M., Mukaratirwa, S., Githigia, S., Saimo, M., Sikasunge, C., Maingi, N., Lubega, G. W., Kassuku, A., Michael, L., Siziya, S., Krecek, R. C., Noormahomed, E., Vilhena, M., Dorny, P. and Willingham 3rd, A. L., 2003. The emergence of *Taenia solium* cysticercosis in eastern and southern Africa as a serious agricultural problem and public health risk. *Acta Trop.* 87, 13–23.

Plank, R. and Dean, D., 2000. Overview of the epidemiology, microbiology, and pathogenesis of *Leptospira* spp. in humans. *Microbes Infect.* 2, 1265–76.

Ploug, T., Holm, S. and Gjerris, M., 2015. The stigmatization dilemma in public health policy–the case of MRSA in Denmark. *BMC Public Health* 15, 640.

Pozio, E., 2007. World distribution of *Trichinella* spp. infections in animals and humans. *Vet. Parasitol.* 149, 3–21.

Pozio, E., La Rosa, G., Serrano, F. J., Barrat, J. and Rossi, L., 1996. Environmental and human influence on the ecology of *Trichinella spiralis* and *Trichinella britovi* in western Europe. *Parasitology* 113 (Pt 6), 527–33.

Pozio, E. and Zarlenga, D. S., 2013. New pieces of the *Trichinella* puzzle. *Int. J. Parasitol.* 43, 983–97.

Pulliam, J. R., Epstein, J. H., Dushoff, J., Rahman, S. A., Bunning, M., Jamaluddin, A. A., Hyatt, A. D., Field, H. E., Dobson, A. P., Daszak, P. and Henipavirus Ecology Research Group (HERG), 2012. Agricultural intensification, priming for persistence and the emergence of Nipah virus: A lethal bat-borne zoonosis. *J. R. Soc. Interface* 9, 89–101.

Purcell, R. H. and Emerson, S. U., 2008. Hepatitis E: An emerging awareness of an old disease. *J. Hepatol.* 48, 494–503.

Rasigade, J. P., Laurent, F., Hubert, P., Vandenesch, F. and Etienne, J., 2011. Lethal necrotizing pneumonia caused by an ST398 *Staphylococcus aureus* strain. *Emerg. Infect. Dis.* 17, 1153.

Rindi, L. and Garzelli, C., 2014. Genetic diversity and phylogeny of *Mycobacterium avium. Infect. Genet. Evol.* 21, 375–83.

Ruiz-Fons, F., 2017. A review of the current status of relevant zoonotic pathogens in wild swine (*Sus scrofa*) populations: Changes modulating the risk of transmission to humans. *Transbound Emerg. Dis.* 64, 68–88.

Saenz, R. A., Hethcote, H. W. and Gray, G. C., 2006. Confined animal feeding operations as amplifiers of influenza. *Vector Borne Zoonotic Dis.* 6, 338–46.

Safdar, N. and Bradley, E. A., 2008. The risk of infection after nasal colonization with *Staphylococcus aureus. Am. J. Med.* 121, 310–15.

Sanchez, J., Dohoo, I. R., Christensen, J. and Rajic, A., 2007. Factors influencing the prevalence of *Salmonella* spp. in swine farms: A meta-analysis approach. *Prev. Vet. Med.* 81, 148–77.

Scallan, E., Hoekstra, R. M., Angulo, F. J., Tauxe, R. V., Widdowson, M. A., Roy, S. L., Jones, J. L. and Griffin, P. M., 2011. Foodborne illness acquired in the United States–major pathogens. *Emerg. Infect. Dis.* 17, 7–15.

Schmidt, C. W., 2009. Swine CAFOs & novel H1N1 flu: Separating facts from fears. *Environ. Health Perspect.* 117, A394–401.

Schultsz, C., Jansen, E., Keijzers, W., Rothkamp, A., Duim, B., Wagenaar, J. A. and van der Ende, A., 2012. Differences in the population structure of invasive *Streptococcus suis* strains isolated from pigs and from humans in The Netherlands. *PLoS One* 7, e33854.

Scientific Opinion of the Panel on Biological Hazards, 2009. Assessment of the public health significance of meticillin resistant *Staphylococcus aureus* (MRSA) in animals and food. *EFSA J.* 993, 1–73.

Sivaraman, K., Venkataraman, N. and Cole, A. M., 2009. Staphylococcus aureus nasal carriage and its contributing factors. *Future Microbiol.* 4, 999–1008.

Smith, T. C., Harper, A. L., Nair, R., Wardyn, S. E., Hanson, B. M., Ferguson, D. D. and Dressler, A. E., 2011. Emerging swine zoonoses. *Vector Borne Zoonotic Dis.* 11, 1225–34.

Sofos, J. N. and Geornaras, I., 2010. Overview of current meat hygiene and safety risks and summary of recent studies on biofilms, and control of *Escherichia coli* O157:H7 in nonintact, and *Listeria monocytogenes* in ready-to-eat, meat products. *Meat Sci.* 86, 2–14.

Stoffregen, W. C., Olsen, S. C., Jack Wheeler, C., Bricker, B. J., Palmer, M. V., Jensen, A. E., Halling, S. M. and Alt, D. P., 2007. Diagnostic characterization of a feral swine herd enzootically infected with *Brucella*. *J. Vet. Diagn. Invest.* 19, 227–37.

Strutzberg-Minder, K. and Kreienbrock, L., 2011. Leptospire infections in pigs: Epidemiology, diagnostics and worldwide occurrence. *Berl. Munch. Tierarztl. Wochenschr.* 124, 345–59.

Sun, J., Yang, M., Sreevatsan, S., Bender, J. B., Singer, R. S., Knutson, T. P., Marthaler, D. G., Davies, P. R., 2017. Longitudinal study of *Staphylococcus aureus* colonization in swine veterinarians in the USA. BMC Infect Dis., 17:690.

Sun, J., Yang, M., Sreevatsan, S. and Davies, P. R., 2015. Prevalence and characterization of *Staphylococcus aureus* in growing pigs in the USA. *PLoS One* 10, e0143670.

Sundstrom, K., Wahlstrom, H., Ivarsson, S. and Sternberg Lewerin, S., 2014. Economic effects of introducing alternative *Salmonella* control strategies in Sweden. *PLoS One* 9, e96446.

Sweileh, W. M., 2017. Global research trends of World Health Organization's top eight emerging pathogens. *Global Health* 13, 9.

Szulowski, K., Iwaniak, W., Weiner, M. and Zlotnicka, J., 2013. *Brucella suis* biovar 2 isolations from cattle in Poland. *Ann. Agric. Environ. Med.* 20, 672–5.

Tan, C. T. and Chua, K. B., 2008. Nipah virus encephalitis. *Curr. Infect. Dis. Rep.* 10, 315–20.

Taylor, L. H., Latham, S. M. and Woolhouse, M. E., 2001. Risk factors for human disease emergence. *Philos. Trans. R. Soc. Lond. B. Biol. Sci.* 356, 983–9.

Teshale, E. H., Denniston, M. M., Drobeniuc, J., Kamili, S., Teo, C. G. and Holmberg, S. D., 2015. Decline in hepatitis E virus antibody prevalence in the United States from 1988–1994 to 2009–2010. *J. Infect. Dis.* 211, 366–73.

Thepault, A., Meric, G., Rivoal, K., Pascoe, B., Mageiros, L., Touzain, F., Rose, V., Beven, V., Chemaly, M. and Sheppard, S. K., 2017. Genome-wide identification of host-segregating epidemiological markers for source attribution in *Campylobacter jejuni*. *Appl. Environ. Microbiol.*

Thiry, D., Mauroy, A., Pavio, N., Purdy, M. A., Rose, N., Thiry, E. and de Oliveira-Filho, E. F., 2017. Hepatitis E virus and related viruses in animals. *Transbound Emerg. Dis.* 64, 37–52.

Thys, S., Mwape, K. E., Lefevre, P., Dorny, P., Phiri, A. M., Marcotty, T., Phiri, I. K. and Gabriel, S., 2016. Why pigs are free-roaming: Communities' perceptions, knowledge and practices regarding pig management and taeniosis/cysticercosis in a *Taenia solium* endemic rural area in eastern Zambia. *Vet. Parasitol.* 225, 33–42.

Torremorell, M., Allerson, M., Corzo, C., Diaz, A. and Gramer, M., 2012. Transmission of influenza A virus in pigs. *Transbound Emerg. Dis.* 59 Suppl 1, 68–84.

Trang do, T., Siembieda, J., Huong, N. T., Hung, P., Ky, V. D., Bandyopahyay, S. and Olowokure, B., 2015. Prioritization of zoonotic diseases of public health significance in Vietnam. *J. Infect. Dev. Ctries* 9, 1315–22.

Trevisan, C., Devleesschauwer, B., Schmidt, V., Winkler, A. S., Harrison, W. and Johansen, M. V., 2016. The societal cost of *Taenia solium* cysticercosis in Tanzania. *Acta Trop.*

Tyson, G. H., Tate, H. P., Abbott, J., Tran, T. T., Kabera, C., Crarey, E., Young, S., McDermott, P. F., Sprague, G., Campbell, M., Adeyemo, O., Browne-Silva, J., Myers, M., Thitaram, S. and Zhao, S., 2016. Molecular subtyping and source attribution of *Campylobacter* isolated from food animals. *J. Food Prot.* 79, 1891–7.

Uddin Khan, S., Atanasova, K. R., Krueger, W. S., Ramirez, A. and Gray, G. C., 2013. Epidemiology, geographical distribution, and economic consequences of swine zoonoses: A narrative review. *Emerg. Microbes Infect.* 2, e92.

Uhlemann, A. C., Hafer, C., Miko, B. A., Sowash, M. G., Sullivan, S. B., Shu, Q. and Lowy, F. D., 2013. Emergence of sequence type 398 as a community- and healthcare-associated methicillin-susceptible *Staphylococcus aureus* in northern Manhattan. *Clin. Infect. Dis.* 57, 700–3.

Uhlemann, A. C., Porcella, S. F., Trivedi, S., Sullivan, S. B., Hafer, C., Kennedy, A. D., Barbian, K. D., McCarthy, A. J., Street, C., Hirschberg, D. L., Lipkin, W. I., Lindsay, J. A., Deleo, F. R. and Lowy, F. D., 2012. Identification of a highly transmissible animal-independent *Staphylococcus aureus* ST398 clone with distinct genomic and cell adhesion properties. *MBio* 3(2):e00027–12. doi:10.1128/mBio.00027-12.

van Cleef, B. A., van Benthem, B. H., Haenen, A. P., Bosch, T., Monen, J. and Kluytmans, J. A., 2013. Low incidence of livestock-associated methicillin-resistant *Staphylococcus aureus* bacteraemia in The Netherlands in 2009. *PLoS One* 8, e73096.

van Cleef, B. A., van Benthem, B. H., Verkade, E. J., van Rijen, M. M., Kluytmans-van den Bergh, M. F., Graveland, H., Bosch, T., Verstappen, K. M., Wagenaar, J. A., Bos, M. E., Heederik, D. and Kluytmans, J. A., 2015. Livestock-associated MRSA in household members of pig farmers: Transmission and dynamics of carriage, a prospective cohort study. *PLoS One* 10, e0127190.

van Cleef, B. A., Van Benthem, B. H., Verkade, E. J., VAN Rijen, M. M., Kluytmans-Van Den Bergh, M. F., Graveland, H., Bosch, T., Verstappen, K. M., Wagenaar, J. A., Heederik, D. and Kluytmans, J. A., 2016. Health and health-related quality of life in pig farmers carrying livestock-associated methicillin-resistant *Staphylococcus aureus*. *Epidemiol. Infect.* 144, 1774–83.

van Cleef, B. A., Verkade, E. J., Wulf, M. W., Buiting, A. G., Voss, A., Huijsdens, X. W., van Pelt, W., Mulders, M. N. and Kluytmans, J. A., 2010. Prevalence of livestock-associated MRSA in communities with high pig-densities in The Netherlands. *PLoS One* 5, e9385.

van Rijen, M. M., Van Keulen, P. H. and Kluytmans, J. A., 2008. Increase in a dutch hospital of methicillin-resistant *Staphylococcus aureus* related to animal farming. *Clin. Infect. Dis.* 46, 261–3.

van Samkar, A., Brouwer, M. C., Schultsz, C., van der Ende, A. and van de Beek, D., 2015. *Streptococcus suis* meningitis in The Netherlands. *J. Infect.* 71, 602–4.

Verkade, E., Kluytmans-van den Bergh, M., van Benthem, B., van Cleef, B., van Rijen, M., Bosch, T., Schouls, L. and Kluytmans, J., 2014. Transmission of methicillin-resistant *Staphylococcus aureus* CC398 from livestock veterinarians to their household members. *PLoS One* 9, e100823.

Verkade, E., van Benthem, B., den Bergh, M. K., van Cleef, B., van Rijen, M., Bosch, T. and Kluytmans, J., 2013. Dynamics and determinants of *Staphylococcus aureus* carriage in livestock veterinarians: A prospective cohort study. *Clin. Infect. Dis.* 57, e11–7.

Viana, M., Mancy, R., Biek, R., Cleaveland, S., Cross, P. C., Lloyd-Smith, J. O. and Haydon, D. T., 2014. Assembling evidence for identifying reservoirs of infection. *Trends Ecol. Evol.* 29, 270–9.

Vincent, A., Awada, L., Brown, I., Chen, H., Claes, F., Dauphin, G., Donis, R., Culhane, M., Hamilton, K., Lewis, N., Mumford, E., Nguyen, T., Parchariyanon, S., Pasick, J., Pavade, G., Pereda, A., Peiris, M., Saito, T., Swenson, S., Van Reeth, K., Webby, R., Wong, F. and Ciacci-Zanella, J., 2014. Review of influenza A virus in swine worldwide: A call for increased surveillance and research. *Zoonoses Public Health* 61, 4–17.

Vluggen, C., Soetaert, K., Duytschaever, L., Denoel, J., Fauville-Dufaux, M., Smeets, F., Bruffaerts, N., Huygen, K., Fretin, D., Rigouts, L., Saegerman, C., Mathys, V., 2016. Genotyping and strain distribution of *Mycobacterium avium* subspecies *hominissuis* isolated from humans and pigs in Belgium, 2011–2013. *Euro Surveill.* 21, 30111.

Voss, A., Loeffen, F., Bakker, J., Klaassen, C. and Wulf, M., 2005. Methicillin-resistant *Staphylococcus aureus* in pig farming. *Emerg. Infect. Dis.* 11, 1965–6.

Wales, A. D., Cook, A. J. and Davies, R. H., 2011. Producing *Salmonella*-free pigs: A review focusing on interventions at weaning. *Vet. Rec.* 168, 267–76.

Wales, A. D. and Davies, R. H., 2017. *Salmonella* vaccination in pigs: A review. *Zoonoses Public Health* 64, 1–13.

Walter, J., Espelage, W., Adlhoch, C., Cuny, C., Schink, S., Jansen, A., Witte, W., Eckmanns, T. and Hermes, J., 2017. Persistence of nasal colonisation with methicillin resistant *Staphylococcus aureus* CC398 among participants of veterinary conferences and occurrence among their household members: A prospective cohort study, Germany 2008–2014. *Vet. Microbiol.* 200, 13–18.

Wasinski, B. and Dutkiewicz, J., 2013. Leptospirosis–current risk factors connected with human activity and the environment. *Ann. Agric. Environ. Med.* 20, 239–44.

Wassenberg, M. W. M., 2010. Costs and effects of MRSA control in Dutch hospitals.

Wassenberg, M. W. M., Bootsma, M. C., Troelstra, A., Kluytmans, J. A. and Bonten, M. J., 2011. Transmissibility of livestock-associated methicillin-resistant *Staphylococcus aureus* (ST398) in Dutch hospitals. *Clin. Microbiol. Infect.* 17, 316–19.

Whitby, M., McLaws, M. L. and Berry, G., 2001. Risk of death from methicillin-resistant *Staphylococcus aureus* bacteraemia: A meta-analysis. *Med. J. Aust.* 175, 264–7.

Wilhelm, B. J., Young, I., Cahill, S., Nakagawa, R., Desmarchelier, P. and Rajic, A., 2016. Rapid systematic review and meta-analysis of the evidence for effectiveness of primary production interventions to control *Salmonella* in beef and pork. *Prev. Vet. Med.*

Willemse, N., Howell, K. J., Weinert, L. A., Heuvelink, A., Pannekoek, Y., Wagenaar, J. A., Smith, H. E., van der Ende, A. and Schultsz, C., 2016. An emerging zoonotic clone in The Netherlands provides clues to virulence and zoonotic potential of *Streptococcus suis*. *Sci. Rep.* 6, 28984.

Winskill, P., Harrison, W. E., French, M. D., Dixon, M. A., Abela-Ridder, B. and Basanez, M. G., 2017. Assessing the impact of intervention strategies against *Taenia solium* cysticercosis using the EPICYST transmission model. *Parasit. Vectors* 10, 73.

Wulf, M. W., Verduin, C. M., van Nes, A., Huijsdens, X. and Voss, A., 2012. Infection and colonization with methicillin resistant *Staphylococcus aureus* ST398 versus other MRSA in an area with a high density of pig farms. *Eur. J. Clin. Microbiol. Infect. Dis.* 31, 61–5.

Wyckoff, A. C., Henke, S. E., Campbell, T. A., Hewitt, D. G. and VerCauteren, K. C., 2009. Feral swine contact with domestic swine: A serologic survey and assessment of potential for disease transmission. *J. Wildl. Dis.* 45, 422–9.

Yang, W. Z., Yu, H. J., Jing, H. Q., Xu, J. G., Chen, Z. H., Zhu, X. P., Wang, H., Liu, X. C., Wang, S. W., Liu, L. G., Zu, R. Q., Luo, L. Z., Xiang, N. J., Liu, H. L., Zhong, W. J., Liu, L., Meng, L., Yuan, H., Gao, Y. J., DU, H. M., Ou, Y. B., Ye, C. Y., Jin, D., Lv, Q., Cui, Z. G., Huang, Y., Zhang, S. Y., An, X. D., Huang, T., Zhou, X. Y., Feng, L., Pang, Q. D., Shu, Y. L. and Wang, Y., 2006. An outbreak of human *Streptococcus suis* serotype 2 infections presenting with toxic shock syndrome in Sichuan, China. *Zhonghua Liu Xing Bing Xue Za Zhi* 27, 185–91.

Young, I., Wilhelm, B. J., Cahill, S., Nakagawa, R., Desmarchelier, P. and Rajic, A., 2016. A rapid systematic review and meta-analysis of the efficacy of slaughter and processing interventions to control non-typhoidal *Salmonella* in beef and pork. *J. Food Prot.* 79, 2196–210.

Yugo, D. M. and Meng, X. J., 2013. Hepatitis E virus: Foodborne, waterborne and zoonotic transmission. *Int. J. Environ. Res. Public Health* 10, 4507–33.

Chapter 2

Salmonella control in pig production

Jan Dahl, Danish Agriculture and Food Council (DAFC), Denmark

1 Introduction

Salmonella is a genus of bacteria comprising more than 2000 different serotypes. *Salmonella* can be divided into several subtypes based on many different typing methods. The reader is referred to microbiological textbooks for further details.

Clinically and epidemiologically *Salmonella* infections can be divided into primary human pathogens, (typhi and paratyphi), primary animal pathogens with some preference for species (*Salmonella* Dublin in cattle, *S.* Choleraesuis, *S.* Pullorum/Gallinarum in poultry) and zoonotic pathogens, although the lines between these groups are not 100% clear. Species-specific types are found in other species and cause infection and disease in humans, although at a relatively low frequency, and the zoonotic types can cause disease in animals.

Salmonella infections of zoonotic origin are one of the most frequent causes of zoonotic infections worldwide. In 2014 the EU recorded 88715 cases of human salmonellosis (European Food Safety Authority, 2015). It is estimated that registered cases only account for somewhere between 5% and 20% of actual cases (Anon., 1994–2015).

http://dx.doi.org/10.19103/AS.2017.0030.02

Pires et al. (2014) reviewed several studies from Denmark, Sweden, regions of the EU, United States, Japan and New Zealand, attributing pork as a source of human salmonellosis. Estimates ranged from less than one per cent of human cases in the United States to 60% of human cases in New Zealand. The differences probably also reflect differences in the methodology, but there are clear indications, that at least in some countries, pork is an important source of human salmonellosis.

The majority of zoonotic *Salmonella* infections in humans are food borne, although water-borne infections and direct contact to live, infected animals also contribute. Human-to-human transmission from people with clinical infection is responsible for a significant proportion of infections within households, although the actual contribution is difficult to quantify, because family members are often exposed to the same food source. Non-heat-treated vegetables can also be a source of infection, but this is generally considered to be caused by either faecal contamination in the field, for example, through the use of contaminated surface water for irrigation, or through cross contamination from infected food of animal origin somewhere in the food chain, including cross contamination in the kitchen at the consumer level.

The major sources of *Salmonella* worldwide are eggs, poultry, pork and beef, but other sources can contribute as well.

This chapter will describe the epidemiology and possible interventions in the pig and pork production chain. *Salmonella* is the 'model' organism, but the framework can be used for other pathogens.

One of the key issues in effective control of *Salmonella* in the pork production chain is that *Salmonella* can be introduced into the chain at all levels of production, and propagate or spread sufficiently to contaminate large quantities of pork ready for consumption. Even at the very last step of production during transport to the slaughterhouse, in lairage, at the slaughter line, during cutting and processing or even in the kitchen, *Salmonella* can propagate, contaminate, be contaminated or cross-contaminate other products. A thorough understanding of the mechanisms at all levels is necessary to make cost-effective interventions. The most cost-effective interventions are specific for a production system and should take into consideration the prevalence in pigs and pork, farming structure, herd sizes and slaughterhouse structure.

2 The production system

To understand the complexity of the issue, a brief introduction to the modern pork production chain is necessary. Backyard production or production in subsistence farming is not specifically covered in this chapter. A key component of these types of production is the use of food scraps as an important part of the feed.

Pigs are fed mixtures of feed ingredients to obtain a diet capable of sustaining effective growth and meat quality. In most cases this feed is a mixture of cereals, feed sources that supply proteins (like soy meal or other types of byproducts from plant oil production) and supplements of vitamins, minerals and amino acids. Many of these ingredients are traded over long distances, across the world.

Modern pig and pork production can be divided into five principal parts: Production and distribution of genes (gilts, boars and semen), production of piglets, production of finishers, slaughter and processing. These parts of the production chain can be geographically separated, or occur at the same locations; each of the parts can be further subdivided, and each level of production can receive pigs or pork from several units from the previous level.

A key issue in understanding the epidemiology is to understand differences between risk factors for introducing *Salmonella* into the system and risk factors for propagating and spreading *Salmonella* within the system.

3 Surveillance and monitoring

No *Salmonella* programme can exist without some sort of monitoring or surveillance. Depending on the situation and the programme in the individual country or production system, two principally different methods serology and microbiology, are used.

Serology is often used in countries where all finisher herds are monitored. In Europe this is the situation in Denmark, Germany and the Netherlands. The United Kingdom stopped a surveillance programme based on serology in 2014.

The advantage of serology is that it is relatively cheap, when large numbers of samples are collected. But experience has shown that some serotypes give a better serological response than others. *Salmonella typhimurium* is more likely to result in a serologically positive pig than most other serotypes in the Danish MIX-ELISA (Kranker and Dahl, 2001). Serology will not produce immediate results. Time to seroconvert can range from seven days to several weeks (Nielsen et al., 1995). It is not possible to differentiate between different serotypes based on serology, although some serological tests are targeting specific serotypes, based on differences between antigen structures of the bacteria. However, these antigens are shared by several *Salmonella* types.

Microbiology is used in a few low-prevalence countries to monitor the situation in primary production. Normally this surveillance does not cover all herds, but only a sample of herds. This is the situation in Sweden. However, monitoring of pork is done using microbiology in countries where an ongoing monitoring of pork is implemented (Anon., 1994–2015).

4 Feed

Feed can introduce *Salmonella* into the system at all levels of primary production. Especially protein sources like soy meal and other byproducts are often contaminated (Anon., 1994–2015). *Salmonella* can even become established and propagate in feed mills, sometimes referred to as a 'house strain'.

This risk can be mitigated by using heat treatment of feed ingredients before they leave the feed mill. Good results have been achieved by ensuring a temperature of 93°C for 90 seconds (European Food Safety Authority, 2008), although shorter periods are often used. This heat treatment is often part of the pelleting process. Good manufacturing practices are essential to avoid cross contamination after heat treatment. Alternatively, acidification, preferably using organic acids, can reduce the level of *Salmonella* in contaminated feed, although with variable results, and not always to levels where the risk is negligible (European Food Safety Authority, 2008).

However, epidemiological studies from Denmark in the 1990s gave conflicting results on the effect of using heat-treated pelleted feed (Dahl, 1997). All whole, compound feed had to be heat treated before leaving the feed mill, and GMP and HACCP programmes had to be in place to ensure a low level of *Salmonella* in the feed. But, surprisingly, studies showed that herds using pelleted feed had a higher *Salmonella* level than herds using feed mixed at the farm (home-mixed feed). This feed is normally not heat treated, and there are high-risk ingredients like imported non-heat-treated byproducts from plant oil production or soy meal in the feed.

At the same time microbiological data showed that serotype distribution in feed and in pigs differed. In feed *S. typhimurium* was rarely found (Anon., 1994–2015). But it was the predominant serotype found in faecal or carcass samples, and many of the serotypes found in feed were rare as a cause of human salmonellosis (Anon., 1994–2015).

This led to the hypothesis that feed could have a dual role in the epidemiology of *Salmonella*. It could introduce *Salmonella* into primary production. But feed could also change the gut flora of the pig, making it more vulnerable to *Salmonella* already present in the herd. A risk assessment by Snary et al. (2016) concluded that in a low-prevalence production system feed is responsible for a large proportion of the *Salmonella* in pigs and pork. But in a high-prevalence production system, there is sufficient *Salmonella* already present, so the introduction of *Salmonella* through feed is of minor importance.

Clinical trials investigating the effect of heat-treated, pelleted feed and non-heat-treated meal feed showed that both the pelleting process and the heat treatment had an effect on *Salmonella* prevalence in pigs. Also the particle size had an effect. Finely ground feed resulted in a higher prevalence than coarsely ground feed (Jørgensen et al., 1999; Kjeldsen and Dahl, 1999; Wilhelm et al., 2012). Microbiological and physico-chemical analyses showed that feed that reduced the *Salmonella* prevalence resulted in more gram-positive bacteria in the gut, a lower pH and a higher level of organic acids, compared to feed that resulted in a higher prevalence (Jørgensen et al., 1999).

Epidemiological studies (Dahl, 1997) also showed that the use of liquid feed resulted in a lower *Salmonella* level than the use of dry feed. A natural fermentation process is occurring in many liquid feed systems, resulting in a low pH and high levels of gram-positive, lactic acid–producing bacteria in the soup. Also contributing to this is that fermented or acidified byproducts like whey are an important part of the feed in these systems. It is generally believed that these factors contribute to the protective effect of liquid feed, although the mechanism is not well documented.

Finally, the use of organic acids in feed has been shown in several clinical trials to reduce the *Salmonella* prevalence, although the results have been variable (Wilhelm et al., 2012).

Although all these epidemiological studies, clinical trials and microbiological studies suggest it is possible to manipulate the gut flora so it is more resistant to *Salmonella* infections, the mechanism is not clear. It is also clear that although these effects have been shown in multiple studies, practical real-life observations have shown that even herds using what seem to be optimal feeding systems can have high *Salmonella* prevalence.

In conclusion, feed can act as a risk factor for introducing *Salmonella* into pig herds at all levels of the production. But feed can also influence the quantitative level in the herd, irrespective of the *Salmonella* level in the feed. And to complicate things, some of the risk-mitigating steps to eliminate *Salmonella* in the feed, like heat treatment, can affect the gut flora negatively, so *Salmonella* already present in the herd can multiply.

5 The environment

The environment, especially wildlife, has traditionally been seen as an important vector of *Salmonella* in the pig herd. *Salmonella* has been isolated from many different species of wildlife, including sea gulls, rodents and other birds and animals that frequently can be found in and around pig herds, especially if biosecurity is lacking.

However, a Danish study (Skov et al., 2008) found that *Salmonella* in wildlife close to pig herds were more likely to be infected from the pig herds than the other way around. This should in no way be seen as an excuse for less optimal biosecurity rules, good rodent control and bird proofing. Of course *Salmonella* can be introduced into the herd by birds and rodents, as can other pathogens, both zoonotic and non-zoonotic. But poor biosecurity will not contribute significantly to the *Salmonella* problem in a production system, where *Salmonella* is introduced by pigs and feed on a regular basis.

6 Replacement animals

Few studies have focused on the role of replacement animals (gilts in sow herds, weaners or growers in finisher herds), although it is a well-established way of introducing pathogens. Dahl (2011) described the effect of introducing gilts from breeder or multiplier herds with different levels of *Salmonella* infection. Negative sow herds had an increased risk of becoming positive, if they received even a relatively low number of positive gilts from the multiplier herd. Dahl concluded that the effect of introduction was at the qualitative level. The herds became positive, but within-herd factors determined the prevalence.

In contrast, sow herds already positive were not affected by the introduction of positive gilts. Stopping the introduction of *Salmonella*-positive gilts into positive herds will not automatically reduce the level in the sow herd, although it is of course important, if the target is eradication. Kranker and Dahl (2001) analysed the association between sow herds and finisher herds and found that especially sow herds that were positive for *S. typhimurium* resulted in positive meat juice samples from finisher herds buying weaners or growers from these herds.

The Scandinavian countries and Ireland are so far the only countries that have implemented *Salmonella* control in the breeding system, although the methodologies are quite different between the countries. Microbiology is used in Sweden, and serology with microbiological follow-up, depending on serological results, is used in Denmark. Ireland uses serology, either based on meat-juice samples from slaughtered pigs or blood samples collected on farm (Anon., 2016b).

Research into factors influencing *Salmonella* in sow herds is limited. Kranker and Dahl (2001) found that herds using home-mixed meal for sows and weaners had a lower level than herds using pelleted feed for sows and weaners. More importantly, they showed that herds sourcing growers from these herds had a higher seroprevalence, if the sow herd was *Salmonella* positive, especially if the herd was infected with *S. typhimurium*. The study could not prove an association at the quantitative level. Sow herds with higher levels of positive pen faecal samples did not pose a greater risk for the finisher herds compared to sow herds with fewer positive samples, although the power of the study was limited.

This is probably due to the fact that the number of pigs introduced into the finisher herd is so high and that the probability that the finisher herd gets infected is very high, even with a low prevalence in the sow herd. Herd factors, like type of feed, hygiene and management in the finisher herd, will then decide the level of prevalence at slaughter.

7 Finisher herds

Finisher phase is the final stage of the primary production part of the chain, and the prevalence of pigs and the serotype distribution amongst positive pigs entering the slaughterhouse is the result of all the preceding parts of the chain.

There are no critical control points in the primary production chain that can ensure that pigs are *Salmonella* negative when entering the slaughterhouse. However, there are possibilities for reduction of *Salmonella* in the finishing period.

Dahl et al. (1997) showed under semi-experimental conditions, in herds positive for *S. typhimurium*, it was possible to wean batches of pigs at 3–4 weeks, transport the weaners directly to clean and disinfected facilities physically separated from other pigs and keep them *Salmonella* negative until slaughter.

Unfortunately practical experience from Denmark has shown that it is difficult to obtain the same success under practical conditions. Epidemiological studies and practical experience have shown all-in all-out management by room and farm to be less effective than the results obtained by Dahl et al. (1997). This might be because it is difficult to obtain the high level of hygiene and biosecurity implemented in the study by Dahl et al. (1997) under practical conditions, or because transfer of *Salmonella* from sow to piglet does occur, but at a sufficiently low rate, so it did not occur in the study.

Use of organic acids in water or feed, or the use of *Salmonella*-reducing feeding principles, like liquid feed or home mixed, coarsely ground feed have been implemented as interventions in high-prevalence herds with varying effects (Wilhelm et al., 2012). Practical experience from Denmark has shown some effect, but also, that the prevalence is not reduced to 'close to zero'.

8 Vaccination as a reduction strategy

Several studies have investigated the effect of vaccination as a reduction strategy in infected pig herds. Wilhelm et al. (2012) did a review on a number of studies on the effect of vaccination and concluded that there are indications of an effect but also that the results were inconsistent. Although some studies have shown promising results, there is still a lack of studies showing an effect that justifies the cost of vaccination, or proves an effect, that reduces the *Salmonella* prevalence sufficiently, except in situations with clinical salmonellosis.

9 The slaughterhouse

Numerous studies have shown high levels of *Salmonella* in lairage in many countries with a moderate to high level in primary production. Some studies have found that the same serotypes found in swab samples from lairage could be found on the carcass afterwards, indicating that pigs from low-prevalence or *Salmonella*-negative herds can become contaminated after transport and before slaughter.

Several studies have evaluated different protocols for cleaning and disinfection and demonstrated that it is difficult to achieve effective removal of *Salmonella* in the lairage environment. Walia et al. (in press) showed that only through high-pressure cleaning including use of detergent, followed by disinfection with an adequate disinfectant and a drying period was one successful in removing *Salmonella* from the lairage pen.

However, none of the studies discussed the effect of an intensive and effective hygiene protocol, given that in most countries, medium to large slaughterhouses will slaughter pigs from many herds during the day, and in most countries a large part of the herds are *Salmonella* positive. Thus lairage will be contaminated again very rapidly after cleaning and disinfection. In most slaughterhouse settings, effective cleaning and disinfection are not possible between batches of pigs. Effective cleaning and disinfection of lairage can only be expected to contribute significantly to *Salmonella* reduction in the slaughterhouse in low-prevalence regions.

Baptista et al. (2010) showed that there is an association between *Salmonella* prevalence in the herd of origin and the probability of a culture positive carcass swab from carcasses from the same herd at slaughter. However, they also found that the carcass prevalence was heavily influenced by pigs from other herds. They demonstrated across several slaughterhouses that the introduction of a relatively low number of *Salmonella*-positive pigs from other herds to the slaughterhouse on the same day increased the carcass prevalence considerably. A likely explanation could be cross contamination at lairage. Also cross contamination during the slaughter process can contribute to this effect.

Sørensen et al. (1999) showed that in certain cases *Salmonella* infections can persist for long periods in the slaughterhouse environment and contaminate carcasses.

After stunning and bleeding, pigs are normally passed through the scalding tank in industrialized slaughter plants. Faecal material and material from the mouth and pharynx of the pig are deposited in the scalding tank. To avoid extensive cross contamination at this step, the temperature of the scalding tank has to reach a sufficiently high temperature. In a review, published by WHO and FAO (Anon., 2016a), experts recommended a scalding/temperature combination of 61°C for 8 minutes, or 70°C for 2–3 minutes.

During the slaughter process there is ample opportunity for contamination from the intestinal channel or cross contamination through slaughterhouse personnel, equipment and meat inspection. Removal of the pluck is a high-risk operation, since fluid from the oral cavity and pharynx can spill onto the carcass. Removal of the intestinal channel is another high-risk operation, especially if the intestinal channel is lacerated, and gut content is spilled onto the carcass. Spillage from the rectum can also be a problem. Different procedures have been applied to control spillage from the rectum. Bunging using plastic bags, inserting plugs of different types or applying a gutter, where the rectum is kept away from the carcass, are frequently used in order to reduce spillage.

The only true *Salmonella*-reducing control point is the singeing/flaming process. Performed correctly, this will effectively reduce the *Salmonella* level on the outer surface of the carcass. But since the high-risk operations of eviscerations occur after singeing, this process alone cannot be considered a true critical control point, since it will not control the contamination that occurs during evisceration.

Decontamination steps after slaughter are implemented in some slaughterhouses. Hot water, steam, organic acids, chlorinated water or other chemicals (Hamilton et al., 2010) have been used with varying results, but generally they seem to have an effect. During the cooling process *Salmonella* levels will also be reduced, and this reduction can be affected by the cooling methodology.

10 *Salmonella* reduction in Danish pig and pork production: a case story

The Danish *Salmonella* programme has been in operation since the mid-1990s. Many adjustments have been made over the years. It is not the author's intention to give a detailed description of the developments over the years. Details are thoroughly described in the annual report of *Zoonosis in Denmark* 1994–2015 (Anon., 1994–2015). Here, only major changes and experiences over more than 20 years are discussed. In 1993 an outbreak of *Salmonella* caused more than 1000 registered clinical cases in humans in Denmark, an incidence of more than 20 per 100 000 Danes (Wegener and Baggesen, 1996). Pork had already been identified as a source of *Salmonella* in previous years, but this outbreak accentuated the need for action.

Data were limited, and not much was known about risk factors and possible interventions. Only Sweden, and to some extent Norway, had surveillance and intervention in pig herds. Sweden had initiated a programme because of an outbreak of *Salmonella* in humans in 1955 (Lundbech et al., 1955). In the Swedish programme, *Salmonella* was eradicated from the herds, when it was found. In most cases, herds were depopulated, cleaned, disinfected and restocked, followed by an intensive test programme.

In a Danish context, this was seen as too rigorous and would be unacceptably costly. The Danish programme was based on a reduction strategy, where it was assumed that at every level of the production, *Salmonella* prevalence should be reduced, leading to a reduced level of introduction into the next phase of the production chain.

A comprehensive programme was initiated in the feed mills, where the intention was that all compound feed delivered to pig herds should be free from *Salmonella*. The most important step is heat treatment of the feed to 93°C during the production process and prevention of cross contamination after heat treatment. Basically, the programme was based on a similar programme for *Salmonella* in feed for poultry. The previous ten years, Denmark had initiated a *Salmonella* programme for poultry and successfully reduced or almost eradicated *Salmonella* in broilers (Anon., 1994–2015).

In Denmark, approximately half the herds used feed delivered as compound feed, and the other half used a home-mixed feed, mainly based on home-grown barley or wheat, and purchased protein sources like soy meal. These ingredients are not necessarily heat treated and are often found to be contaminated (Anon., 1994–2015).

In primary production, a surveillance programme was initiated in breeding and multiplying herds in 1994. Each month 20 blood samples were collected from pigs between 4 and 7 months of age. Later statistical evaluations showed that ten samples were sufficient to establish a reliable measure of the *Salmonella* level. Blood samples were analysed using the Danish *Salmonella* Mix-ELISA (Nielsen et al., 1995). For each herd an index was calculated, based on samples from the last three months. Details of this can be found in annual report of *Zoonosis in Denmark* (Anon., 1994–2015).

Initially breeding and multiplying herds with an index above a certain level would be banned for sale of gilts. Later on this was modified several times, but the general principle was always that herds with a high *Salmonella* index were a greater risk for sow herds buying gilts than herds with a low level or *Salmonella*-negative herds.

Sow herds were not regularly monitored, partly because it is costly to monitor many thousands of herds, when it cannot be done at a slaughterhouse (sows are delivered to slaughterhouses, but the number of sows is much lower, and in Denmark some sows are exported and slaughtered in Germany, making a slaughterhouse-based sampling scheme ineffective).

Finisher herds were monitored using meat juice, collected at the slaughterhouse, and analysed using the Danish *Salmonella* Mix-ELISA (Nielsen et al., 1995).

Initially the result was a reduction in positive meat juice samples from 1995 to 1998. However, from 2000 it became evident that despite the comprehensive programme, and introduction of economic incentives in both breeding/multiplying herds and in finisher herds, more of the finisher herds showed signs of *Salmonella* infection in the monitoring system. Over the years, more herds would have one or more positive meat juice samples, indicating an ongoing spread of *Salmonella* between herds.

To investigate this effect, Dahl (2013) modelled the development of serological prevalence in Danish pig herds from 1998 to 2012 and found that the number of herds positive for *Salmonella* had increased from 30% in 1998 to more than 60% in 2012. But the prevalence in positive herds had remained stable.

Furthermore, a study showed that the probability of introduction of *Salmonella* from breeding/multiplying herds into sow herds was not dependent on the quantitative level of *Salmonella* in the breeding/multiplying herd. Introducing gilts from positive, but low-prevalence breeding/multiplying herds was just as risky as introducing gilts from high-prevalence herds (Dahl, 2011). Only introduction of gilts from negative breeding/multiplying herds into negative sow herds reduced the risk. Similar results were found when analysing the association between sow herds and finisher herds. It is not the quantitative level in the sow herd that determines the risk for the finisher herd. So the established paradigm that effective *Salmonella* control in primary production can be achieved by reducing the quantitative level in all parts of the primary production was proven wrong.

This has been investigated as described on Danish data. But, it is supported by observations from other countries, who have established a serological *Salmonella* surveillance programme. There is, to the author's best knowledge, currently no country with a medium to high level of *Salmonella* in the pig production that has achieved a reduction in primary production using this approach.

The only countries that have achieved and maintained a successful reduction and control of *Salmonella* in primary production are some of the other Scandinavian countries where herd-level eradication programmes are enforced, but at a high cost.

The key lesson from the Danish experience is that it is imperative to avoid introduction of *Salmonella* from positive herds to negative herds. Otherwise more and more herds will become positive.

The Danish *Salmonella* programme was changed after these results (Anon., 1994–2015). All Danish herds have an official *Salmonella* status. This includes a declaration of the serotypes cultured in samples from the herd.

The other key intervention point has been improved slaughterhouse hygiene. When the programme was initiated in the early 1990s, swab samples were taken from a large variety of cuts at all slaughterhouses and were analysed bacteriologically for *Salmonella*. But this made standardization of sampling methodology difficult, and it was difficult to compare results between slaughterhouses. In 2000 the slaughterhouse sampling method

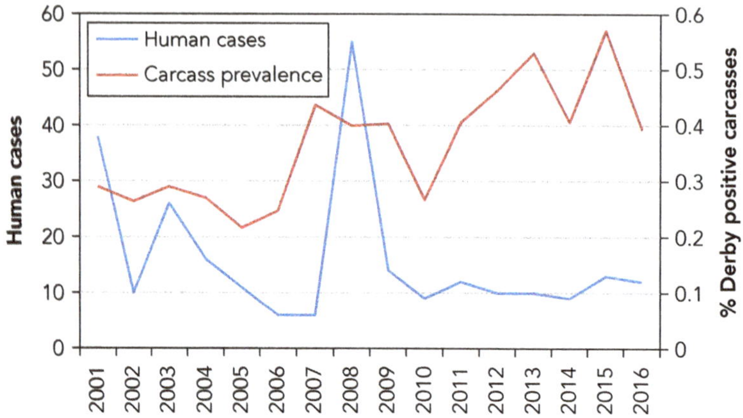

Figure 1 Human cases of *Salmonella* Derby from 2001 to 2016 (Anon., 2017) and proportion of Derby positive pig carcasses from large Danish slaughterhouses (extracted January 2017, Jan Dahl).

was revised and replaced by a standardized carcass swab sampling (Sørensen et al., 2001). Results of the sampling are collected at the industry organizations and sent to the Danish Veterinary and Food Administration.

The results are used by the slaughterhouses as part of the surveillance of the hygiene, in combination with results from the ongoing surveillance of quantitative levels of coliforms and visual surveillance of the hygiene. Over the years interpretation of the results has been refined, and guidelines have been established for when the slaughterhouse has an unacceptably high level (Sørensen and Møgelmose, 2005).

The target in the current plan is to achieve a carcass prevalence based on the swab samples at or below 1% positive carcasses. Despite the increase in the number of positive herds and the increase in meat-juice prevalence, a strong focus on slaughterhouse hygiene has resulted in a reduction in positive swab samples from carcasses from 1.5% to 1% (Anon., 1994–2015). In the same period, serotype distribution has changed and *Salmonella* Derby levels increased from 18% in the positive samples in 2001 to 40% in 2015 (based on data collected from slaughterhouses, that are members of the Danish Agriculture and Food Council). Fortunately, *Salmonella* Derby is rarely the cause of human salmonellosis. In 2015 there were 13 registered cases of human salmonellosis attributed to *S.* Derby (Fig. 1). The peak in human cases in 2008 was part of a larger outbreak, where the source remained unknown.

The Danish National Food Institute calculates each year the number of registered human cases of salmonellosis attributed to different sources of *Salmonella*. In 1993, more than 1000 cases of human salmonellosis were attributed to Danish pork. In 2015 this number reduced to 35.

11 Conclusion

In conclusion, the Danish *Salmonella* control programme has been very successful in reducing the number of human cases from Danish pork. However, the reduction has been achieved through improved hygiene in slaughterhouses. The *Salmonella* level in primary production has increased through continued spread in the production pyramid.

12 Where to look for further information

A comprehensive review of interventions for the control of *Salmonella* in pigs and pork was presented by a group of experts for the Food and Agriculture Organization and the World Health Organization and published in 2016 (Anon., 2016a).

13 References

Anon. (1994–2015). Annual report of *Zoonosis in Denmark* 1994–2015. http://www.food.dtu.dk/english/Publications/Disease-causing-microorganisms/Zoonosis-annual-reports.

Anon. (2016a). Interventions for the control of non-typhoidal *Salmonella* spp. in beef and pork. Meeting report and systematic review. FAO and WHO. http://apps.who.int/iris/bitstream/10665/249529/1/9789241565240-eng.pdf?ua=1.

Anon. (2016b). National pig *Salmonella* control programme. https://www.agriculture.gov.ie/farmingsectors/pigs/pigsalmonellacontrolprogramme/.

Anon. (2017). Data extracted from the public database from SSI. http://www.ssi.dk/Smitteberedskab/Sygdomsovervaagning/Sygdomsdata.aspx?sygdomskode=SALM&xaxis=Aar&show=Graph&datatype=Laboratory&extendedfilters=False#HeaderText.

Baptista, F. M., Dahl, J. and Nielsen, L. R. (2010). Factors influencing *Salmonella* carcass prevalence in Danish pig abattoirs. *Prev. Vet. Med.* 95(3–4): 231–8. doi:10.1016/j.prevetmed.2010.04.007. Epub 26 May 2010.

Dahl, J. (1997). Cross-sectional epidemiological analysis of the relations between different herd factors and *salmonella*-seropositivity. In: *International Symposia on Veterinary Epidemiology and Economics Proceedings, ISVEE 8: Proceedings of the 8th Symposium of the International Society for Veterinary Epidemiology and Economics*, Paris, France (published as Epidemiologie et Sante Animale, Issues 31–32), Veterinary public health session, p. 04. 23 July 1997.

Dahl, J. (2011). Association between serological *salmonella* monitoring in breeding herds and meat-juice prevalence in sow herds with production of finishers. In: *International Conference on the Epidemiology and Control of Biological, Chemical and Physical Hazards in Pigs and Pork*. http://lib.dr.iastate.edu/.

Dahl, J. (2013). Development in with-in herd prevalence, between herd prevalence and carcass prevalence in Danish pigs and pork compared to number of attributable human cases from 1995 to 2012. In: *Proceedings from International Conference on the Epidemiology and Control of Biological, Chemical and Physical Hazards in Pigs and Pork*. http://lib.dr.iastate.edu/.

Dahl, J., Wingstrand, A., Nielsen, B. and Baggesen, D. L. (1997). Elimination of *Salmonella typhimurium* infection by the strategic movement of pigs. *Vet. Rec.* 140(26): 679–81

European Food Safety Authority (2008). Microbiological risk assessment in feedingstuffs for food-producing animals. *The EFSA Journal* 720: 1–84

European Food Safety Authority (2015). The European Union summary report on trends and sources of zoonoses, zoonotic agents and food-borne outbreaks in 2014. *EFSA Journal: EFSA Journal* 13(12): 4329 (191pp.).

Hamilton, D., Holds, G., Lorimer, M., Kiermeier, A., Kidd, C., Slade, J. and Pointon A. (2010). Slaughterfloor decontamination of pork carcases with hot water or acidified sodium chlorite - a comparison in two Australian abattoirs. *Zoonoses Public Health* 57(Suppl. 1): 16–22.

Jørgensen, L., Dahl, J. and Wingstrand, A. (1999). The effect of feeding pellets, meal and heat treatment on the *Salmonella*-prevalence of finishing pigs. In: *International Conference on the Epidemiology and Control of Biological, Chemical and Physical Hazards in Pigs and Pork*. http://lib.dr.iastate.edu/

Kjeldsen, N. J. and Dahl, J. (1999). The effect of feeding non-heat treated, non-pelleted feed compared to feeding pelleted, heat-treated feed on the *Salmonella*-prevalence of finishing pigs. In: *International Conference on the Epidemiology and Control of Biological, Chemical and Physical Hazards in Pigs and Pork*. http://lib.dr.iastate.edu/.

Kranker, S. and Dahl, J. (2001). Bacteriological and serological examination and risk factor analysis of *Salmonella* occurrence in sow herds, including risk factors for high *Salmonella* seroprevalence in receiver finishing herds. In: *International Conference on the Epidemiology and Control of Biological, Chemical and Physical Hazards in Pigs and Pork*. http://lib.dr.iastate.edu/do/search/?q=Kranker&start=0&context=1759512.

Lundbeck, H., Plazikowski, U. and Silverstolpe, L. (1955). The Swedish *Salmonella* outbreak of 1953. *J. Appl. Bacteriol.*, 18: 535–48.

Nielsen, B., Baggesen, D., Bager, F., Haugegaard, J. and Lind, P. (1995). The serological response to *Salmonella* serovars typhimurium and infantis in experimentally infected pigs. The time course followed with an indirect anti-LPS ELISA and bacteriological examinations. *Vet. Microbiol.* 47, 205-18.

Pires, S. M., Vieira, A. R., Hald, T. and Cole, D. (2014). Source attribution of human salmonellosis: An overview of methods and estimates. *Foodborne Pathog. Dis.*, 11(9), 667–76.

Skov, M. N., Madsen, J. J., Rahbek, C., Lodal, J., Jespersen, J. B., Jørgensen, J. C., Dietz, H. H., Chriél, M. and Baggesen, D. L. (2008). Transmission of *Salmonella* between wildlife and meat-production animals in Denmark. *J. Appl. Microbiol.* 105(5): 1558–68.

Snary, E. L., Swart, A. N., Simons, R. R., Domingues, A. R., Vigre, H., Evers, E. G., Hald, T. and Hill, A. A. (2016). A quantitative microbiological risk assessment for *Salmonella* in pigs for the European Union. *Risk Anal.* 36(3): 437–49.

Sørensen, L. L. and Møgelmose, V. (2005). The intensified control program for *salmonella* in Danish pork. In: *SafePork 2005: 6th International Symposium on the Epidemiology and Control of Foodborne Pathogens in Pork*. http://lib.dr.iastate.edu/.

Sørensen, L. L., Sørensen, R., Klint, K. and Nielsen, B. (1999). Persistent environmental strains of *Salmonella* infantis at two Danish slaughterhouses, two case-stories. In: *International Conference on the Epidemiology and Control of Biological, Chemical and Physical Hazards in Pigs and Pork*. http://lib.dr.iastate.edu/.

Sørensen, L. L., Wachmann, H., Dahl, J. and Nielsen, B. (2001) The new Danish *Salmonella* surveillance on fresh pig carcasses based on pooled swab samples including compatibility with levels of the former system. In: *International Conference on the Epidemiology and Control of Biological, Chemical and Physical Hazards in Pigs and Pork*. http://lib.dr.iastate.edu/.

Walia, K. (In press). The efficacy of different cleaning and disinfection procedures to reduce *Salmonella* and *Enterobacteriaceae* in the lairage environment of a pig abattoir. *Int. J. Food Microbiol.* (accepted).

Wegener, H. C. and Baggesen, D. L. (1996). Investigation of an outbreak of human salmonellosis caused by *Salmonella* enterica ssp. enterica serovar Infantis by use of pulsed field gel electrophoresis. *Int J. Food Microbiol.* 32(1–2): 125–31.

Wilhelm, B., Rajić, A., Parker, J., Waddell, L., Sanchez, J., Fazil, F., Wilkins, W. and McEwen, S. A. (2012). Assessment of the efficacy and quality of evidence for five on-farm interventions for *Salmonella* reduction in grow-finish swine: A systematic review and meta-analysis. *Prev. Vet. Med.* 107(1–2): 1–20.

Chapter 3

Dealing with the challenge of antibiotic resistance in pig production

Paul D. Ebner and Yingying Hong, Purdue University, USA

1 Introduction

The use of antibiotics is among the most researched, yet most controversial practices in US food animal production. Several questions surround the practice, such as:

- Why are antibiotics used with otherwise healthy animals?
- Do bacteria become resistant to antibiotics? If so, do these bacteria impact human health?
- What impact does antibiotic resistance have on animal health?
- What happens to unabsorbed antibiotics? What is the environmental impact?
- What are the alternatives?

Confounding the matter, there is little scientific consensus on the answers to some of these questions. When that lack of consensus is coupled with different approaches to mitigating risk, countries with otherwise similar livestock production practices and industry structures can have drastically different antibiotic use policies.

Until recently, science was somewhat limited in its ability to accurately address some of the key questions surrounding the antibiotic use in food animals, especially those dealing with bacteria that become resistant on the farm and whether these bacteria impact human health (and if so, to what extent). The advances in DNA sequencing technologies and

http://dx.doi.org/10.19103/AS.2017.0030.03

the use of these technologies in quickly and comprehensively characterizing the related microbial communities, however, offer a substantial amount of hope.

Here, we will cover the practice of antibiotic use in food animal production and the various questions this practice poses. We will start with a brief history of antibiotic use in livestock production prior to diving into less clear waters. We will also look at the current state of research focused on developing non-antibiotic means of controlling bacterial infections in livestock. Finally, we will examine research in phage therapy as a case study of a (re)emerging technology that could be utilized in biocontrol of bacterial pathogens in agriculture.

2 Historical background

Since their widespread introduction to human medicine, antibiotics have been integral in treating diverse bacterial infections. Sulphonamides (technically synthetic antibacterials) were introduced in the 1930s and primarily used initially to treat *Streptococcus* infections. In the 1940s, the use of penicillin became popular in human medicine and the use of antibiotics to control bacterial infections in veterinary medicine soon followed.

In 1946, however, food animal production began its history of using antibiotics not only to treat infections, but as a prophylaxis and, in some cases, to promote growth. Moore et al. (1946) first showed how inclusion of streptomycin in chicken feed led to improved growth in the birds. Soon after, another group of animal nutritionists utilizing fermentation extracts as a source of B12 in animal feeds reported that improved growth was not from B12, but from the crude chlortetracycline, also found in the fermentation extract (Stokstad et al., 1949). Thus started in motion the practice of administering antibiotics to otherwise healthy animals for the explicit purpose of improving efficiency [see Gustafson and Bowen (1997) for a historical overview]. Today, antibiotics are used in food animal production in three primary ways: 1) to improve efficiency, 2) to prevent infections and 3) to treat established infections.

Among those three types of uses, the use of antibiotics for growth promotion is the most controversial. While the practice is often disparaged, it should be acknowledged that under some conditions, antibiotics used in this manner can provide an effective management tool and can reduce costs. In terms of actual performance gains, in 2002, Cromwell, through a semi-meta-analysis, estimated that including antibiotics in the diet of weaned pigs improves growth and feed efficiency by roughly 16% and 7%, respectively (Cromwell, 2002).

There currently is, however, substantial discussion as to whether these types of gains are occurring currently. Most studies measuring antibiotic-associated improvements in efficiencies are quite dated. Livestock and poultry production systems have changed substantially with almost all (by volume) swine and poultry production in the United States taking place indoors. Facilities are, in general, cleaner and producers have more vaccines and other health aids at their disposal, so microbial exposures of animals are quite different in the present than in the past decades.

3 Unintended consequences of antibiotic use

Most concerns regarding the use of antibiotics in livestock production derive from one main issue: antibiotic resistance. Antibiotic resistance in this context usually refers to a bacterium acquiring in some manner the means to withstand the lethal effects of an antibiotic to which the bacterium was previously susceptible. It is an unintended consequence of almost all antibiotic therapies. In any given bacterial population, there are likely small subpopulations of bacteria which are resistant to a given antibiotic. Exposure of that bacterial community to the antibiotic can remove susceptible bacteria allowing resistant bacteria to become greater percentages of the total population. The matter is complicated, however, by the ability of some bacteria to then transfer the genetic material conferring antibiotic resistance not only vertically (to their offspring), but horizontally (to unrelated bacteria present in the environment) as well. Thus, one bacterium can, in many cases, share the genetic information necessary for antibiotic resistance with unrelated, often highly dissimilar, neighbouring bacteria [see Blair et al. (2015) for a detailed review].

The phenomenon of antibiotic resistance was witnessed almost immediately following the widespread use of antibiotics in human medicine. It was not long after penicillin was introduced to human medicine that clinicians saw significant increases in the number of infections associated with penicillin-resistant bacteria (Davies and Davies, 2010). Thus started what has metaphorically been an arms race of developing new antibiotics to treat previously susceptible bacterial infections, followed shortly thereafter by widespread resistance to the new drug. In some cases, it is clear that antibiotic development is on the losing end of this race as the treatment of some diseases may be rapidly approaching what is essentially a post-antibiotic era (e.g. *Neisseria ghonorrhoeae*; Unemo and Shafer, 2014).

Antibiotic resistance often develops regardless of how the drugs are used, whether it is for growth promotion or treatment of an infection. As such, similar scenarios of antibiotic resistance have played out in livestock production. Review of any meta-analysis will show significant increases in the isolation of bacteria resistant to different drugs over the past six decades. While this increase in antibiotic resistance as a result of antibiotic use in livestock production is hardly debatable, the bigger question remains of whether bacteria that develop resistance on the farm impact human health and, if so, to what extent. Considerable effort has gone into answering these questions, with little consensus thus far. It is clear that bacteria that develop antibiotic resistance on the farm make their way to human bacterial populations through contaminated food, the environment or other vehicles. The roles these bacteria play in human diseases are not clear. In some cases, antibiotic-resistant infections can be traced directly to farms. In most cases, the complexity of microbial ecology makes such sleuthing and the identification of a clear culprit impossible.

Arguably, the use of antibiotics (judiciously or less judiciously) in human medicine likely has a much greater direct impact on antibiotic resistance as it affects public health. The human medicine factor, however, should not absolve concerns about potential similar effects of veterinary use of antibiotics. At this point, quantifying the impact of veterinary use antibiotics on antibiotic resistance as it relates to human health may be impossible. According to Frank Aarestrup, a professor at the National Food Institute in Denmark and a leading antibiotic resistance scientist, 'the actual importance of the animal reservoir of antimicrobial resistance for human health is an area of great controversy and unfortunately, no precise estimates exist. Estimates have ranged from almost zero to a major contribution to the human health burden' (Aarestrup, 2015). All agree, however, that the issue of antibiotic resistance, in general, is critical with serious consequences. As such, reductions in antibiotic use that do not compromise animal health and well-being should be encouraged (as many livestock organizations currently do).

4 Changes in antibiotic use and availability

4.1 FDA policies

In 2014, the US Food and Drug Administration (USFDA) introduced new guidelines that represent what is arguably the largest shift in US antibiotic use policy as it pertains to livestock production since the practice was introduced (USFDA, 2015a). In short, the FDA's new guidelines called for US drug manufacturers to phase out the marketing of certain antibiotics for performance in livestock. The antibiotics targeted are from a list of drugs the FDA previously determined as 'medically important'. Almost all major classes of antibiotics (e.g. tetracyclines, ß-lactams, aminoglycosides, etc.) are included in this list. To date, all drug manufacturers in the United States have indicated their compliance with this guideline and, over a three-year period, have removed performance indications for these drugs. It is important to note, however, that many of the antibiotics no longer approved for performance still have prophylactic or treatment indications and can still be used thusly. Additionally, ionophores are classified as antibiotics by the FDA, but are not included in the list of medically important drugs. Ionophores are used extensively in poultry and beef production and by volume may comprise approximately 30% of the total amount of antibiotics used in livestock production (USFDA, 2015b). Their use in this manner will presumably remain possible. It is also worth noting that many of the negative consequences associated with antibiotic use, namely the development of resistance, are not seen with ionophore use on comparative scales.

4.2 Drug discovery

The development of new antibiotics is a very expensive and time-consuming endeavour. Bringing a new antibiotic to market can cost an estimated 2.8 billion USD (DiMasi et al., 2016). Coupled with the fact that other types of drugs may generate higher, more continuous revenue, there is currently much less interest and/or investment in traditional antibiotic development than in past decades, and only a few global companies with R&D efforts geared towards new antibiotic discoveries [see Luepke et al. (2016) for a review]. Over the past three decades, only one truly new antibiotic has been discovered with all other new drugs essentially being derivatives of compounds already in use (Ling et al., 2015).

4.3 Antibiotic use in livestock production outside of the United States

Most European Union (EU) countries, and eventually the EU itself, made similar reductions in the types of antibiotic use approved for livestock production in the 2000s. Scandinavian countries, namely Denmark and Sweden, were the first to remove the use of what is termed in the EU as 'antibiotic growth promoters' in livestock production. In effect, the change in policy discontinued most in-feed antibiotic use, which is more comprehensive than the new FDA policy described earlier. In 2006, the EU followed suit prohibiting the inclusion of antibiotics in animal feeds.

In each case, these policy changes were made in an attempt to improve public health and reduce the threat of antibiotic-resistant bacteria. The impact these prohibitions have had specifically on public health has been, and is currently, the subject of much debate. Several EU countries collect and publish yearly reports on the antibiotic use in veterinary and human medicine and results of antibiotic resistance surveillance programmes, with Denmark recently celebrating their 20th year of doing so (DANMAP, 2016). What is clear is that the EU policies have led to an overall decrease in the amount of antibiotics used in livestock production. It is also clear that while the prohibitions may have initially resulted in higher levels of some bacterial infections in livestock (namely scouring or diarrhoea in pigs), these issues have seemingly been largely overcome. In 2015, in Denmark, for example, pig production has increased since 1995, when the country saw a reduction in the use of antibiotics (DANMAP, 2016). The overall structure of Danish pork production, however, has changed significantly over that time period with decreases in the number of farms, increases in the size of remaining or new farms and increases in the weaned pig exports. At this time, it is unclear as to how much increases in pork production numbers can be attributed to the changes in the structure of Danish pork production versus the implementation of management practices to overcome any impact of antibiotic withdrawal.

5 Antibiotic alternatives

Taken together, the increasing threat of antibiotic resistance coupled with large-scale changes in antibiotic use policy points to a clear need to identify compounds or therapies that may be effective alternatives to antibiotics in controlling bacterial infections. The likelihood of finding wholesale substitutions for antibiotics, however, may be exceedingly low. Antibiotics are exceptionally effective drugs with, in most cases, minimal toxicity. Additionally, the drugs have traditionally had multiple wide-ranging uses from the treatment of specific infections to improving feed efficiency as noted earlier. Searching for compounds that accomplish all these objectives affordably and with minimal side effects may be, at this point, a fool's errand. Nevertheless, it is quite reasonable to predict that some combination of available or emerging alternatives together could mirror the benefits of antibiotic use without the concomitant risk to public health. What follows are short summaries of some of the most researched antibiotic alternatives focusing on their modes of action and potential uses and benefits.

5.1 Probiotics, prebiotics and synbiotics

Perhaps the most often cited antibiotic alternatives are probiotics. Probiotics are usually defined as live microorganisms, either bacteria or yeast, that when taken in adequate amount can provide the animal host with beneficial effects, such as resistance to enteric pathogen infections, improvement of overall gut health or growth promotion. While the exact mechanisms of probiotics are not completely understood (probiotics are also not monolithic), in general, probiotics elicit beneficial effects through 1) competitive exclusion of pathogens by competing for nutrients and colonization sites, 2) production of antimicrobial compounds, either from probiotics or from probiotic-stimulated host cells, 3) host immunomodulation, including the stimulation of the gut immune system and reduction/promotion of anti-inflammation and 4) enhancement of nutrient digestion and adsorption. A handful of studies have shown the potential of using probiotics as feed additives for disease prevention, growth promotion and food safety. For example, beef cattle fed a *Lactobacillus acidophilus*-based probiotic had reduced prevalence of *Escherichia coli* O157:H7 in faeces and on hides (Brashears et al., 2003; Younts-Dahl et al., 2004; Stephens et al., 2007). This probiotic may also increase growth efficiency and is therefore widely used in cattle feedlots in both the United States and Canada. Similarly, feeding lactic acid bacteria complex to early weaned piglets significantly improved pig growth performance and digestibility of crude protein and crude fibre (Giang et al., 2010). A separate study also showed that feeding weaned pigs with *Lactobacillus rhamnosus* prior to a challenge with *E. coli* K88 enhanced host production of secretory IgA, attenuated *E. coli* K88-induced proinflammatory cytokine IL6 elevation and reduced diarrhoea (Zhang et al., 2010b). In poultry, various probiotics have shown efficacy in reducing colonization and shedding of both *Salmonella* and *Campylobacter* (Lutful Kabir, 2009).

Markets for feed-based probiotics have grown rapidly in recent years. Global probiotic sales in 2010 were estimated at $180 million (Cheng et al., 2014), but are predicted to reach $4.71 billion by 2021 (Markets and Markets, 2016). Despite the wide use of probiotics in food animal production, probiotics are not without challenges; inconsistency is likely foremost. A probiotic that shows disease prevention or growth promotion effects in one animal model may fail to elicit similar effects in another study with animals of different age or species, at different growth stages, with different health statuses or under different husbandry practices. Lower efficacy could be partially due to poor adaption to or colonization of the complex host intestinal microbial environment. Currently, there is a significant focus on utilizing 16s rRNA sequencing and various '-omics' techniques (e.g. metagenomics and metatranscriptomics) to better define what constitutes a healthy gut microbiome and how probiotics and indigenous gut microorganisms interact to improve or maintain a health microbiome (Rebollar et al., 2016; Valeriano et al., 2016). Such studies could improve probiotic selection and optimization.

Variation in probiotic efficacy may also be the result from variation in dosing or the ability of adequate amounts of the probiotic to actually reach areas in the animal where they would be of most benefit. Adverse gut environments (i.e. stomach acid), feed processing, transportation and storage can all negatively impact probiotic activity and more process control studies could improve probiotic survivability under various conditions [see Amalaradjou and Bhunia (2012) for more details].

Prebiotics are compounds that generally cannot be digested by host; instead, they promote animal health through selective metabolism in the gut microbial environment (Gibson et al., 2004). A wide range of prebiotics have been studied for their capacity to select for or promote the growth of beneficial gut microorganisms. Several groups have shown that feeding xylooligosaccharides, arabinoxylooligosaccharides and galactooligosaccharides to chickens can increase *Bifidobacterium* (Courtin et al., 2008; Jung et al., 2008), which is a commensal bacterium known to act antagonistically against pathogens. Conversely, mannan-oligosaccharides, such as β-galactomannan, can effectively bind to type I fimbriae of *Salmonella* and prevent *Salmonella* colonization of the gut (Badia et al., 2013). Similar to probiotics, there are some inconsistencies in the efficacy observed with prebiotics, which may be partially due to poor persistence of prebiotic selected bacteria in gut microbial environment. Again, the use of aforementioned techniques could enhance our understanding of the prebiotic effects on gut microbiome as well.

It is important to note that probiotics and prebiotics could be used together for synergistic effects. Such combinations, also called synbiotics, have been studied in swine (Modesto et al., 2009; Gaggia et al., 2010), poultry (Dibaji et al., 2014; Sarangi et al., 2016) and cattle (Yasuda et al., 2007; Hasunuma et al., 2011) with mixed results. The interactions between probiotics and prebiotics before and after animal administration are still largely understudied, which limits the ability to select for synergistic effects or to at least confirm the absence of antagonistic effects during development. As a result, the use of synbiotics in agriculture is still limited.

5.2 Plant extracts

It has long been known that some plant extracts have antimicrobial, anti-inflammatory and antioxidative properties. For example, acillin in garlic extract has a broad-spectrum antimicrobial activity against several Gram-positive and Gram-negative pathogenic bacteria *in vitro*. These compounds, also referred to as phytochemicals, are mainly secondary metabolites belonging to the groups of terpenoids/essential oils, phenolics, alkaloids and lectins/polypeptides (Windisch et al., 2008), and have been used in disease prevention and treatment in traditional and nontraditional human medicine.

The use of plant extracts in agricultural animals, on the other hand, was much less studied until recently as scientists are actively searching for antibiotic alternatives for agriculture use. To date, a wide variety of plant extracts have been evaluated as feed additives for their antimicrobial and growth promotion potentials in livestock and poultry. In general, studies in swine and poultry thus far have not demonstrated consistent antimicrobial effects of plant extracts against pathogen colonization or shedding *in vivo* (Mitsch, 2004; Robyn et al., 2013). The failure to deliver an *in vivo* antimicrobial effect could be partly due to the fact that the concentrations of active antimicrobial phytochemicals in feed did not reach minimum inhibitory concentrations against pathogens (Windisch et al., 2008). Depending on the phytochemicals of interest, increasing plant extract concentrations in feed to a certain extent can alter feed palatability and nutrition profile. In addition, the complex intestinal environment may also interfere with some phytochemical activities. In a study that failed to demonstrate the efficacy of allicin, a garlic phytochemical, to control *Campylobacter* in chickens, the authors explained that intestinal mucus may interfere with the antimicrobial efficacy of allicin *in vivo* (Robyn et al., 2013), as mucus could prevent *Campylobacter* from directly interacting with allicin and allicin could also be inactivated by cysteine in cysteine-rich regions of mucus.

Similarly, it is inconclusive as to whether plant extracts can promote animal growth performance. In studies where growth promotion via plant extracts was observed, such effects may be attributed to appetite stimulation, modification of intestinal morphology or selection of beneficial gut microorganisms. Li et al. (2012) showed that compared to pigs fed a control diet, weanling pigs fed a diet containing an essential oil blend (18% thymol and cinnamaldehyde) at 100 ppm had higher weight gains and improved faecal scores. Growth improvements were coupled with significantly higher dry matter and protein digestibility, plasma total oxidative capacity, villus height/crypt depth in the jejunum, improved *Lactobacilli/E. coli* ratios in the colon, significantly lower intestinal *E. coli* and lower total aerobic bacterial concentrations in the rectum.

One unique advantage of using plant extracts compared to other antibiotic alternatives is the potential antioxidative benefit to meat quality. Several studies have demonstrated improved oxidative stability of meat products when animals were fed with plant extracts containing phenolic phytochemicals (Hashemi and Davoodi, 2011).

As most studies use mixes of bioactive compounds, it is sometimes difficult to determine the mechanism of action, toxicity and stability of individual phytochemicals. This is further confounded by the fact that the phytochemical composition of plant extract can change greatly due to variations in plants (different plant parts, growing conditions, etc.), extraction methods and storage conditions. *In vivo* studies with purified phytochemicals of high bioactivity would be of benefit in more rationally utilizing phytochemicals in animal production.

5.3 Bacteriocins

Bacteriocins are a group of ribosomally synthesized antimicrobial peptides naturally produced by certain bacteria in order to gain a competitive advantage over other cohabiting bacteria. A wide range of bacteria, both Gram-positive and Gram-negative (e.g. *Lactobacillus*, *Lactococcus*, *Pediococcus*, *Carnobacterium*, *Enterococcus*, *Escherichia*, *Bacillus*, *Paenibacillus*, *Staphylococcus*, *Pseudomonas* and *Clostridium*) can produce bacteriocins (Svetoch and Stern, 2010). As a group, bacteriocins largely function by inhibiting bacterial cell wall synthesis (Piper et al., 2009), forming pores on cell walls and interfering with DNA (Parks, 2007), RNA (Metlitskaya et al., 2006) and protein metabolism (Bagley et al., 2005). It is notable that many bacteriocins act distinctly from antibiotics. For example, both nisin and vancomycin function by targeting the peptidoglycan precursor lipid II, and hence inhibit cell wall synthesis. However, nisin binds to sites different from vancomycin-binding sites and therefore remains effective against many vancomycin-resistant Gram-positive bacteria (Piper et al., 2009). In general, the development of antibiotic resistance does not lead to cross resistance to bacteriocins (Severina et al., 1998).

Aside of antimicrobial mechanisms, bacteriocins also have relatively low toxicity, mostly narrow spectra, and can be produced *in situ* by many probiotics (Cotter et al., 2013). To date, various bacteriocins have been studied, some extensively, in both humans and animals. Animal studies have mostly focused on controlling pathogens important to both animal health and food safety (Ogunbanwo et al., 2004; Cutler et al., 2007; Grilli et al., 2009; Teixeira et al., 2013). Stahl et al. (2004) purified two colicins that showed an inhibitory effect on F4 and F18 positive *E. coli* at concentrations of 0.25 μg/ml to 25 μg/ml *in vitro*. One of the colicins, colicin E1, was evaluated in preventing post-weaning diarrhoea in pigs artificially challenged with F18 *E. coli* (Cutler et al., 2007). While the levels of targeted *E. coli* in faecal samples were significantly lower in colicin-treated pigs than non-treated pigs only at day 1 post-*E. coli* challenge, pigs receiving colicin E1 at a high dose of 16.5 mg/kg had significantly lower incidences and severity of diarrhoea compared to non-treated pigs throughout the 4-day experiment.

In terms of food safety, several bacteriocins have shown promising results in reducing *Campylobacter* and *Salmonella* in poultry. Stern et al. (2006) challenged chickens with 10^8 CFU of *Campylobacter jejuni* and reported greater than 10^6 log CFU reduction of *C. jejuni* in the caeca after a 3-day treatment of bacteriocin OR-7 at 250 mg/kg in feed compared to non-treatment groups. In a different study, bacteriocin E-760 reduced naturally acquired *Campylobacter* to undetectable levels in chickens when administered at 125 mg/kg in feed, whereas *Campylobacter* levels in untreated chickens increased to 6.17 log CFU/g in the caeca over the course of the study (Line et al., 2008). While bacteriocins have shown promising antimicrobial efficacies, as with most antimicrobials, the development of resistance appears inevitable. Currently, studies on bacteriocin resistance are mostly limited to laboratory conditions likely due to less extensive use of bacteriocins in practice. Nevertheless, bacteriocin resistance in different bacteria, such as *Listeria monocytogenes*, *Campylobacter*, *Staphylococcus aureus*, *Clostridium botulinum* and *Bacillus cereus*, has been reported. Bacteria, however, may not develop effective resistance mechanisms to natural bacteriocins (Peschel and Sahl, 2006; Hoang et al., 2011). This could be at least partially due to bacteriocins having multiple targets in bacteria. In addition, there are currently limited data indicating horizontal transfer of bacteriocin resistance among bacteria (Mantovani et al., 2011). At this stage, further risk assessment studies regarding bacteriocin resistance are still needed for applications under different conditions, which could allow the development of resistance mitigation strategies if necessary.

5.4 Virulence inhibitors

Many pathogenic bacteria encode virulence factors to facilitate cell mobility, adhesion to host cells, toxin production, cell-to-cell communication (quorum sensing) and biofilm formation, among other processes. Improved understanding of bacterial pathogenicity and characterization of virulence factors may provide a new antibiotic alternative approach by inhibiting essential virulence factors without directly killing or inhibiting bacterial growth. For example, the type III secretion system (T3SS) plays an important role in the adhesion of many Gram-negative bacteria to mammalian cells. In the *Salmonella enterica* Typhimurium infection model, T3SS secretes a set of Sip proteins, which can alter the permeability of host cell membrane and facilitate the translocation of other T3SS-secreted virulence proteins into host cells (Galán, 2001). Several acyl salicylaldehyde compounds have previously been identified as T3SS inhibitors. Hudson et al. (2007) demonstrated that acyl salicylaldehyde compounds could reduce the T3SS secretion of virulence proteins, hence reducing *Salmonella* haemolysis, and *Salmonella* internalization *in vitro*. By using a bovine intestinal ligated loop model, the same study also showed that pre-incubation of *Salmonella* with T3SS inhibitors suppressed intestinal secretory and inflammatory responses during *Salmonella* infection *in vivo*. Besides T3SS, other groups have explored inhibitors of several other virulence factors, such as the bacterial adrenergic sensors that regulate bacterial pathogenicity when the host is under stress, various toxins and pilus formation (Defoirdt, 2013; Cheng et al., 2014).

As virulence inhibitors generally do not kill or inhibit bacteria growth, they are less likely to select for resistant bacteria, which could be a unique advantage of these compounds. It should be noted, however, that among the antibiotic alternatives discussed in this chapter, the application of virulence inhibitors in preventing and treating bacterial infections in agriculture animals is the least studied. Results of many of the *in vivo* studies assessing antibacterial efficacy of virulence inhibitors were limited to mouse models and thus may not reflect applications in agriculture animals with more complex biological and environmental conditions. The safety of virulence inhibitors has also not yet been thoroughly studied. Safety evaluations should include not only the toxicity of individual compounds, but also the impact of inhibitors on similar pathways or factors in commensal bacteria (Rasko and Sperandio, 2010).

6 Case study: phage therapy

Prior to the widespread introduction of antibiotics in human medicine, a treatment called 'phage therapy' received a great deal of attention and research. Bacteriophages are viruses that infect and, in many cases, destroy bacterial cells in a process that predates *Homo sapiens*. Bacteriophages are specific for bacteria, with most bacteriophages only able to infect certain bacterial species or serovars within a species.

Phage therapy is the use of bacteriophages, specifically bacteriophage-mediated destruction of bacterial cells, to treat bacterial infections. The practice dates to the early twentieth century shortly after the actual discovery of bacteriophages themselves. Research into phage therapy, like almost all non-antibiotic antibacterial therapies, was largely neglected as antibiotic use became popular and efforts focused almost exclusively on developing or identifying new antibiotics. The increasing threat of antibiotic resistance coupled with the lower-tech appeal of phage therapy has led to a renewed interest in phage therapy development as an alternative in treating bacterial infections.

In recent years, a great deal of applied phage therapy research has focused on agriculture and food production. Numerous groups both in the United States and abroad are currently engaged in developing phage-based interventions to prevent pathogen transmission, treat bacterial infections and reduce bacterial contamination of food, among others (O'Flaherty et al., 2009).

Our lab began examining potential uses of phage therapy in livestock production in the late 2000s. Our initial phage research was in response to the increases in foodborne pathogen shedding seen in pigs following transportation and lairage. These increases in pathogen shedding are also common in poultry and cattle and are thought to result from recurring previous infections, rapid infections from contaminated transport vehicles and equipment or holding pens, the stress associated with transport and relocation or a combination thereof. Regardless of the exact mechanism, several groups have shown that the number of animals shedding foodborne pathogens can increase significantly following short transportation and after lairage (Hurd et al., 2002; Larsen et al., 2004).

In terms of food safety, this is arguably the most inopportune time for increases in foodborne pathogen shedding as it can serve to increase the number of animals bringing such pathogens into the processing facility. The issue is confounded by the fact that producers have very few approved antibacterial options this close to processing due to required withdrawal times and other limitations. In this respect, phage therapy could be ideal as a means of preventing increases in pathogen shedding as the treatment is 'natural' and rapid. Additionally, the treatment only needs to be effective for a short period as the animals are soon processed and there would be little concern or time for rebound infections.

In our earliest studies, we challenged market weight pigs with *Salmonella enterica* Typhimurium and allowed these 'seeder' pigs to contaminate a facsimile-holding pen. We then treated larger groups of market weight pigs with an anti-*Salmonella* phage cocktail by oral gavage (another group of untreated market weight pigs served as controls) before co-mingling phage-treated pigs with seeder pigs in the contaminated environment. After six hours, we measured *Salmonella* concentrations in both phage-treated and untreated pigs. Phage-treated pigs had significantly lower concentrations of *Salmonella* in the ileum, caecum and faeces compared to untreated pigs indicating that the phage treatment could reduce the increases in foodborne pathogen shedding associated with transport and lairage (Wall et al., 2010; Zhang et al., 2010a).

We have since shown that phages can be simply added to the feed of pigs prior to transportation to produce similar reductions in *Salmonella* colonization in treated pigs (Saez et al., 2011). We expanded our research focus to food matrices and have shown that bacterial contamination of ground pork, ground beef (raw and undercooked), eggs and spinach can be significantly reduced with the application of anti-*Salmonella* or anti-*E. coli* O157:H7 phages (Hong et al., 2014, 2016a).

Our most recent research has focused on the impact of phage treatment on the host in effort to better measure the safety of the treatment or the impact of the treatment on non-targeted bacteria. Similar to other groups, we found no detectable deleterious acute immune response to phage treatment in mice. We compared the impact of phage treatment on the gut microbiome to that of antibiotic treatment as well. Our results indicate that compared to antibiotic treatment, phage therapy has little impact on the gut microbiome (Hong et al., 2016b).

Taken together, our results and those of other groups indicate that phage therapy has great potential as an antibacterial, especially in certain settings. To date, there are some phage-based products approved for use in the United States as food additives or food processing aids (the author has no affiliation with these companies). At the time of this writing, however, no phage-based products have been approved and marketed for the treatment of specific infections in live animals in the United States.

7 Future trends and conclusion

Antibiotic use in livestock production remains controversial, namely due to the development of antibiotic-resistant bacteria that often results from any type of antibiotic therapy. While there continues to be a debate as to the impact bacteria that develop resistance on the farm have on human health, most agree that alternatives to antibiotics that can effectively control bacterial infections in livestock production are needed. Currently, most of the research on antibiotic alternatives focuses on probiotics, prebiotics and plant extracts. There are also several promising emerging technologies including the use of bacteriocins, manipulation of bacterial virulence factors and the possibility of phage-based infection control. The advancement of DNA sequencing technologies and various '–omics' could facilitate the identification of the most effective technologies that limit bacterial pathogens while maintaining or enhancing gut health and the associated gut microbiome.

8 Where to look for further information

As more challenges related to antibiotic resistance development emerge, the need for alternative antibacterial compounds and therapies becomes more apparent. Proposed alternatives (including many described here) must not only be effective, but safe. There may not be single compound that can fully replace the wide range of benefits seen with antibiotics (e.g., improved growth, disease prevention, infection elimination, etc.). It is more likely, at least in the short term, that these benefits could be provided by a collection of compounds, each with the capacity to provide one or two of the benefits associated with antibiotics. Developing such co-therapies for both animals and humans is the subject of labs throughout the globe. More information can be found in the following references, but also by visiting the various governmental agencies responsible for regulation of antibiotic use, the numerous large-scale antibiotic resistance monitoring groups (e.g., DANMAP, NARMS, etc.) and professional organizations for health care providers responsible for overseeing the use of different compounds.

9 References

Aarestrup, F. M. 2015. The livestock reservoir for antimicrobial resistance: A personal view on changing patterns of risks, effects of interventions and the way forward. *Philos. Trans. R. Soc. Lond.* 370:20140085.

Amalaradjou, M. A. and Bhunia, A. K. 2012. Modern approaches in probiotics research to control foodborne pathogens. *Adv. Food Nutr. Res.* 67:185–239.

Badia, R., Lizardo, R., Martinez, P. and Brufau, J. 2013. Oligosaccharide structure determines prebiotic role of B-galactomannan against *Salmonella enterica* ser. Typhimurium *in vitro.*. *Gut. Microbes.* 4(1): 72–5.

Bagley, M. C., Dale, J. W., Merritt, E. A. and Xiong, X. 2005. Thiopeptide antibiotics. *Chem. Rev.* 105:685–714.

Blair, J. M., Webber, M. A., Baylay, A. J., Ogbolu, D. O. and Piddock, L. J. 2015. Molecular mechanisms of antibiotic resistance. *Nat. Rev. Microb.* 13:42–51.

Brashears, M. M., Galyean, M. L., Loneragan, G. H., Mann, J. E. and Killinger-Mann, K. 2003. Prevalence of *Escherichia coli* O157:H7 and performance by beef feedlot cattle given *Lactobacillus* direct-fed microbials. *J. Food Prot.* 66:748–54.

Cheng, G., Hao, H., Xie, S., Wang, X., Dai, M., Huang, L. and Yuan, Z. 2014. Antibiotic alternatives: The substitution of antibiotics in animal husbandry? *Front. Microb.* 5:217.

Cotter, P. D., Ross, R. P. and Hill, C. 2013. Bacteriocins – a viable alternative to antibiotics? *Nat. Rev. Microbial.* 11(2):95–105.

Courtin, C. M., Swennen, K., Broekaert, W. F., Swennen, Q., Buyse, J., Decuypere, E., Michiels, C. W., De Ketelaere, B. and Delcour, J. A. 2008. Effects of dietary inclusion of xylooligosaccharides, arabinoxylooligosaccharides and soluble arabinoxylan on the microbial composition of caecal contents of chickens. *J. Sci. Food Agric.* 88:2517–22.

Cromwell, G. L. 2002. Why and how antibiotics are used in swine production. *Anim. Biotechnol.* 13(1):7–27.

Cutler, S. A., Lonergan, S. M., Cornick, N., Johnson, A. K. and Stahl, C. H. 2007. Dietary conclusion of Colicin E1 Is effective in preventing postweaning diarrhea caused by F18-positive *Escherichia coli* in pigs. *Antimicrob. Ag. Chemother.* 51(11):3830–5.

Danish Integrated Antimicrobial Resistance Monitoring and Research Programme [DANMAP] 2016. *DANMAP 2015*. The Danish Integrated Antimicrobial Resistance Monitoring and Research Programme. (Ed.) Høg, B. B., Korsgaard, H., Bygade, M., Sönksen, U. W. Copenhagen, Denmark.

Davies, J. and Davies, D. 2010. Origins and evolution of antibiotic resistance. *Microbiol. Mol. Biol. Rev.* 3:417–33.

Defoirdt, T. 2013. Antivirulence therapy for animal production: Filling an arsenal with novel weapons for sustainable disease control. *PLoS Pathog.* 9(10):e1003603.

Dibaji, S. M., Seidavi, A., Asadpour, L. and da Silva, F. M. 2014. Effect of a synbiotic on the intestinal microflora of chickens. *J. App. Poult. Res.* 23(1):1–6.

DiMasi, J. A., Grabowski, H. G. and Hansen, R. W. 2016. Innovation in the pharmaceutical industry: New estimates of R&D costs. *J. Health Econ.* 47:20–33.

Gaggia, F., Mattarelli, P. and Biavati, B. 2010. Probiotics and prebiotics in animal feeding for safe food production. *Int. J. Food Microbiol.* 141(Suppl. 1):S15–S28.

Galán, J. E. 2001. *Salmonella* interactions with host cells: Type III secretion at work. *Annu. Rev. Cell Dev. Biol.* 17:53–86.

Giang, H. H., Viet, T. Q., Ogle, B. and Lindberg, J. E. 2010. Growth performance, digestibility, gut environment and health status in weaned piglets fed a diet supplemented with potentially probiotic complexes of lactic acid bacteria. *Livest. Sci.* 129:95–103.

Gibson, G. R., Probert, H. M., Loo, J. V., Rastall, R. A. and Roberfroid, M. B. 2004. Dietary modulation of the human colonic microbiota: Updating the concept of prebiotics. *Nutr. Res. Rev.* 17:259–75.

Grilli, E., Messina, M. R., Catelli, E., Morlacchini, M. and Piva, A. 2009. Pediocin A improves growth performance of broilers challenged with *Clostridium perfringens*. *Poult. Sci.* 88:2152–8.

Gustafson R. H. and Bowen R. E. 1997. Antibiotic use in animal agriculture. *J. App. Microb.* 83:531–41.

Hashemi, S. R. and Davoodi, H. 2011. Herbal plants and their derivatives as growth and health promoters in animal nutrition. *Vet. Res. Comm.* (2011) 35: 169–80.

Hasunuma, T., Kawashima, K., Nakayama, H., Murakami, T., Kanagawa, H., Ishii, T., Akiyama, K., Yasuda, K., Terada, F. and Kushibiki, S. 2011. Effect of cellooligosaccharide or synbiotic feeding on growth performance,fecal condition and hormone concentrations in Holstein calves. *Anim. Sci. J.* 82(4):543–8.

Hoang, V. K., Stern, N. J. and Lin, J. 2011. Development and stability of bacteriocin resistance in *Campylobacter* spp. *J. App. Microb.* 111(6):1544–50.

Hong, Y., Pan, Y. and Ebner, P. D. 2014. Development of bacteriophage treatments to reduce *E. coli* O157:H7 contamination of beef products and produce. *J. Anim Sci.* 92:1366–77.

Hong, Y., Schmidt, K., Marks, D., Hatter, S., Marshall, A., Albino, L. and Ebner, P. 2016a. Treatment of *Salmonella*-Contaminated Eggs and Pork with a Broad-Spectrum, Single Bacteriophage: Assessment of Efficacy and Resistance Development. *Foodborne Pathog. Dis.* 13(12): 679–88.

Hong, Y., Thimmapuram, J., Zhang, J., Collings, C., Bhide, K. and Schmidt, Ebner P. D. 2016b. The impact of orally administered phages on host immune response and surrounding microbial communities. *Bacteriophage* 6(3):1–10.

Hudson, D. L., Layton, A. N., Field, T. R., Bowen, A. J., Wolf-Watz, H., Elofsson, M., Stevens, M. P. and Galyov, E. E. 2007. Inhibition of type III secretion in *Salmonella enterica* serovar Typhimurium by small-molecule Inhibitors. *Antimicrob. Ag. Chemother.* 51(7):2631–5.

Hurd, H. S., McKean, J. D., Griffith, R. W., Wesley, I. V. and Rostagno, M. H. 2002. *Salmonella enterica* infections in market swine with and without transport and holding. *Appl. Environ. Microbiol.* 68:2376–81.

Jung, S. J., Houde, R., Baurhoo, B., Zhao, X. and Lee, B. H. 2008. Effects of galacto-oligosaccharides and a *Bifidobacteria lactis*-based probiotic strain on the growth performance and fecal microflora of broiler chickens. *Poult. Sci.* 87:1694–9.

Larsen, S. T., Hurd, H. S., McKean, J. D., Griffith, R. W. and Wesle, I. V. 2004. Effect of short-term lairage on the prevalence of *Salmonella enterica* in cull sows. *J. Food Prot.* 67:1489–93.

Li, P., Piao, X., Ru, Y., Han, X., Xue, L. and Zhang, H. 2012. Effects of adding essential oil to the diet of weaned pigs on performance, nutrient utilization, immune response and intestinal health. *Asian-Aust. J. Anim. Sci.* 25:1617–26.

Line, J. E., Svetoch, E. A., Eruslanov, B. V., Perelygin, V. V., Mitsevich, E. V., Mitsevich, I. P., Levchuk, V. P., Svetoch, O. E., Seal, B. S., Siragusa, G. R. and Stern, N. J. 2008. Isolation and purification of enterocin E-760 with broad antimicrobial activity against Gram-positive and Gram-negative Bacteria. *Antimicrob. Ag. Chemother.* 52(3):1094–100.

Ling, L. L., Schneider, T., Peoples, A. J., Spoering, A. L., Engels, I., Conlon, B. P., Mueller, A., Schäberle, T. F., Hughes, D. E., Epstein, S., Jones, M., Lazarides, L., Steadman, V. A., Cohen, D. L., Felix, C. R., Fetterman, K. A., Millett, W. P., Nitti, A. G., Zullo, A. M., Chen, C. and Lewis K. 2015. A new antibiotic kills pathogens without detectable resistance. *Nature* 517:455–9.

Luepke, K. H., Suda, K. J., Boucher, H., Russo, R. L., Bonney, M. W., Hunt, T. D., Mohr III, J. F. 2016. Past, present, and future of antibacterial economics: Increasing bacterial resistance, limited antibiotic pipeline, and societal complications. *Pharmacotherapy* 37(1):71–84.

Lutful Kabir S. M. 2009. The role of probiotics in the poultry industry. *Int. J. Mol. Sci.* 10(8):3531–46.

Mantovani H. C., Cruz A. M. O. and Paiva, A. D. 2011. *Bacteriocin Activity and Resistance in Livestock Pathogens*. A. Méndez-Vilas (Ed.) Formatex Research Center, Badajoz, Spain, pp. 53–63.

Markets and Markets. 2016. Probiotics in animal feed market by bacteria (*Lactobacilli*, *Streptococcus Thermophiles*, and *Bifidobacteria)*, livestock (cattle, poultry, swine, and aquaculture), form (dry and liquid), function (yield, immunity, and productivity), and by region - global forecast to 2021. Available at: http://www.marketsandmarkets.com/Market-Reports/probiotics-animal-feed-market-85832335.html.

Metlitskaya, A., Kazakov, T., Kommer, A., Pavlova, O., Praetorius-Ibba, M., Ibba, M., Krasheninnikov, I., Kolb, V., Khmel, I. and Severinov, K. 2006. Aspartyl-tRNA synthetase is the target of peptide nucleotide antibiotic microcin C. *J. Biol. Chem.* 281:18033–42.

Mitsch, P., Zitterl-Eglseer, K., Kohler, B., Gabler, C., Losa, R. and Zimpernik, I. 2004. The effect of two different blends of essential oil components on the proliferation of *Clostridium perfringens* in the intestines of broiler chickens. *Pout. Sci.* 83:669–75.

Modesto, M., D'Aimmo, M. R., Stefanini, I., Trevisi, P., De Filippi, S., Casini, L., Mazzoni, M., Bosi, P. and Biavati, B. 2009. A novel strategy to select *Bifidobacterium* strains and prebiotics as natural growth promoters in newly weaned pigs. *Livest. Sci.* 122:248–58.

Moore, P. R., Evenson, A., Luckey, T. D., McCoy, E., Elveh-jem, E. A. and Hart, E. B. 1946. Use of sulphasuccidine, streptothricin and streptomycin in nutrition studies with the chicken. *J. Biol. Chem.* 165:437–41.

O'Flaherty, S., Ross, R. P. and Coffey, A. 2009. Bacteriophage and their lysins for elimination of infectious bacteria. *FEMS Microbiol. Rev.* 33(4):801–19.

Ogunbanwo, S. T., Sanni A. I. and Onilude, A. A. 2004. Influence of bacteriocin in the control of *Escherichia coli* infection of broiler chickens in Nigeria. *World J. Microbiol. Biotechnol.* 20:51–6.

Parks, W. M., Bottrill, A. R., Pierrat, O. A., Durrant, M. C. and Maxwell, A. 2007. The action of the bacterial toxin, microcin B17, on DNA gyrase. *Biochimie* 89:500–7.

Peschel, A. and Sahl, H. G. 2006. The co-evolution of host cationic antimicrobial peptides and microbial resistance. *Nat. Rev. Microb.* 4(7):529–36.

Piper, C., Draper, L. A., Cotter, P. D., Ross, R. P. and Hill, C. 2009. A comparison of the activities of lacticin 3147 and nisin against drug-resistant Staphylococcus aureus and Enterococcus species. *J. Antimicrob. Chemother.* 64:546–51.

Rasko, D. A., Sperandio V. 2010. Anti-virulence strategies to combat bacteria-mediated disease. *Nat. Rev. Drug Disc.* 9(2):117–28.

Rebollar, E. A., Antwis, R. E., Becker, M. H., Belden, L. K., Bletz, M. C., Brucker, R. M., Harrison, X. A., Hughey, M. C., Kueneman, J. G., Loudon, A. H., McKenzie, V., Medina, D., Minbiole, K. P., Rollins-Smith, L. A., Walke, J. B., Weiss, S., Woodhams, D. C. and Harris, R. N. 2016. Using 'omics' and integrated multi-omics approaches to guide probiotic selection to mitigate chytridiomycosis and other emerging infectious diseases. *Front. Microb.* 7:68.

Robyn, J., Rasschaert, G., Hermans, D., Pasmans, F. and Heyndrickx, M. 2013. Is allicin able to reduce *Campylobacter jejuni* colonization in broilers when added to drinking water? *Poult. Sci.* 92(5):1408–18.

Saez, A. C., Zhang, J., Rostagno, M. H. and Ebner, P. D. 2011. Direct feeding of microencapsulated bacteriophages to reduce *Salmonella* colonization in pigs. *Foodborne Path. Dis.* 12:1269–74.

Sarangi, N. R., Babu, L. K., Kumar, A., Pradhan, C. R., Pati, P. K. and Mishra, J. P. 2016. Effect of dietary supplementation of prebiotic, probiotic, and synbiotic on growth performance and carcass characteristics of broiler chickens. *Vet. World* 9(3):313–19.

Severina, E., Severin, A. and Tomasz, A. 1998. Antibacterial efficacy of nisin against multidrug-resistant Gram-positive pathogens. *J. Antimicrob. Chemother.* 41:341–7.

Stahl, C. H., Callaway, T. R., Lincoln, L. M., Lonergan, S. M. and Genovese, K. J. 2004. Inhibitory activities of colicins against *Escherichia coli* strains responsible for postweaning diarrhea and edema disease in swine. *Antimicrob. Ag. Chemoth.* 48(8):3119–21.

Stephens, T. P., Loneragan, G. H., Karunasena, E. and Brashears, M. M. 2007. Reduction of *Escherichia coli* O157 and *Salmonella* in feces and on hides of feedlot cattle using various doses of a direct-fed microbial. *J. Food Prot.* 70:2386–91.

Stern, N. J., Svetoch, E. A., Eruslanov, B. V., Perelygin, V. V., Mitsevich, E. V., Mitsevich, I. P., Pokhilenko, V. D., Levchuk, V. P., Svetoch, O. E. and Seal, B. S. 2006. Isolation of a *Lactobacillus salivarius* strain and purification of its Bacteriocin, which is inhibitory to *Campylobacter jejuni* in the chicken gastrointestinal system. *Antimicrob. Ag. Chemother.* 50(9):3111–16.

Stokstad, E. L.R, Jukes, T. H., Pierce, J., Page, A. C. and Franklin, A. L. 1949. The multiple nature of the animal protein factor. *J. Biol. Chem.* 180:647–54.

Svetoch, E. A. and Stern, N. J. 2010. Bacteriocins to control *Campylobacter* spp. In poultry-A review. *Poult. Sci.* 89(8):1763–8.

Teixeira, M. L., Dalla Rosa, A. and Brandelli, A. 2013. Characterization of an antimicrobial peptide produced by *Bacillus subtilis* subsp. *spizezinii* showing inhibitory activity towards *Haemophilus parasuis*. *Microbiology*. 159:980–8.

Unemo, M. and Shafer, W. 2014. Antimicrobial resistance in *Neisseria gonorrhoeae* in the 21st century: Past, evolution, and future. *Clin. Microbiol. Rev.* 27(3): 587–613.

United States Food and Drug Administration [USFDA]. 2015a. Guidance for Industry (GFI) #213: New Animal Drugs and New Animal Drug Combination Products Administered in or on Medicated Feed or Drinking Water of Food- Producing Animals: Recommendations for Drug Sponsors for Voluntarily Aligning Product Use Conditions with GFI #209. Retrieved from: http://www.fda.gov/downloads/AnimalVeterinary/GuidanceComplianceEnforcement/GuidanceforIndustry/UCM299624.pdf.

United States Food and Drug Administration [USFDA]. 2015b. 2014 Summary report on Antimicrobials Sold or Distributed for Use in Food-Producing Animals. Retrieved from: https://www.fda.gov/downloads/ForIndustry/UserFees/AnimalDrugUserFeeActADUFA/UCM476258.pdf.

Valeriano, V. D. V., Balolong, P. M. and Kang, D. K. 2016. Probiotic roles of *Lactobacillus* spp in swine: Insights from gut microbiota.. *J. App. Microb.* 54(6):110–19.

Wall, S. K., Zhang, J., Rostagno, M. H. and Ebner, P. D. 2010. Use of phage therapy to reduce lairage associated *Salmonella* infections under production-like conditions. *App. Env. Microb.* 76:48–53.

Windisch, W., Schedle, K., Plitzner, C. and Kroismayr, A. 2008. Use of phytogenic products as feed additives for swine and poultry. *J. Anim. Sci.* 86(suppl14):E140–8.

Yasuda, K., Hashikawa, S., Sakamoto, H., Tomita, Y., Shibata, S. and Fukata, T. 2007. A new synbiotic consisting of *Lactobacillus casei* subsp. casei and dextran improves milk production in Holstein dairy cows. *J. Vet. Med. Sci.* 69(2):205–8.

Younts-Dahl, S. M., Galyean, M. L., Loneragan, G. H., Elam, N. A. and Brashears, M. M. 2004. Dietary supplementation with *Lactobacillus*- and *Propionibacterium*-based direct-fed microbials and prevalence of *Escherichia coli* O157 in beef feedlot cattle and on hides at harvest. *J. Food Prot.* 67:889–93.

Zhang, J., Kraft, B. K., Pan, Y., Wall, S. K., Saez, A. and Ebner, P. D. 2010a. Development of an anti-*Salmonella* phage cocktail with increased host range. *Food. Path. Dis.* 7:1415–20.

Zhang, L., Xu, Y. Q., Liu, H. Y., Lai, T., Ma, J. L., Wang, J. F. and Zhu, Y. H. 2010b. Evaluation of *Lactobacillus rhamnosus* GG using an *Escherichia coli* K88 model of piglet diarrhoea: Effects on diarrhoea incidence, faecal microflora and immune responses. *Vet. Microbiol.* 141:142–8.

Chapter 4

Detecting veterinary drug residues in pork

Amy-Lynn Hall, United States Food and Drug Administration, USA

1 Introduction

Pork is one of the preferred meats in the world and is a billion dollar industry for producers worldwide (Giamalva et al., 2014). According to statistics from the US Department of Agriculture (USDA) Foreign Agricultural Service (2016), in 2015, *per capita* pork consumption averaged approximately 90 pounds in the European Union and China and 50 pounds in the United States. There are a variety of veterinary drugs approved for use in companion and food-producing animals, including swine. Therefore, consumers rely on regulatory authorities to ensure that edible products obtained from animals treated with a veterinary drug product do not contain drug-derived residues that might constitute a human health hazard.

A residue is defined as any compound present in edible tissues that results from the use of a drug, and includes the drug, its metabolites and any other substance formed in or on food because of the drug's use (Anon., 2012). The definition is broad enough to include any substance formed in or on food including resistant bacteria. The US Food and Drug Administration (FDA) considers edible tissues to include muscle, liver, kidney, fat/skin, milk, eggs and honey (United States Food and Drug Administration, 2016). Residues are considered violative if they are above the tolerance (FDA's safety number). There are two types of violative residues (United States Food and Drug Administration Compliance Program Guidance Manual, 2005): the first type of violative residue results from the use of approved drugs, when residues are above the established tolerances. Residues are considered safe if they are below the established tolerance. The second type of violative residue results from the use of unapproved drugs. If a residue from an unapproved drug is detected at any concentration, it is considered violative.

http://dx.doi.org/10.19103/AS.2017.0030.04
Published by Burleigh Dodds Science Publishing Limited, 2018.

The FDA's Center for Veterinary Medicine (CVM) is directed to regulate new animal drugs for food-producing animals by the Food, Drug and Cosmetic Act. CVM's mission is to protect human and animal health, which includes ensuring that safe and effective new animal drugs reach the market through the drug approval process. FDA-CVM evaluates new animal drugs to ensure that they are safe and effective for their intended use. New animal drugs are evaluated for effectiveness, target of animal safety, environmental impact, chemistry, manufacturing and controls, human food safety and labelling. The purpose of this chapter is to describe the risk assessment-based approach that the FDA uses to determine the human food safety of new animal drugs before they are approved for use.

2 Human food safety evaluation of new animal drugs

Most food-producing animals, including swine, are treated with drugs at some point during their lifetime. The following are some common swine ailments: *Escherichia coli* diarrhoea (Fairbrother et al., 2005), swine respiratory disease (Harms et al., 2002) and swine dysentery (Alvarez-Ordonez et al., 2013). There are a variety of new animal drugs approved for use in swine to treat such diseases (United States Food and Drug Administration, 1998, 2005, 2015).

Because some residues from drugs remain in the edible tissues of food-producing animals after administration, it is important that the FDA regulate the amount of residues that remain in the edible tissues after treatment. The human food safety evaluation of new animal drugs in food-producing animals helps ensure that food derived from treated animals is safe for human consumption. The FDA conducts a human food safety assessment of new animal drugs for use in food-producing animals through hazard identification, hazard characterization and mitigation to reduce human exposure to residues in food derived from treated animals to meet the mandated safety standard of 'a reasonable certainty of no harm'. The assessment looks at the microbial food safety, toxicology and residue chemistry aspects of the new animal drug (United States Food and Drug Administration, 2003, 2016).

2.1 Microbial food safety evaluation

The FDA evaluates the effects of antimicrobial exposure on the emergence or selection of resistant bacterial pathogens of human health concern. FDA considers an antimicrobial new animal drug to be 'safe' if it concludes that there is reasonable certainty of no harm to human health from the proposed use of the drug in food-producing animals. Two potential approaches to address the risk associated with new animal drugs possessing antimicrobial activity are 1) a hazard characterization and 2) a complete, qualitative resistance risk assessment. The hazard characterization provides information regarding the chemical, biochemical, microbiological and physical properties of the antimicrobial new animal drug that bear on characterizing the downstream effects of the drug. The qualitative risk assessment approach is comprised of a release assessment, an exposure assessment, a consequence assessment and a risk estimation. The release assessment estimates the probability that the proposed use of the antimicrobial new animal drug in food-producing animals will result in the emergence or selection of resistant bacteria in the animal. The release assessment estimates whether each factor would have a high, medium or low likelihood of favouring resistance emergence. The exposure assessment describes the likelihood of human exposure to food-borne bacteria of human health concern through particular exposure pathways. The consequence assessment describes the probability that human exposure to resistant bacteria will result in an adverse health consequence. Drugs are ranked as important, highly important or critically important to human health. The risk estimate integrates the results from the release, exposure and consequence assessments into an overall risk estimation associated with the proposed conditions of use for the drug. The risk estimation ranks drugs as high, medium or low and represents the potential for human health to be adversely impacted by the selection or emergence of antimicrobial food-borne bacteria associated with the use of the drug in food-producing animals. Additional information regarding the microbial food safety assessment can be found in Guidance for Industry #152, 'Evaluating the Safety of Antimicrobial New Animal Drugs with Regard to their Microbiological Effects on Bacteria of Human Health Concern' (United States Food and Drug Administration, 2003).

For antimicrobial drugs, studies also might need to be conducted to evaluate the effect of the drug residues on the human intestinal microflora (e.g. development of resistance, alteration of the barrier effect and overgrowth of potentially pathogenic microorganisms). New animal drugs with antimicrobial activity may be assigned a microbiological acceptable daily intake (ADI). We will discuss this term in the next section.

2.2 Toxicology evaluation

CVM is concerned with both intermittent and chronic exposure of people to residues. The standard battery of toxicity studies are designed to evaluate the oral toxicity to humans who may be exposed to residues through the consumption of food derived from animals treated with a new animal drug. The toxicity tests attempt to predict the effects of drugs over a lifetime of oral exposure by extrapolating from the toxicological model species to humans and from short-term exposure to long-term exposure. The basic toxicology battery includes 90-day feeding studies in a rodent and a non-rodent mammalian species to test the oral toxicity of the drug in the toxicologic species (United States Food and Drug Administration, 2006d), repeat-dose (chronic) toxicity studies to define toxic effects based on long-term exposures to the compound and/or its metabolites (United States Food and Drug Administration, 2006c), developmental toxicity studies to identify any potential effects on human prenatal development (United States Food and Drug Administration, 2006b), multi-generation reproduction study to detect any effect of the parent drug or its metabolites on mammalian reproduction (United States Food and Drug Administration, 2006e), genotoxicity studies (United States Food and Drug Administration, 2012a) and carcinogenicity studies (United States Food and Drug Administration, 2006a).

For compounds that are non-carcinogens, the toxicity testing allows for the calculation of the ADI for the total residues of the drug in the edible tissues. The ADI is defined as the amount of drug residues that can be consumed by a person every day for a lifetime without adverse health effects (United States Food and Drug Administration, 2016). The toxicological ADI is based on the new animal drug's toxicological or pharmacological properties as determined through the toxicology studies. A weight of evidence approach is used for this hazard characterization, meaning that all data from available studies and other applicable scientific information are considered. Traditional toxicology studies help to determine if the new animal drug causes an adverse effect in a biological test system and identify the highest dose of the new animal drug that produces no observable effect, which may be a no-observed-effects dose (NOEL), a no-observed-adverse-effects dose

(NOAEL) or a benchmark dose (BMD). The toxicological ADI is calculated by dividing the NOEL/NOAEL or BMD by an appropriate safety factor. The safety factor used in the calculation of the toxicological ADI reflects uncertainties associated with the extrapolation of data and information from toxicology studies to humans, including extrapolation of long-term, chronic effects from laboratory studies with shorter-term exposures, extrapolation of animal data to humans and variability in sensitivity to the toxicity of the new animal drug among humans. Some new animal drugs have the potential to cause acute toxicity to the human consumer following consumption of a single meal or consumption of food over a single day (United States Food and Drug Administration, 2016). In these cases, the ADI is not the appropriate safe intake value for quantifying the dose above which exposure from a single meal or over a single day can produce acute adverse effects; instead determining an acute reference dose (ARfD) is the most appropriate approach. The ARfD is an estimate of the amount of residues, expressed on a body weight basis, which can be ingested in a period of 24 hours or less without adverse effects to the consumer (United States Food and Drug Administration, 2016).

New animal drugs with antimicrobial activity may be assigned both a toxicological ADI and a microbiological ADI. A harmonized step-wise approach is used to determine if residues of antimicrobial new animal drugs, their metabolites or any excipients reaching the human colon remain microbiologically active, affect intestinal bacteria and whether a microbiological ADI is needed. The two microbiological endpoints of current public health concern to be considered when establishing a microbiological ADI are 1) an increase of the population(s) of resistant bacteria and 2) the disruption of the colonization barrier. The final assigned ADI (toxicological or microbiological) is generally determined to be the lower of the toxicological or microbiological ADIs. If a new animal drug is determined not to present a hazard to human health, an ADI may not need to be established (United States Food and Drug Administration, 2016).

The final ADI is used to calculate the safe concentration of total residues. The safe concentration is the amount of total residues of a new animal drug that can be consumed from each edible tissue every day for the lifetime of a human without exposing the human to residues in excess of the ADI. In calculating the safe concentration of residues in edible tissues, the ADI is adjusted for the anticipated dietary consumption of animal-derived food products (United States Food and Drug Administration, 2016). The safe concentration is calculated by multiplying the ADI by the average weight of a human (60 kg), and dividing by a food consumption value (Table 1). The food consumption values are assigned to

Table 1 Food consumption values

Edible tissue	Quantity consumed per day
Muscle	300 g
Liver	100 g
Kidney	50 g
Fat	50 g
Milk	1500 mL
Eggs	100 g
Honey	20 g

each of the edible commodities based on the portion of the diet they comprise. The consumption values assume that a person will not eat a full serving of all edible tissues in a single day. In other words, a person generally will not eat a full serving of kidney at lunch and a full serving of liver at dinner. The consumption value for milk is high enough to account for babies and young children where milk is the major food product they are consuming (Friedlander et al., 1999). Because milk, eggs, honey and meat all can be consumed in a single day, drug sponsors who seek approval for drugs in lactating dairy cattle, laying hens or honey must partition the ADI for meat and non-meat (i.e. milk, eggs and honey). The drug sponsors must specify what percentage of the total ADI is to be used for meat and what portion is reserved for milk, eggs and honey (i.e. 60% for meat and 20% for milk, 10% eggs and 10% for honey). The portion of the ADI reserved for milk, eggs and honey is not available for meats, even in animal species that do not produce milk or eggs for human consumption.

2.3 Residue chemistry evaluation

The purpose of the residue chemistry studies is to assess the quantity and nature of residues in tissues derived from animals treated with new animal drugs. Results from the residue chemistry studies allow the CVM to control the exposure to the compound so that the final ADI is not exceeded when humans consume edible tissues from animals treated with the approved new animal drug. The CVM controls the exposure through the assignment of tolerances and withdrawal periods and/or milk discard times. Residue chemistry studies generally submitted to the CVM include total residue and metabolism studies, comparative metabolism studies and residue depletion studies. Where appropriate, an analytical method for monitoring residues will be designated (United States Food and Drug Administration, 2016).

The total residue and metabolism study is conducted with the radiolabelled drug in the target animal species. This study is intended to show where the administered drug is deposited in the target species, how it is metabolized and its pattern of elimination. Metabolism of the proposed drug is evaluated to separate the total residue into all of its metabolites. A major metabolite is considered to be any radiolabelled compound, including the parent drug, that comprises at least 10% of the total residue (Friedlander et al., 1999). By measuring the major radiolabelled components of the drug at various time points following the end of treatment, it is possible to identify the drug residues that persist the longest and in which edible tissues drug elimination occurs most slowly. A target tissue and marker residue can be determined from the results of the total residue and metabolism study. The target tissue is generally the edible tissue from which residues deplete most slowly. The target tissue is often liver or kidney and rarely, muscle or fat (Friedlander et al., 1999). Because it is not possible to follow all residues throughout the body, one residue that has a known relationship to the total residues is chosen for monitoring. This residue is called the marker residue. The marker residue could be the parent drug, a metabolite or a combination of residues. The FDA establishes the ratio between the marker residue and the total amount of residues in the target tissue (M:T ratio). The M:T ratio establishes the relationship between the marker residue and all of the drug residues present in the target tissue, including the parent drug and marker residue itself. The M:T ratio is established at a time when the amount of total residues in the target tissue is equal to or less than the target tissue's safe concentration (Fig. 1). At this time point, the concentration of the marker residue is established as the tolerance

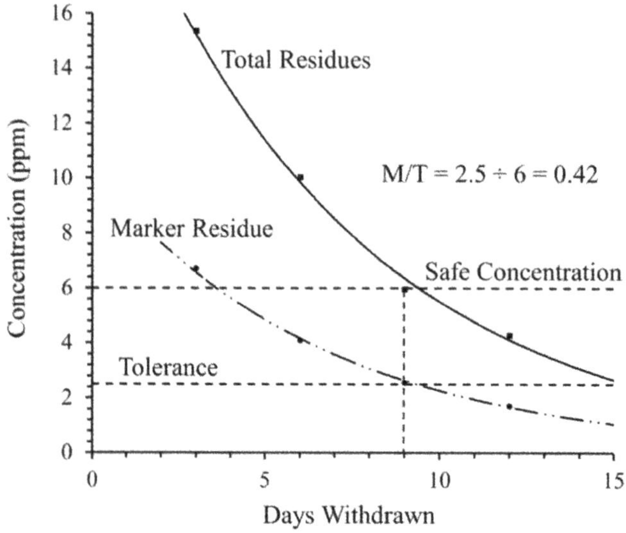

Figure 1 Relationship between the safe concentration, concentration of total residues, concentration of the marker residue and the tolerance. The information depicted is from the Freedom of Information Summary for New Animal Drug Application (NADA) 141–206.

so that the absence of the marker residue above the tolerance confirms that the total residue concentration is at or below the safe concentration (United States Food and Drug Administration, 2016). An indicator of the total residues, the marker residue, is needed because the ADI is based on the evaluation of the health effects of total residues during the microbial food safety and toxicology evaluations. This approach ensures that humans are not exposed to total residues that exceed the ADI when they consume edible tissues from animals treated with drugs. Tolerance is defined as the maximum concentration of a marker residue, or other residue indicated for monitoring, that can legally remain in a specific edible tissue of a treated animal (Anon., 2012). The tolerance provides a link to the ADI and establishes the safety of the edible tissues. If residues are below the tolerance, the tissue is considered safe to consume. If residues are above the tolerance, the tissue is not safe for consumption. Tolerances, when available, are published in the Code of Federal Regulations (specifically, 21 CFR 556).

Comparative metabolism studies are conducted in the toxicological species to confirm that the animals used in the toxicity studies were exposed to the same metabolites present in food products derived from treated animals. The metabolic profiles in the toxicological and target species are qualitatively compared to assess whether the same major metabolites are present in both species. If both species present with the same major metabolites, it is concluded that the toxicological species was autoexposed to the same compounds to which humans will be exposed when eating food derived from treated animals (United States Food and Drug Administration, 2016).

Residue depletion studies are conducted using the final product under field conditions at the highest dose for the longest labelled duration to provide data for calculation of the withdrawal period and/or milk discard time. These studies should demonstrate the depletion of the marker residue post-treatment to below the tolerance. The withdrawal period and milk discard time are defined as the time interval between the last administration of a new animal drug and when the animal can be safely slaughtered for food or the milk can be safely consumed. The withdrawal period and milk discard time are calculated using the upper 99th percentile statistical tolerance limit with 95% confidence. This means that when animals treated with the new animal drug according to labelled indications are withheld from slaughter for the assigned withdrawal period, there is only a 5% probability that, by chance, 1 animal in 100 will have residues of the marker compound in the target tissue that exceed the established tolerance (Friedlander et al., 1999). This statistical approach ensures that if the drug is used according to the approved labelling, there is a very small chance of producing a violative residue. The withdrawal period and milk discard time appear on the product label and in the CFR (specifically 21 CFR 520, 522, 524, 526, 528, 529 and 558).

2.4 Analytical methods for residues

Drug sponsors must provide a practicable method for analysing tissue residues as described in the CFR (specifically, 21 CFR 514.1(b)(7)) except when it is not reasonable to expect the new animal drug to become a component of food at concentrations considered unsafe. A method for analysing the marker residue in the target tissue consists of the determinative (quantitative) and confirmatory procedures. The determinative procedure quantifies the amount of new animal drug residue present in the tested matrices. The confirmatory procedure unequivocally identifies the marker residue present in the tested matrices. The following performance criteria are evaluated: calibration range and linearity, bias (systematic error), precision and accuracy, limit of detection and lower limit of quantitation, robustness and ruggedness, specificity and selectivity, parallelism, stability and system suitability. Because the tolerance for the marker residue is determined using the determinative method, it is directly tied to that method. The analytical method should perform optimally from one-half the tolerance to two times the tolerance (United States Food and Drug Administration, 2016). When a method is revised, it is necessary to confirm the relationship between the original and the revised methods. The USDA uses the official regulatory method to monitor animal-derived foods for residues of the new animal drugs (Anon., 2016).

3 Human food safety evaluation of carcinogens

The Delaney Clause of the Federal Food, Drug and Cosmetic Act prohibits the use of compounds 'found to induce cancer when ingested by man or animal, or if it is found, after tests which are appropriate for the evaluation of the safety of the food additive, to induce cancer in man or animal' (section 409(c)(3)(A)). An exception to the Delaney Clause exists for new animal drugs when 'it is determined by methods of examination prescribed or approved by the Secretary…that no residue of that compound will be found in the food produced from those animals under conditions of use reasonable certain to be followed in practice' (21 CFR 500.80(a)). To implement this provision of the Delaney Clause, the FDA uses quantitative risk assessment procedures to determine when the amount of residue of carcinogenic concern in the edible tissues produces an incremental estimated increase in cancer risk for a lifetime of one in one million. This level functionally defines 'no residue' (Friedlander et al., 1999). Following a determination of the 'no residue' level, residue and metabolism studies as described previously are conducted.

4 Violative residues exceeding established tolerances

The National Residue Program (NRP) is a collaborative interagency programme established to protect the public from exposure to harmful levels of chemical residues in meat, poultry, milk and egg products produced domestically or imported into the United States (Anon., 2016). A violation occurs when the Food Safety and Inspection Service (FSIS) laboratory detects concentrations of a chemical compound that is above its established tolerance or action level in a sample. The goal of the NRP is to keep violative residues out of the food supply. When residues are below the tolerance, the edible tissues from an animal treated with a drug are safe to consume. Because of the multiple safety factors and levels of conservatism used in the assignment of the ADI, tolerances and withdrawal periods, if a violation occurs from the use of an approved drug, it is most likely caused by misuse or by not following label directions. Examples of not following label directions include using a different dose than what is prescribed on the label, following a different route of administration than that which is approved, treating an animal for a different duration (e.g. treating an animal for 7 days when the drug is only meant to be used for 5 days) and treating a different (unapproved) species or a different disease. Violative residues also may be caused by the use of unapproved drugs. Any amount of residues from the use of an unapproved drug is considered a violation.

The level of concern regarding a violative residue depends on the information known about the drug residue. When the drug residue is found in the species and slaughter class for which the drug is approved, but the residue exceeds the established tolerance, this is considered Concern Level 1. There is greater concern for Concern Level 2 violations, when the drug is approved in a food-producing species but the drug residue is found in a species or slaughter class for which the drug is not approved. The greatest concern arises for Concern Level 3 violations, when the drug is not approved for use in any food animal in the United States.

There are three major concerns for people consuming food with violative residues (Barton, 2000). The first concern is that consumption of violative residues may lead to an increase in microorganisms developing resistance because of higher drug residues in the tissues from the treated animals. The second concern is for an increased risk of people having allergic reactions after consuming tissues from treated animals. The third concern is for acute or chronic health effects from consuming drug residues that are violative. Examples of drugs that can cause acute health effects in humans are beta-agonists, such as clenbuterol, and penicillin and sulpha drugs. Consumption of excessive amounts of clenbuterol can cause increased heart rate, muscle tremors, headache, nausea, dizziness and fever (Martinez-Navarro, 1990; Paige et al., 1997; Smith, 2000). Consumption of excessive amounts of penicillin (Dewdney et al., 1991) and sulpha drugs (Baynes et al., 2016, Choquet-Kastylevsky et al., 2002) can cause allergenic reactions including swelling, shock, skin rashes, asthma and fever. Avermectins and nitrofurans can cause chronic human health effects. The health effects in humans are extrapolated from the effects demonstrated in animals. Consumption of excessive amounts of avermectins can cause neurotoxic effects, central nervous system depression, ataxia and developmental effects to male reproductive organs (Fisher and Mrozik, 1992). Nitrofurans are classified as carcinogens (National Research Council, 1999), and are not approved for use in food-producing animals.

5 Future trends: judicious use of medically important antimicrobial drugs in food-producing animals

Antimicrobial resistance, and the resulting failure of antimicrobial therapies in humans, is a mounting public health problem of global significance. The FDA issued a guidance document in 2012, Guidance for Industry #209 (United States Food and Drug Administration, 2012b), addressing the judicious use of medically important antimicrobial drugs in food-producing animals. This guidance document summarizes some of the key reports and scientific literature related to the use of antimicrobial drugs in animal agriculture and outlines FDA's current thinking on strategies for assuring that medically important antimicrobial drugs are used judiciously in food-producing animals in order to help minimize antimicrobial resistance development. The FDA believes the use of medically important antimicrobial drugs in food-producing animals for production purposes (e.g. to promote growth or improve feed efficiency) represents an injudicious use. In addition, the use of medically important antimicrobial drugs in food-producing animals should be limited to those uses that include veterinary oversight or consultation. In order for drug sponsors to align their approved new animal drugs voluntarily with Guidance for Industry #209, the FDA issued another guidance document in 2013, Guidance for Industry #213 (United States Food and Drug Administration, 2013). This guidance document provides drug sponsors with specific recommendations on how to modify the use conditions of their medically important antimicrobial drug products. The voluntary process would help phase out the use of medically important antimicrobial drugs for production purposes and phase in veterinary oversight of therapeutic uses of these drugs starting 1 January 2017.

6 Where to find further information

Federal Food, Drug, and Cosmetic Act: http://www.fda.gov/RegulatoryInformation/Legislation/FederalFoodDrugandCosmeticActFDCAct/default.htm

For more information on FDA approved new animal drugs, visit the website, Animal Drugs @ FDA, located at: https://animaldrugsatfda.fda.gov/adafda/views/#/search

Summaries of the safety and effectiveness data used to support the approval of new animal drugs are available in the Freedom of Information Summaries located at: http://www.fda.gov/AnimalVeterinary/Products/ApprovedAnimalDrugProducts/FOIADrugSummaries/default.htm

Guidance for Industry #3 (General Principles for Evaluating the Human Food Safety of New Animal Drugs Used in Food-Producing Animals): http://www.fda.gov/AnimalVeterinary/Products/ApprovedAnimalDrugProducts/FOIADrugSummaries/default.htm

Electronic CFR: http://www.ecfr.gov/cgi-bin/ECFR?page=browse

Test methods used by USDA FSIS laboratories to support FDA's inspection programme, ensuring that meat, poultry and egg products are safe, wholesome and accurately labelled can be found at: http://www.fsis.usda.gov/wps/portal/fsis/topics/science/laboratories-and-procedures/guidebooks-and-methods/chemistry-laboratory-guidebook

Link to more information regarding the National Residue Program: https://www.fsis.usda.gov/wps/portal/fsis/newsroom/!ut/p/a0/04_Sj9CPykssy0xPLMnMz0vMAfGjzOINAg3MDC2dDbwMDIHQ08842MTDy8_YwNtMvyDbUREAzbjixQ!!/

?1dmy&current=true&urile=wcm%3Apath%3A%2Ffsis-content%2Finternet%2Fmain%2Ftopics%2Fdata-collection-and-reports%2Fchemistry%2Fresidue-chemistry

7 References

Alvarez-Ordonez, A., Martinez-Lobo, F. J., Arguello, H., Carvajal, A. and Rubio, P. (2013), 'Swine dysentery: aetiology, pathogenicity, determinants of transmission and the fight against the disease', *International Journal of Environmental Research and Public Health*, 10, (5) 1927–47.

Anon. (2012), 'New Animal Drugs; Updating Tolerances for Residues of New Animal Drugs in Food', *US Federal Register*, 77 FR 72254, 72254–68, https://www.federalregister.gov/documents/2012/12/05/2012-29322/new-animal-drugs-updating-tolerances-for-residues-of-new-animal-drugs-in-food (accessed 5 January 2017).

Anon. (2016), '2016 Residue Sampling Plans', *United States Department of Agriculture, Food Safety and Inspection Service and Office of Public Health Science*, https://www.fsis.usda.gov/wps/wcm/connect/04d78d46-c519-428c-a856-fe6416ae9e18/2016-Blue-Book.pdf?MOD=AJPERES (accessed 5 January 2017).

Barton, M. D. (2000), 'Antibiotic use in animal feed and its impact on human health', *Nutrition Research Reviews*, 13, 279–99.

Baynes, R. E., Dedonder, K., Kissell, L., Mzyk, D., Marmulak, T., Smith, G., Tell, L., Gehring, R., Davis, J. and Riviere, J. E. (2016), 'Health concerns and management of select veterinary drug residues', *Food and Chemical Toxicology*, 88, 112–22.

Choquet-Kastylevsky, G., Vial, T. and Descotes, J. (2002). 'Allergic adverse reactions to sulfonamides', *Current Allergy and Asthma Reports*, 2, 16–25.

Dewdney, J. M., Maes, L., Raynaud, J. P., Blanc, F., Scheid, J. P., Jackson, T., Lens, S. and Verschueren, C. (1991), 'Risk assessment of antibiotic residues of beta-lactams and macrolides in food products with regard to their immune-allergic potential', *Food Chemistry and Toxicology*, 29, 477–83.

Fairbrother, J. M., Nadeau, E. and Gyles, C. L. (2005), 'Escherichia coli in postweaning diarrhea in pigs: an update on bacterial types, pathogenesis, and prevention strategies', *Animal Health Research Reviews*, 6, (1) 17–39.

Fisher, M. H. and Mrozik, H. (1992). 'The chemistry and pharmacology of avermectins', *Annual Review of Pharmacology and Toxicology*, 32, 537–53.

Friedlander, L. G., Brynes, S. D. and Fernandez, A. H. (1999), 'The human food safety evaluation of new animal drugs', *Chemical Food Borne Hazards and Their Control*, 15(1), 1–11. https://doi.org/10.1016/s0749-0720(15)30203-6

Giamalva, J., Boone, P., Hausman, P., Lundy, D., Holmes, C., Smith, D. and Wilson, S. (2014), 'Pork and Swine Industry & Trade Summary', United States International Trade Commission, Publication ITS-11, 1–84.

Harms P. A., Halbur, P. G. and Sorden, S. D. (2002), 'Three cases of porcine respiratory disease complex associated with porcine circovirus type 2 infection', *Journal of Swine Health and Production*, 10, 27–30.

Martinez-Navarro, J. F. (1990), 'Food poisoning related to consumption of illicit beta-agonist in liver', *Lancet*, 336, 1311.

National Research Council (US) Committee on Drug Use in Food Animals. (1999), The Use of Drugs in Food Animals: Benefits and Risks, National Academies Press , Washington, D.C.

Paige J. C., Tollefson, L. and Miller, M. (1997), 'Public health impact of drug residues in animal tissues', *Veterinary and Human Toxicology*, 39, 162–9.

Smith D. J. (2000), 'Total radioactive residues and clenbuterol residues in swine after dietary administration of [^{14}C] clenbuterol for seven days and preslaughter withdrawal periods of zero, three, or seven days', *Journal of Animal Science*, 78, 2903–12.

United States Food and Drug Administration. (1998), 'Freedom of Information Summary Supplemental New Animal Drug Application (NADA) 97–505',
United States Food and Drug Administration. (2002), 'Freedom of Information Summary Original New Animal Drug Application (NADA) 141–206', http://www.fda.gov/downloads/Animal Veterinary/Products/ApprovedAnimalDrugProducts/FOIADrugSummaries/ucm117750.pdf (accessed 1 October 2016).
United States Food and Drug Administration. (2003), 'Evaluating the Safety of Antimicrobial New Animal Drugs with Regard to Their Microbiological Effects on Bacteria of Human Health Concern (CVM Guidance for Industry #152), http://www.fda.gov/downloads/AnimalVeterinary/GuidanceComplianceEnforcement/GuidanceforIndustry/UCM052519.pdf (accessed 1 October 2016).
United States Food and Drug Administration Compliance Program Guidance Manual. (2005), 'Illegal Residues in Meat, Poultry, Seafood, and Other Animal Derived Foods', http://www.fda.gov/downloads/AnimalVeterinary/GuidanceComplianceEnforcement/ComplianceEnforcement/UCM113433.pdf (accessed 5 January 2017).
United States Food and Drug Administration. (2005), 'Freedom of Information Summary Original New Animal Drug Application (NADA) 141–244', http://www.fda.gov/downloads/AnimalVeterinary/Products/ApprovedAnimalDrugProducts/FOIADrugSummaries/ucm118061.pdf, (accessed 5 January 2017).
United States Food and Drug Administration. (2006a), 'Studies to Evaluate the Safety of Residues of Veterinary Drugs in Human Food: Carcinogenicity Testing (VICH GL28)', http://www.fda.gov/downloads/AnimalVeterinary/GuidanceComplianceEnforcement/GuidanceforIndustry/UCM052527.pdf (accessed 5 January 2017).
United States Food and Drug Administration. (2006b), 'Studies to Evaluate the Safety of Residues of Veterinary Drugs in Human Food: Developmental Toxicity Testing (GL32)', http://www.fda.gov/downloads/AnimalVeterinary/GuidanceComplianceEnforcement/GuidanceforIndustry/UCM052522.pdf (accessed 1 October 2016).
United States Food and Drug Administration. (2006c), 'Studies to Evaluate the Safety of Residues of Veterinary Drugs in Human Food: Repeat-Dose (Chronic) Toxicity Testing (GL37)', http://www.fda.gov/downloads/AnimalVeterinary/GuidanceComplianceEnforcement/GuidanceforIndustry/UCM052505.pdf (accessed 1 October 2016).
United States Food and Drug Administration. (2006d), 'Studies to Evaluate the Safety of Residues of Veterinary Drugs in Human Food: Repeat-Dose (90-Day) Toxicity Testing (GL31)', http://www.fda.gov/downloads/AnimalVeterinary/GuidanceComplianceEnforcement/GuidanceforIndustry/UCM052523.pdf (accessed 1 October 2016).
United States Food and Drug Administration. (2006e), 'Studies to Evaluate the Safety of Residues of Veterinary Drugs in Human Food: Reproduction Toxicity Testing (GL22)', http://www.fda.gov/downloads/AnimalVeterinary/GuidanceComplianceEnforcement/GuidanceforIndustry/UCM052655.pdf (accessed 1 October 2016).
United States Food and Drug Administration. (2012a), 'Studies to Evaluate the Safety of Residues of Veterinary Drugs in Human Food: Genotoxicity Testing (GL23)', http://www.fda.gov/downloads/AnimalVeterinary/GuidanceComplianceEnforcement/GuidanceforIndustry/UCM052656.pdf (accessed 1 October 2016).
United States Food and Drug Administration. (2012b), 'The Judicious Use of Medically Important Antimicrobial Drugs in Food-Producing Animals (GFI #209)', http://www.fda.gov/downloads/AnimalVeterinary/GuidanceComplianceEnforcement/GuidanceforIndustry/UCM216936.pdf (accessed 8 October 2016).
United States Food and Drug Administration. (2013), 'New Animal Drugs and New Animal Drug Combination Products Administered In or On Feed or Drinking Water of Food-Producing Animals: Recommendations for Drug Sponsors for Voluntarily Aligning Product Use Conditions with GFI #209 (GFI #213)', http://www.fda.gov/downloads/AnimalVeterinary/GuidanceComplianceEnforcement/GuidanceforIndustry/UCM299624.pdf (accessed 8 October 2016).

United States Food and Drug Administration. (2015), 'Freedom of Information Summary Original New Animal Drug Application (NADA) 141-438', http://www.fda.gov/downloads/AnimalVeterinary/Products/ApprovedAnimalDrugProducts/FOIADrugSummaries/ucm064900.pdf (accessed 5 January 2017).

United States Food and Drug Administration. (2016), 'General Principles for Evaluating the Human Food Safety of New Animal Drugs Used in Food-Producing Animals (CVM Guidance for Industry #3)', http://www.fda.gov/downloads/AnimalVeterinary/GuidanceComplianceEnforcement/GuidanceforIndustry/UCM052180.pdf (accessed 8 October 2016).

USDA Office of Global Analysis/Foreign Agricultural Service. (2016), 'Livestock and Poultry: World Markets and Trade', https://apps.fas.usda.gov/psdonline/circulars/livestock_poultry.pdf (accessed 4 January 2017).

Part 2

Quality

Chapter 5

Producing consistent quality meat from the modern pig

R. D. Warner and F. R. Dunshea, University of Melbourne, Australia; and H. A. Channon, University of Melbourne and Australian Pork Limited, Australia

1 Introduction

The key issues that present challenges to the pig industry in the production of high-quality pork are (1) efficient production of lean carcasses, (2) occurrence of quality aberrations particularly those associated with post-mortem muscle metabolism resulting in either, rapid pH fall and low pH or high final muscle pH (pale, soft, exudative (PSE) and dark, firm, dry (DFD) pork, respectively) and (3) producing excellent quality pork that is suitable to be sold as fresh and processed pork products.

1.1 Efficient production of lean carcasses

Due to rapidly increasing global human population, there is increasing demand for food, and in particular animal protein. For the pig industry, this has been achieved through genetics and animal selection, targeted nutrition and the use of entire (uncastrated) males, and more recently, metabolic modifiers. Continued selection for lean and efficient growth in pigs will not, by itself, achieve the realisation of efficiencies required to meet the global demand for high-quality meat (Dunshea et al., 2016). Thus, the use of entire pigs and metabolic modifiers are key strategies that can be implemented in order to achieve optimal levels of efficiency. However, the utilisation of entire males and of metabolic modifiers may have some impacts on meat quality, which include objectionable flavour, toughness and dryness (lack of juiciness).

http://dx.doi.org/10.19103/AS.2017.0030.05

Pork producers aim to produce pork efficiently and sustainably, in order to satisfy consumer demand for lean pork and remain competitive. This continuing drive to improve efficiency and meet customer requirements has led to modifications in on-farm management practices to reduce costs of production for pork producers, whilst producing lean carcasses, at lighter carcass weights. In many countries, this has reduced intramuscular fat (IMF) content of pork, negatively affecting flavour and juiciness. The effect of low fat content on eating quality is exacerbated by consumers often overcooking pork. Consumer knowledge of how to prepare and cook pork that satisfies taste and quality expectations therefore requires ongoing effort.

1.2 Aberrations in quality due to muscle metabolism

PSE and DFD are quality conditions which are a result of rapid glycolytic metabolism and limited glycolytic metabolism, respectively, post-mortem. The occurrence of PSE in a pig carcass is quantified by measuring the rate of pH and temperature fall in the early period (pre-rigor) post-mortem as well as by low muscle ultimate pH. DFD in a pig carcass is quantified by measuring the ultimate pH of the muscles once rigor mortis has been established. Genetics, handling and feeding on-farm as well as stress factors are implicated in the occurrence of both PSE and DFD.

1.3 Producing excellent meat quality

Visual appearance – Consumers buy meat after examining visually, hence the appearance of the pork cut on display in the supermarket or retail shop is very important. If a muscle displaying the quality condition of PSE is adjacent to a muscle displaying DFD, the contrast can be unacceptable (see Fig. 1). For some consumers, pale and dark pork (due to being PSE and DFD) are also unacceptable, although other consumers can be non-selective for colour at the point of purchase. Also of visual (and sensory) importance is the amount of weep/drip/water in the tray and the level of marbling or IMF.

Sensory quality – Tenderness, juiciness and flavour are the key eating quality attributes of pork that influence consumer appreciation of pork. There is clear evidence to show that some consumers are prepared to pay more for meat with superior eating quality. Consumers expect premium value, as well as premium quality, from meat (Pethick et al., 2006) and the quality can be influenced by both environmental and genetic factors.

Processing quality – PSE pork that was avoided in the fresh state by a customer because of its unattractive appearance is just as likely to be rejected by a meat processor. A good use for PSE pork in processed products has not been found even in dry sausage, where the use of PSE pork enables shorter drying times. Use of PSE pork results in deterioration to a soft, crumbly texture in the finished product. When PSE pork is used in processing, it must be mixed with quantities of normal pork sufficient to maintain product quality.

For the modern Western consumer, quality meat encompasses not only visual and sensory traits, but also more intangible credence traits. These intangible traits that are equally, and increasingly, important to consumers include concerns around animal welfare, sustainability and environmental management (e.g. clean/green). The importance of these more intangible quality traits are acknowledged, but outside the scope of this chapter.

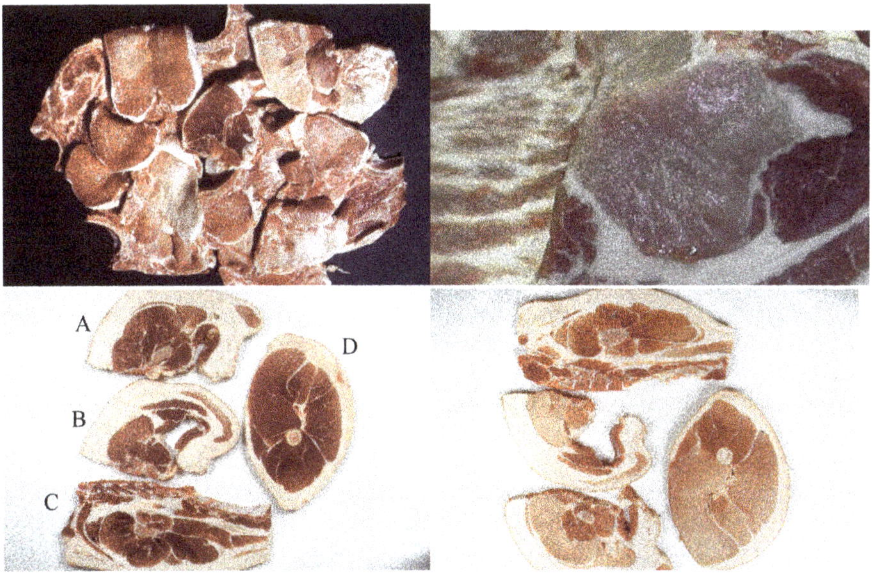

Figure 1 Typical variation in pork quality. Top left: Pork chops purchased from a supermarket and displayed together to show the variation in pork quality (source: Robert Kauffman, University of Wisconsin-Madison, USA). Top right: Two toning between two muscles in an exposed surface of the lumbar region of the pork carcass (source: Robyn Warner, University of Melbourne, Australia). The muscle on the left is *longissimus lumborum* and exhibits PSE. The muscle on the right is the *spinalis dorsi* and exhibits DFD. Bottom left: cross sections through a DFD carcass (source: Robert Kauffman). Regions of the carcass being; A rump, B loin, C shoulder, D leg. Bottom right: Cross sections through a PSE carcass (source: Robert Kauffman). Same regions of the pig carcass as for bottom left.

2 Issues/challenges for control of pig meat quality

2.1 Efficient production of lean carcasses

Extensive research has been conducted on the challenges and quality risks associated with efficient production of lean pork.

2.1.1 Entire male pigs and boar taint

In some countries, male pigs are not castrated (i.e. left 'entire') due to leanness and growth rate advantages compared to castrated male pigs. Castration can result in fatter carcasses and reduced growth performance, compared to entire male pigs (Campbell and Taverner, 1988). As entire males approach sexual maturity, this results in high circulating levels of the hormones androstenone and testosterone. Testes development and function is controlled by gonadotropin-releasing hormone (GnRH) that triggers the release of luteinising hormone and follicle-stimulating hormone from the pituitary gland, which then regulate the secretion of testicular steroids, including testosterone and androstenone. Boar taint is characterised by an unpleasant taste or odour in the meat from intact male pigs and is primarily attributed to increased concentrations of androstenone (5α-androst-16-en-3-one) and skatole (3-methylindole) (Grindflek et al., 2011). Testicular steroids reduce liver clearance of skatole in entire male pigs, and it therefore accumulates in fat. Skatole is a breakdown product of tryptophan in the hindgut of the pig and is metabolised by hepatic cytochrome P450 enzymes and sulphotransferase. The non-metabolised part accumulates in adipose tissue, causing an intense faecal-like odour in the cooked meat (Babol and Squires, 1995; Claus et al., 1994). Skatole and androstenone, the two major causative compounds implicated in boar taint, behave synergistically in their contribution to boar taint. Androstenone is a pheromonal steroid synthesised in the testes and metabolised in the liver. Part of androstenone accumulates in adipose tissue causing a urine-like odour. Androstenone is lipophilic and hydrophobic and is stored in the lipid component of tissues (Gower, 1972). Skatole is preferentially deposited in fatty tissue and is lipophilic and hydrophilic (Claus et al., 1994). The concentration at which untrained consumers can detect these compounds is 1.0 and 0.2 μg/g of fat for androstenone and skatole (Bonneau, 1998), respectively, and the levels in entire male pigs destined for slaughter are shown in Fig. 2. In a survey of four piggeries in Australia and New Zealand, the incidence of pig carcasses with levels of androstenone and skatole above the threshold was estimated to be 6–18% (Hennessy et al., 1997). Anecdotally, women and people of Asian origin are more sensitive to boar taint, which may explain the low threshold levels for skatole of 0.l03 μg/g of fat in Asian consumers (Leong et al., 2011) and 0.028 μg/g in Singaporean consumers (Ying, 2011). Some entire male pigs can have unacceptably high levels of androstenone and skatole at a lighter liveweight (D'Souza et al., 2011), suggesting that carcass weight at slaughter is not an appropriate strategy to manage boar taint.

In general, skatole is more highly correlated with boar taint and overall negative consumer sensory scores for taste, than androstenone. The concentration of skatole in pork muscle explains a large proportion of the variation in consumer scores for odour of cooked pork from entire male and female pigs, and less of the variation in consumer odour scores is explained by muscle androstenone concentration (Lundstrom et al., 1988). Although skatole has a greater effect on odour than androstenone, both compounds contribute similarly to overall flavour (note: flavour comprises both odour and taste). Carcass rejection rates for skatole (>0.2 μg/g) and for androstenone (>1 μg/g) were 9–26%

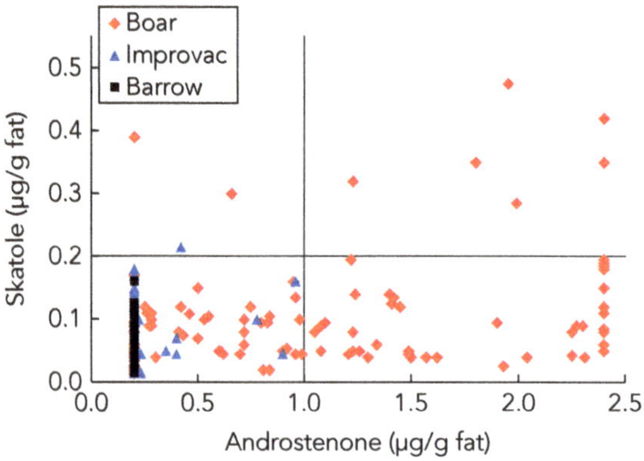

Figure 2 Effect of boars, barrows (castrates) and Improvac® on levels of skatole and androstenone in the pig fat. The horizontal line for skatole (=0.2) and the vertical line for androstenone (=1.0) indicate cutoffs for acceptability. Derived from Dunshea et al. (2001).

and 32–66%, respectively, across six studies (Thomsen et al., 2015) and hence boar taint is a cause for concern. Indole is a compound which may also contribute to boar taint, being a degradation product of tryptophan similar to skatole (Zadinová et al., 2016).

As skatole can be absorbed through the skin of pigs, faecal soiling of pigs and dirty pens may contribute to the incidence of high levels of skatole. Hansen et al. (1995) and Hansen et al. (1994) showed that keeping pigs dirty with faeces and urine and on a concrete floor increased skatole concentration in the backfat relative to keeping them clean and on slatted floor, with particularly higher levels of skatole at higher environmental temperatures and higher stocking density. Subsequent studies showed no effect of cleanliness of the pen or of the pig on skatole concentration (Aluwé et al., 2011; Thomsen et al., 2015). Indole is generally considered to not be under genetic control but is also related to pen cleanliness and feed mixture composition, mostly just intensifying the unpleasant odour caused by skatole (Zadinová et al., 2016).

Whilst male pigs may be castrated to prevent boar taint, this reduces feed efficiency and lean content and has a negative impact on animal welfare (Squires and Schenkel, 2010). Some countries, such as Australia, have raised entire male pigs in order to take advantage of the 30% improvement in pork production profitability through not castrating soon after birth (D'Souza et al., 2003a). Immunocastration has been developed as a method to reduce boar taint, and is discussed under Section 4.1.1.

2.1.2 Metabolic modifiers

The need to improve on-farm feed efficiency that impelled genetic selection, and nutritional strategies to increase carcass leanness, has together also driven the development and industry adoption of exogenous metabolic modifiers such as ractopamine (a β-agonist) and porcine somatotropin (pST). As a result of these technological advances, producers have benefited because of improved production efficiencies while meat processors have improved efficiencies due to increased lean meat yield. Also, there have been improved environmental impacts of pork production because of reduced effluent and greenhouse gas production (Woods et al., 2011). While ostensibly the consumer has also benefited because meat is leaner and less expensive, there have been some concerns that the focus on increasing production efficiency and lean meat yield has been to the detriment of meat quality (Dikeman, 2007; Dunshea et al., 2005, 2016; Parr et al., 2016). For example, using a meta-analysis approach, Dunshea et al. (2005) found that pST decreased IMF (−12%), increased shear force (+9%) and reduced drip loss (−6%) in pork loin muscle with no effect on consumer scores for odour, flavour or juiciness. Similarly, ractopamine increased shear force (+12%) and the limited published data on consumer preferences at that time suggested that ractopamine caused a decrease in tenderness (−6%), a negligible decrease in flavour (−1%) with no effect on juiciness. Unlike for pST, ractopamine had no effect on IMF or marbling. The magnitude of the effect of both technologies on shear force was approximately 0.4 to 0.5 kg (ca. 9–12%) which is close to the ability of consumers to detect a difference in shear force (Miller et al., 2001). More recently, using a Monte Carlo simulation approach, Channon et al. (2016b) indicated that pST and ractopamine increased shear force by 12 and 8% whereas tenderness was reduced by 7 and 5%, respectively.

Some of the impacts of metabolic modifiers on tenderness may be mediated via effects on muscle fibre types or proteolytic systems. Proteolysis is well-known to be involved in meat tenderisation whereas the relationship between fibre type and tenderness is less well understood and also more complex. The majority of studies have reported no effect of pST on fibre type distribution (Aalhus et al., 1997; Beermann et al., 1990; Oksbjerg et al., 1995; Ono et al., 1995; Parr et al., 2016). However, some authors have reported an increase in the proportion of fast oxidative glycolytic fibres and a decrease in proportion of fast glycolytic fibres in the *longissimus* muscle (Solomon et al., 1990, 1991). In contrast, a decrease in the proportion of fast oxidative glycolytic fibres and an increase in the proportion of fast glycolytic fibres were found in another study (Whipple et al., 1992). Also, Solomon et al. (1990) reported that pST administration increased muscle fibre size and subsequent shear force. The use of β-agonists (including ractopamine) in pigs and other species causes well-documented changes in muscle fibre distribution with an increase in the proportion of fast glycolytic fibres (Parr et al., 2016). The use of pST has also been reported to reduce calcium-activated proteolysis in the longissimus muscle, thereby preventing improvements in tenderness during the ageing process (Weikard et al., 1992). It has been hypothesised that a failure of pork to improve in tenderness with ageing may be related to genetic selection and feeding for leanness (Channon et al., 2016a) which has reduced the muscle proteolytic pathways and if this is the case metabolic modifiers may exacerbate these effects.

Therefore, it appears that conservative use of pST and ractopamine will cause a small increase in shear force with a similar reduction in perception of tenderness. However, it should be borne in mind that the magnitude of these increases is similar to those observed with equivalent increases in carcass leanness obtained through breeding or feeding for leanness and may be an inherent consequence of the production of leaner meat (see below).

2.1.3 Breeding and feeding for leanness

Dietary manipulation (see Section 3.5.1) and selection for leanness in pigs can result in an increase in shear force and a consumer perception of increased toughness (Karlsson et al., 1993). In addition, as carcasses become leaner, the content of IMF decreases, which may negatively impact the eating quality of pork, particularly when combined with poor cooking techniques of consumers. IMF consists of the lipid present in the perimysial connective tissue surrounding the muscle fibre bundles and influences eating quality by providing a smoother mouthfeel, reducing water loss during cooking, promoting saliva flow and improving flavour. IMF can also reduce the force required to shear myofibrils and aid the separation of muscle fibre bundles during eating (Essén-Gustavsson et al., 1994).

IMF typically constitutes 0.5–2.5% of muscle wet weight in the pork *M. longissimus* muscle. Positive effects on sensory attributes of pork of IMF content in excess of 2% have been reported in Denmark (Bejerholm and Barton-Gade, 1986; Fernandez et al., 1999; Touraille et al., 1989) and in France (Fernandez et al., 1999; Touraille et al., 1989). In the United States, Devol et al. (1988) suggested a threshold value for IMF content of pork muscle of 2.5–3.0% for optimum tenderness and reported that IMF content was most highly related to tenderness and shear force compared with other evaluated muscle characteristics (including flavour, juiciness, firmness, colour and connective tissue content). However, Fernandez et al. (1999) stated that IMF levels above 3.5% were associated with a high rejection score by French consumers due to negative effects on texture, taste and visual appearance. In Australia, reflective of consumer preferences for pork loin (longissimus) with low levels of marbling, Channon et al. (2001) found that the average IMF content of pork sourced from five large abattoirs was 0.98 ± 0.50%, with 74% of pork loins having IMF levels ranging from 0.5 to 1.4%.

In addition to genetic selection for leanness resulting in the inadvertent selection for reduced IMF, and genetic defects such as the Hal or Rendement Napole (RN) genes (see Section 3.2), there has also been a change in muscle fibre type towards fast glycolytic fibres and perhaps a reduction in the proteolytic systems in pork skeletal muscles (Joo et al., 2013; Karlsson et al., 1993; Parr et al., 2016). All of these together, and separately, can have effects on the visual and sensory quality traits of pork.

2.2 Aberrations in quality due to muscle pH

Irreversible anaerobic glycolysis occurs in the muscle after death due to the removal of oxygen. The concentration of glycogen present in the muscle at slaughter can influence the amount of lactate produced after death as well as the extent of the fall in muscle pH post-slaughter. The pH of living muscle is about 7.0 and declines in well-fed, rested animals to an ultimate pH of about 5.5–5.7 at 24 hours post-slaughter. Many studies have investigated the influence of the rate, and extent, of pH decline on PSE and DFD conditions rather than on pork eating quality *per se*. PSE meat results from a rapid rate of pH decline post-slaughter whilst muscle temperature is still high (>35°C) and causes protein denaturation and cell membrane leakage. This causes denaturation of myofibrillar proteins resulting in a loss in protein solubility and in water-holding capacity (WHC) whilst sarcoplasmic protein denaturation causes a loss in intensity of muscle pigment colour (Offer and Knight, 1988). PSE pork is paler and has been defined as pork with a muscle lightness (L* value) >50 and a drip loss exceeding 5% and an ultimate pH <5.4 (Warner et al., 1997).

The incidence of PSE pork can be minimised by reducing pre-slaughter stress, removing pigs carrying the halothane gene from the herd and incorporating rapid chilling systems into pig processing plants (see Section 4.2). Aaslyng et al. (2007) reported, as part of a study to determine the effect of raw meat quality on consumer preferences for pork eating quality, that the odour and flavour of pork of low pH (<5.4) was rated as more sour and less juicy compared with pork with a pH of >5.8. Furthermore, as juiciness was shown to have a positive influence on consumer preference, and sour taste had a small negative influence, it can be inferred that quality issues can occur as a result of low pH muscles in pork.

DFD pork results from low glycogen reserves in the muscle at the time of slaughter resulting in ultimate pH greater than 6.0. Reduced glycogen stores are often associated with long lairage times (>24 hours), mixing of unfamiliar and/or entire male pigs pre-slaughter, prolonged periods of transportation, extended time off feed, fighting, prolonged stress (D'Souza et al., 1998a,1999a) and over-activity (Deng et al., 2002) (see Section 4.2). The darker colour of DFD pork can be attributed to a higher proportion of reduced myoglobin because at a higher ultimate pH, oxygen uptake and utilisation is greater and less oxymyoglobin is formed (Deng et al., 2002; Fox, 1987) as well as less scattering of light within the muscle structure (Hughes et al., 2014). The dry surface appearance of DFD pork can be attributed to the water being tightly bound to myofibrillar proteins as they are highly charged, given that the isoelectric point (pI) of myofibrillar protein (pI of myofibrillar proteins is ~5–5.2) is lower than the ultimate pH (Judge et al., 1989). DFD pork is defined as exhibiting a surface lightness <42, a drip loss <5% and an ultimate pH >6.0 (Warner, 1994). DFD pork has a shorter shelf life as high pH meat is more susceptible to microbial spoilage than normal pork (Newton and Gill, 1981); however, DFD pork is generally more tender after cooking (Kauffman and Marsh, 1987).

2.3 Producing high-quality meat

2.3.1 Visual quality

Consumers are influenced by what they see when making purchasing decisions, hence the appearance of the pork cut on display in the supermarket or retail butcher shop is very important. Two-toning can occur in pig carcasses (see Fig. 1) due to differences in pH and muscle metabolism either within a muscle or between two adjacent muscles. Thus if a muscle displaying the quality condition of PSE is adjacent to a muscle displaying a dark colour due to the quality condition of high ultimate pH and DFD, the contrast can be unacceptable. For some consumers, pale and/or dark pork (due to being PSE or DFD, respectively) are also unacceptable, although other consumers may be non-selective for colour. Also of visual (and sensory) importance is the amount of weep/drip/water in the tray. Although the use of soaker pads has reduced the amount of drip visible in retail trays, PSE pork will produce large amounts of weep in the tray, due to overloading of the soaker pads, which is visually unappealing.

2.3.2 Sensory quality

Tenderness, juiciness and flavour are the key eating quality attributes of pork that influence consumer appreciation of pork, but there is still no simple 'online' low-cost tool available to industry to grade pork carcasses for these traits. Whilst much research has determined and documented the importance of a number of pre- and post-slaughter management factors in determining pork eating quality, this information has not yet been integrated into an eating quality system.

There is clear evidence to show that some consumers are prepared to pay more for meat with superior eating quality. Consumers expect premium value, as well as premium quality, from meat (Pethick et al., 2006) and quality can be influenced by both environmental and genetic factors. Hughes (2008) showed that 41% of consumers in the EU are prepared to pay more for high-quality food. Furthermore, consumers of beef have been reported to be prepared to pay as much as two to three times more for quality (Lyford et al., 2010). The Australian MSA grading system for beef results in premiums for eating quality which flow to the producer, processor and retailer, and the development of a similar system for pork would be advantageous (D' Souza et al., 2011). Hence the need for a pork eating quality assurance system, that the consumer can reply on, that not only assures eating quality, but also differentiates pork of superior eating quality from pork of everyday or inferior eating quality.

2.3.3 Processing quality

Myofibrillar proteins are very important in meat processing because they largely provide functionality to the meat system (Acton and Dick, 1984, 1996). Water-binding phenomena in processed meats depend largely on the extraction of contractile proteins from myofilaments followed by gelation of these proteins during thermal processing (Asghar et al., 1985). Thus, protein denaturation in PSE pork is important not only because of its effects on fresh pork quality but also because of its implications for the functionality of the myofibrillar proteins during processing. Camou and Sebranek (1991) reported that protein from PSE muscle is in a less functional form than protein from 'normal' muscle, even after solubilisation with salt. Gel strength of extracted salt-soluble (myofibrillar) proteins was reduced by 45% from PSE muscle. The major functional properties of muscle constituents for which the myofibrillar proteins play a predominant role include WHC, fat emulsification and protein gelation (particle-to-particle binding ability) (Acton and Dick, 1984, 1996).

A vital component of making good meat products is to achieve a good structure or texture. If pork has been minced or cut to make a processed product, the 'pieces' must stick to each other in the finished product – whether it is a sausage or a formed ham. This 'stickiness' is predominantly provided by myosin solubilised by salt in the brine or in the emulsion. In PSE pork, myosin does not easily solubilise. In addition, water is needed to create juiciness, but the water must be within the meat fragments and not between them. Pale pork with a soft structure and fluid exuding from it produces a very poor processed product. Processed pork products made from PSE pork will be unattractively pale in colour, fall apart and have excess fluid in the pack on the shelves. Sausages made with 50–100% PSE pork have lower sensory texture, flavour and overall acceptability scores than control samples (Kuo and Chu, 2003). PSE pork also has higher lipid oxidation during storage (Kuo and Chu, 2003) and retail display and hence processed products made with PSE pork will have a shorter shelf life due to development of oxidative rancidity.

3 Influencing factors for the challenges

3.1 Genetic influences on pig meat quality

In pigs, there is considerable variation between and within lines for carcass composition traits (Sosnicki and Newman, 2010). In contrast, only moderate variation exists for meat quality traits such as tenderness, drip loss and colour (Sosnicki and Newman 2010), although IMF content can range from low to high. Even less opportunity exists to improve eating quality traits such as flavour or juiciness through genetic selection (Barbut et al., 2008). The heritability of tenderness in pork, using subjective and objective determinations, is estimated at 0.25–0.30 (Sellier, 1998). Given the heritability, there is certainly opportunity for genetic improvement in tenderness in pigs through selection. There is a distinct lack of market signals in place for eating quality of pork, thus likely preventing implementation of genetic improvement due to lack of reward for the producer.

3.1.1 Intramuscular fat

The heritability of IMF content is relatively high in pigs ($h^2 = 0.26$ to 0.86, mean of 0.50) (Hocquette et al., 2010). Considerable efforts have been made to identify quantitative trait loci (QTL) for IMF in pigs and the existence of a major gene with a great effect on IMF content has been postulated by segregation analysis (Janss et al., 1997). The recessive form of the gene, named *MI*, is associated with increased IMF and appears to originate from the Chinese Meishan pig breed (Janss et al., 1997; Sanchez et al., 2007). In the Duroc pig population, Sanchez et al. (2007) showed that the major genes coding for IMF correspond to two QTL on two different chromosomes, including a recessive Duroc allele and a favourable and dominant allele, surprisingly, in a Large White population. It is evident that research to identify the QTL(s) for IMF in pigs will continue.

3.1.2 Boar taint

Boar taint is considered as a classic example of an undesirable flavour in pork, and is caused by the presence of skatole and androstenone in the fat (Hansson et al., 1980).

The accumulation of both androstenone and skatole, but not indole, in fat is influenced by genetic factors, and distinct breed differences have been identified (Zamaratskaia et al., 2008). In general, Duroc pigs tend to have higher levels of androstenone and Landrace and other white breeds of pigs tend to have high levels of skatole (Rowe et al., 2014; Squires, 1999). The h^2 for androstenone ranges from 0.5 to 0.7 and for skatole from 0.3 to 0.5, in different pig breeds and crosses (Grindflek et al., 2011; Große-Brinkhaus et al., 2015). Genetic relationships between androstenone and skatole are estimated to be 0.3–0.4 (Große-Brinkhaus et al., 2015). Conversely, because indole and skatole are both associated with tryptophan metabolism, the genetic correlation is reported to be 0.71–0.78 (Große-Brinkhaus et al., 2015).

Squires and Schenkel (2010) estimated that the application of markers, to produce pigs that were homozygous for the favourable alleles for boar taint, could reduce average fat skatole and androstenone levels by as much as 20–54% and 26–61%, respectively. The Illumina Porcine 60K+SNP Beadchip (IP60B) was genotyped on 987 pigs divergent for androstenone concentration in a Duroc-based sire line (Duijvesteijn et al., 2010). Significant effects of 37 SNPs on two chromosomes (SSC1 and SSC6; SSC= *sus scrofa* chromosome) were found, which together explained only 2–6% of the variation in androstenone (Duijvesteijn et al., 2010). A genome-wide association analysis (GWAS) for androstenone, skatole and indole in Pietrain sire lines found 33 SNPs associated with at least one of the boar taint components (Große-Brinkhaus et al., 2015). They concluded that all three boar taint compounds are influenced by a large number of loci, which have an additive effect on all traits. In Danish Landrace boars, and using GWAS and IP60B, significant SNP associations were found for skatole on SSC14 and for androstenone on SSC5 (Rowe et al., 2014). Several genes and their SNPs have been identified for both androstenone and skatole and other candidate genes associated with boar taint are yet to be confirmed [see review for list of genes and SNPs; (Zadinová et al., 2016)]. Importantly, it is likely that there is an interaction between the candidate genes as well as with production and reproduction traits. In the future, it is therefore appropriate to focus on genes and SNPs which can lower androstenone and skatole without affecting animal performance (Zadinová et al., 2016).

3.2 Specific genes and gene marker effects

3.2.1 HAL gene

The best known and described gene affecting meat quality in pigs is the HAL gene (Rothschild et al., 2004) which was originally called the Porcine Stress Syndrome (PSS) gene (Harrison, 1979). Animals with the PSS condition have abnormal calcium homeostasis due to a substitution of thymine for cytosine at nucleotide 1843 in the cDNA which codes for the protein responsible for the 'ryanodine receptor', thus now correctly called ryanodine receptor gene (Fuji et al., 1991). This leads to alteration in the amino acid sequence from arginine to cysteine at position 615 of the protein (Fuji et al., 1991). The ryanodine receptor protein is a calcium release channel in the sarcoplasmic reticulum of muscle cells. The mutation is an inherited syndrome, transmitted by a single autosomal recessive gene and two copies (alleles) of the gene are present in affected animals. The gene maps to chromosome 6 (SSC6) and a DNA test is patented (HAL 1843tm; Fuji et al., 1991) and is used worldwide. PSE pork is usually associated with carcasses from pigs which are homozygous positive for the ryanodine receptor (ryr1) gene mutation (nn). Pigs which carry the gene (Nn) respond adversely to pre-slaughter stressors and can exhibit very high levels of PSE. Channon et al. (2000) showed that halothane carriers (known to be heterozygote, Nn, for the ryr1 gene) reacted adversely to negative handling pre-slaughter and also to CO_2 stunning, exhibiting much higher levels of PSE in their muscles, as shown in Fig. 3. Meta-analysis has been used to show that pigs that are homozygous positive for the gene (nn) have loin muscles with 1.7% higher drip loss, 3.4 units lighter meat, 0.5 lower pH_{45} and 0.06 lower ultimate pH, relative to pigs homozygous negative for the gene (Salmi et al., 2010).

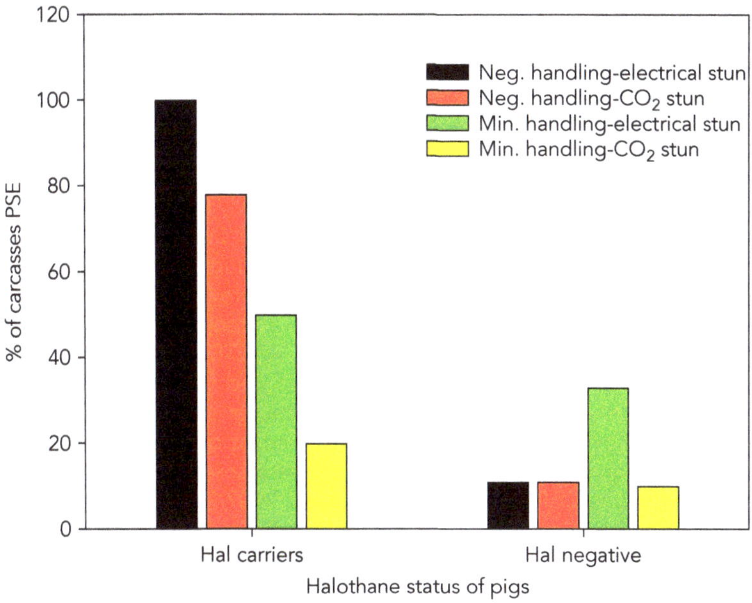

Figure 3 Effect of halothane gene (Hal Carrier, heterozygote for halothane gene, Nn vs Hal Negative, homozygous negative for gene, NN), handling pre-slaughter (Neg. handling, negative handling using electric goads vs Min. handling, minimal handling with no use of electric goads) and stunning (electrical stunning vs CO_2 stunning) on PSE incidence (%) in the loin. Adapted from Channon et al. (2000).

Worldwide, the frequency of the defective allele has been decreased as a result of genetic selection, although some companies maintain the gene in order to take advantage of the increased lean produced from heterozygous pigs (Rothschild et al., 2004).

3.2.2 RN gene

The second major meat quality gene found in pig populations is the 'acid meat gene' (Rothschild et al., 2004) which has also been called the 'Hampshire' effect (Naveau et al., 1996) due to its prevalence in Hampshire pigs. It has now been named the RN (Rendement Napole) gene (Naveau et al., 1996) and is associated with meat that has an extremely low ultimate pH and results in a much lower level of yield in cured and cooked hams (see below) due to excessive drip loss. The RN locus maps to chromosome 15 in the pig (Rothschild et al., 2004). The gene is part of a family that codes for one of the regulatory subunits of the AMP-activated protein kinase complex (named PRKAG3) (Milan et al., 2000). The abnormality manifests phenotypically as an increase in glycogen content in glycolytic fibres (Estrade et al., 1993). In genetic terms, inheritance of the RN gene is at a single locus and there are 2 alleles: one dominant mutant allele (RN-) and one recessive normal allele (rn+). The dominant RN- allele manifests in low ultimate pH muscles. Polymorphisms in the PRKAG3 gene have been found to be associated with variation in pH and colour, independent of the RN gene (Ciobanu et al., 2004). The test for the RN mutation has been used to remove the defect primarily from Hampshire-based lines and the genetic test is an important tool for pig breeders to improve meat quality.

3.2.3 Markers for tenderness

The skeletal muscle protease calpain and its inhibitor calpastatin have been clearly shown to be the main proteases driving tenderisation during ageing. Polymorphisms in the genes for calpastatin (CAST) have been reported, but Ciobanu et al. (2004) concluded that they did not appear to be associated with tenderness. In contrast, during a scan for QTL affecting pork quality in a Berkshire Duroc resource population, QTL with large effects on both shear force and taste panel tenderness were detected at essentially the same position on the q arm of porcine chromosome 2 [SSC2q; (Stearns et al., 2005)]. Subsequent fine-mapping of the tenderness QTL within the same Berkshire Duroc population strengthened support for possible CAST involvement. Within-family linkage analyses revealed three individual half-sib families with most likely QTL positions for one or both tenderness traits that coincide with the map position of CAST (Meyers et al., 2007). A QTL affecting pork tenderness was detected within the Illinois Meat Quality Pedigree and fine-mapped to the region of porcine chromosome 2 (SSC2) harbouring the functional candidate gene calpastatin (CAST) (Meyers and Beever, 2008). Hence, the effectiveness of CAST polymorphisms for marker-assisted selection of pork tenderness can now be assessed. In addition, SNPs have been reported for porcine CAPN1 (calpain) (Yang et al., 2007).

3.3 Breeding for a desirable ultimate pH

The breeding for a 'desirable' ultimate pH (pHu) in order to achieve pork quality has been incorporated into pig genetic breeding programmes (Barbut et al., 2008). The wide-ranging effects of pHu on pork quality and the relative ease of collecting phenotypic data have made this an ideal trait for selection purposes (Sosnicki and Newman, 2010). Ultimate pH has been shown to be genetically and phenotypically correlated with many economically important traits such as meat colour, meat tenderness, WHC and sensory qualities, with the pHu target being >5.7 (Sosnicki and Newman, 2010). It is recognised that although high pHu meat has positive WHC and processing traits, continued selection to increase the pHu above 6.1 may have detrimental effects on shelf life and flavour. For this reason, it is recommended that pig breeding programmes target both an upper and lower threshold for ultimate pH of 6.1 and 5.7, respectively, in order to optimise pork quality.

3.4 Specific breeds

3.4.1 Duroc

Pork from Duroc pigs has been reported to be more tender (Bunter et al., 2008; Candek-Potokar et al., 1998; Thornton et al., 1968), more palatable (Jeremiah et al., 1999; Wood et al., 1996), of better flavour (Candek-Potokar et al., 1998; Martel et al., 1988; McGloughlin et al., 1988) and of higher pork flavour intensity (Jeremiah et al., 1999; Wood et al., 1996) and juicier (Cameron and Enser, 1991; Cameron et al., 1990; Channon et al., 2004) than Large White, Landrace and their crosses. However, other studies have found no differences between Duroc, Large White and Landrace sired pigs for tenderness (Channon et al., 2004; Edwards et al., 1992; Lo et al., 1992a; Martel et al., 1988; McGloughlin et al., 1988) and juiciness (Candek-Potokar et al., 1998; Edwards et al., 1992; Lo et al., 1992b; Martel et al., 1988; McGloughlin et al., 1988; Purchas et al., 1990). Bunter et al. (2008) reported that 25% Durocs have similar IMF levels to Large White pigs and lower IMF levels than 50% Durocs. They also identified that significant variation exists in sire progeny groups within sire for desirable meat quality characteristics as well as between sire groups. Selection through choice of sire breed may not therefore be accurate due to within-breed sire differences, so selection of pigs on the basis of percentage of Duroc content alone may not necessarily result in the production of pork of superior eating quality. In general, the higher IMF content found in Durocs in Australia is not phenotypically associated with improved meat quality (Hermesch, 1997).

3.4.2 Hampshires

Pork from the Hampshire breed is reported to be juicier and more tender than pork from other white breeds of pigs (Fjelkner-Modig and Persson, 1986; Jeremiah et al., 1999; Purchas et al., 1990), in spite of low ultimate pH. IMF content of Hampshire pigs was reported to be 0.3 and 0.7% units higher than Landrace and Yorkshire breeds, respectively (Fjelkner-Modig and Persson, 1986). Hampshire pigs generally have an elevated muscle glycogen compared with other breeds (Sayre et al., 1963) which is thought to explain the superior eating quality. As described above, pigs with the dominant RN^- allele possess a mutation in the PRKAG3 gene that encodes the AMP-activated protein kinase (AMPK) γ3 subunit. AMPK is a major energy sensor in skeletal muscle influencing enzyme activity, gene and protein expression, fibre type and mitochondrial biogenesis (Estrade et al., 1993). This can detrimentally affect the WHC of both fresh and processed meat. High cooking losses associated with the RN gene is caused by a very low ultimate pH, but is not accompanied with a rapid fall in muscle pH or protein denaturation, seen in the development of PSE pork. Any selection programmes that incorporate the Hampshire breed should ensure that the wild-type allele (rn^+) is present.

3.4.3 Berkshires

Pork from Berkshire pigs is anecdotally known for being tender and of excellent quality and has become a niche product in some markets (Honeyman et al., 2006). Berkshires are often reported to produce better quality pork than the typical 'white' pig breeds such as Landrace, Large White and Yorkshire. Berkshires generally have a slower rate of pH decline post-mortem and hence less drip loss, lower cooking loss and less pale colour (Lee et al., 2012; Ryu et al., 2008), that is, lower prevalence of PSE-type pork. Berkshires are also reported to have a higher incidence of type I oxidative fibres (Ryu et al., 2008) which are associated with a higher lipid content within the muscle cell. Pork from Berkshire pigs has been found to have lower shear force (Brewer et al., 2002; Crawford et al., 2010; Lee et al., 2012) and trained sensory panels have scored the pork from Berkshires as either the same as 'White' breeds or higher, for important traits such as juiciness and tenderness (Brewer et al., 2002; Lee et al., 2012; Stoller et al., 2003).

In agreement with the data for Duroc and Berkshires above, meta-analysis of 43 studies has shown that the Landrace x Large White and crosses have higher drip loss and pure Large White and pure Landrace have lower scores for tenderness and juiciness (Trefan et al., 2013).

3.5 Nutritional, management and environmental influences on pork quality

3.5.1 IMF

Generally, IMF levels are strongly linked to subcutaneous and other fat depots, thus nutritional regimens to increase IMF levels are associated with higher overall carcass fatness.

The most successful nutritional strategy in non-ruminants, to dissociate fat accumulation in muscle from that in other parts of the body, has been demonstrated with subtle protein-deficient diets (Hocquette et al., 2010). More specifically, reducing the lysine to energy ratio increases the IMF content and improves the sensory scores for pork (D'Souza et al., 2008). In addition, feeding a vitamin A-restricted diet to pigs has been shown to increase the IMF content (D'Souza et al., 2003b).

3.5.2 Boar taint

One solution for boar taint is to use genetic selection to reduce boar taint, which is discussed in Section 3.1.2. Another promising method for controlling boar taint is to use immunocastration, which works by applying two vaccinations to entire male pigs pre-slaughter (see section 4.1.1 for description). The vaccination applied will stimulate the production of antibodies against GnRH (Squires, 1999). Immunocastration has been shown to be effective in reducing levels of androstenone and skatole to low levels without affecting the superior leanness and feed conversion efficiency of the pigs (Dunshea et al., 2001) (see Fig. 2). Immunocastration of pigs is delivered using Improvac® (or Improvest® in the United States), produced by Zoetis.

Androstenone levels are mostly determined by genetic factors and stage of puberty, whereas skatole levels, in addition to being influenced by genetic background and hormonal status of the pigs, are also controlled by nutritional and environmental factors. Thus skatole may be managed through nutrition such as feeding fermentable carbohydrate (inulin, sugar beet pulp) or bicarbonate (which increases gut pH) (Squires, 1999). As discussed in Section 2.1.1, management factors can also affect skatole levels, particularly pig and pen cleanliness. Thus, manure removal from pens reduces the absorption of skatole by pigs from the environment (Squires, 1999).

Positive effects of certain feed components on skatole levels such as chicory roots, inulin (Hansen et al., 2006; Jensen and Hansen, 2006) and potato starch (Claus et al., 2003; Zamaratskaia et al., 2005) have been reported on the aroma and flavour of pork. Conversely, some diets appear to increase levels of skatole in the pig fat. Garlic essential oil added to both a plant- and animal-based diet and fed to pigs resulted in a substantial increase in skatole concentration in the backfat of pig carcasses (Ling, 2011). Pigs fed with animal by-products like meat and bonemeal had higher scores for unacceptable flavour and odour and the formation of skatole was favoured (Lane and Fraser, 1999; Tuomola et al., 1996).

Danish abattoirs have implemented a colorimetric assay for online sampling for skatole (Hagdrup, 2009; Hagdrup et al., 2007). As it is well recognised that boar taint can arise from androstenone as well as skatole, many laboratories are investigating methods which may be developed to be rapid, and cheap, for online detection of boar taint [see Haugen et al. (2012) for review].

3.5.3 Magnesium and tryptophan

A reduction in the incidence of PSE in pig carcasses has been observed when supplementation of the pigs' diet occurs for two to five days pre-slaughter with magnesium, in the form of magnesium aspartate (D'Souza et al., 1998d; D'Souza et al., 2000), as shown in Fig. 4, magnesium sulphate or magnesium chloride (D'Souza et al., 1999b; D'Souza et al., 2000). A reduction in PSE incidence was also found when the pigs' diet was supplemented with 5 g of tryptophan/kg of feed for 5 days pre-slaughter (Warner et al., 1998).

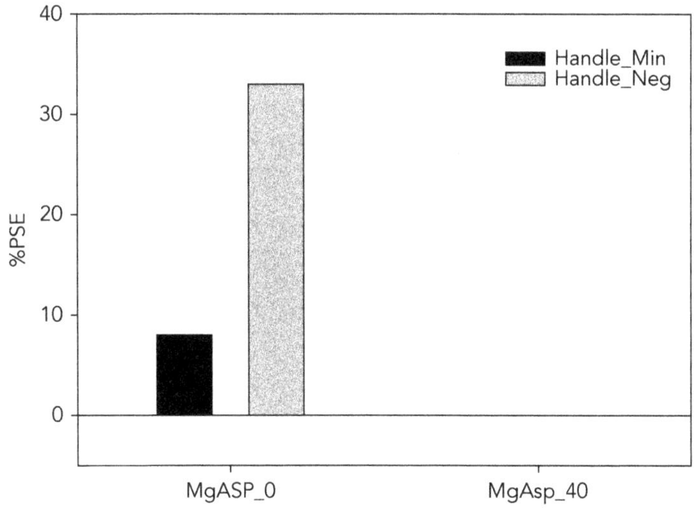

Figure 4 Effect of pre-slaughter handling (Handle_Min, minimal handling and no force used; Handle_Neg, negative handling, including use of electric goads) and supplementation of pigs' diet for 5 days pre-slaughter with magnesium aspartate (MgAsp; 0 vs 40 g per day) on the incidence of PSE (%PSE). Note that the %PSE for the pigs supplemented with 40 g magnesium aspartate was zero. Derived from D'Souza et al. (1998d).

Dietary magnesium supplementation of pigs appears to reduce the stress response by reducing plasma cortisol and catecholamine concentrations (D'Souza et al., 1998d). The incidence of PSE in pigs negatively handled pre-slaughter was reduced from 33% to 0% by feeding 40 g of magnesium aspartate for 5 days pre-slaughter (D'Souza et al., 1998d) (see Fig. 4). Conversely, the effect of dietary tryptophan appears to be through eliciting an increase in brain serotonin synthesis and this has been shown to have a sedative effect (Adeola and Ball, 1992).

3.5.4 Vitamin E

Oxidation of meat during storage, distribution and heating is problematic due to the effects of oxidation on sensory quality, particularly tenderness, flavour, odour and warmed over flavour in re-heated meat, and visual appearance due to premature browning and reduced shelf life. Oxidation also causes a reduction in protein and lipid functionality and nutrient loss. Furthermore, the product quality of processed pork products made from pork that has previously been frozen can be impacted by oxidation. Dietary vitamin E supplementation (*all-rac*-tocopherol acetate) of animal diets is well-known to improve meat quality by reducing post-mortem lipid, protein and myoglobin oxidation. Generally, vitamin E supplementation at doses of 100–200 mg *all-rac*-tocopherol acetate/kg of pig feed for 84–130 days pre-slaughter can result in lower thiobarbituric acid reactive substances (TBARS – measure of lipid oxidation) in fresh pork chops and processed pork patties and bacon (Asghar et al., 1991; Buckley and Connolly, 1980; Lanari et al., 1995). Of the studies where colour and drip loss were measured, the colour was generally better (less browning, longer shelf life) and the drip loss was lower when supplementation occurred at 200 mg/kg, rather than at 100 mg/kg (Asghar et al., 1991; Lanari et al., 1995; Monahan et al., 1992). Meta-analyses have shown that vitamin E supplementation of pigs' diets results in a large reduction in TBARS (lipid oxidation) values of pork loins displayed for 5 days (0.7 vs 0.3 mg/kg tissue) (Trefan et al., 2011).

3.5.5 Fishmeal and supplements high in polyunsaturated fatty acids (PUFAs)

It is well known that high levels of fish products in pig feed give rise to undesirable off-flavours in fresh, processed and frozen pork and pork products (Hertzman et al., 1988). Furthermore, the addition of fish oil to the pig's diet, which is known to result in increased PUFAs in the muscle and fat of pork, increases the intensity of off-flavours and -odours (Jonsdottir et al., 2003). For these reasons, the addition of fishmeal is limited to 5% for growing and finishing pigs in the Australian pig population, with the stipulation that no fishmeal can be included in the diet for a period of five to seven weeks before slaughter (www.daf.qld.gov.au). Similarly, any compounds added to the pigs' diet with the aim of increasing the PUFA content of pork products will likely have an effect on flavour, odour, shelf life and lipid stability of the pork products and hence are not recommended.

3.6 Importance of on-farm and pre-slaughter handling

Pigs are exposed to a number of potential stressors in the period from leaving the production unit up to slaughter at the abattoir. The handling of pigs prior to slaughter can cause a 'stress' response in pigs and ultimately affect the meat quality. However, few studies have investigated the effects of on-farm handling of pigs on meat quality. Negative handling of pigs by stock handlers (i.e. use of electric prodders) on-farm can result in marked reduction of growth and reproductive performance (Hemsworth and Barnett, 1991; Hemsworth et al., 1986, 1987). Also, if a particular husbandry treatment is aversive, the animal may learn to pair the 'punishment' of that particular treatment with the presence of humans (D'Souza et al., 1998c). This pairing of aversive handling with the human handler could lead to further 'stressing' of the pigs at the abattoir due to the difficulty associated with moving pigs from the lairage pen to the stunning area and thus could ultimately influence the meat quality.

Pigs subjected to negative handling on the farm result in a higher incidence of PSE pork (D'Souza et al., 1998c), as shown in Fig. 5. Furthermore, pigs handled with electric goads pre-slaughter had a much higher incidence of PSE compared to pigs that were minimally handled pre-slaughter without the use of electric goads (33–41%, 8–9%, respectively) (D'Souza et al., 1998a,d) (see Fig. 5). Finally, negative handling pre-slaughter had particularly negative effects on the PSE incidence in carcasses of pigs carrying the halothane gene, as shown in Fig. 3. This research informed the Australian Model Code of Practice for the Welfare of Pigs, which states that 'Electric prodders must not be used on pigs, except during loading, transport or unloading, and only when: individual pigs weigh 60 kg (liveweight) or more; and there is reasonable risk to the safety of the driver or the pig(s); and other reasonable action to cause movement has failed' (Primary Industries Standing Committee, 2008). Further, Standard 5 of the voluntary 'Australian National Animal Welfare Standards for Livestock Processing Establishments – preparing meat for human consumption' states that 'Implements used to handle livestock are appropriate for the species and are used judiciously to minimise stress and injury in livestock'.

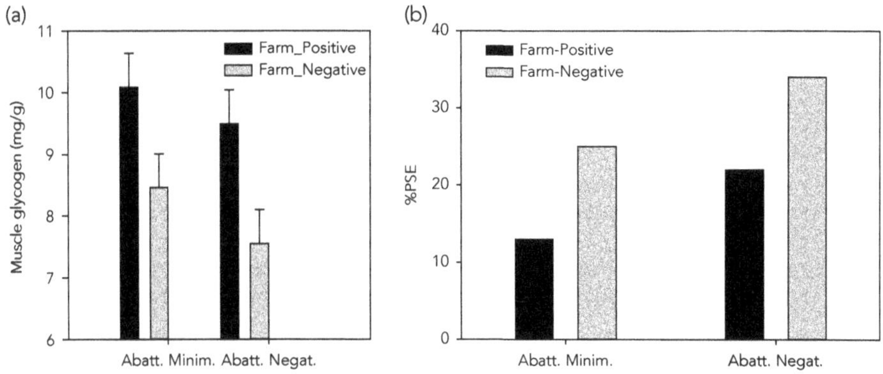

Figure 5 Effect of on-farm handling for 5 days/week for 5 weeks (Farm_positive, when pigs individually approached the experimenter they were gently patted; Farm_negative, when pigs approached the experimenter they were given an electric shock) and pre-slaughter handling (Abatt. Minim., minimal handling and no force used; Abatt. Negat., negative handling, including use of electric goads) on (a) muscle glycogen and (b) %PSE. Derived from D'Souza et al. (1998c).

3.7 Importance of stunning at slaughter

The method used to stun pigs can influence meat and carcass quality attributes of pork (Barton-Gade, 1997; Channon et al., 1997, 2000, 2002, 2003b). Stunning pigs with carbon dioxide can reduce the incidence of ecchymosis and bone fractures, improve meat quality (lower incidence of PSE pork) and improve worker safety compared with electrical stunning (Barton-Gade, 1997; Channon et al., 1997, 2000, 2002, 2003b). In the 1990s, the common method of stunning pigs at slaughter was electrical stunning, and electric goads were commonly used pre-slaughter to drive the pigs up the race to slaughter. The high occurrence of PSE pig meat created problems in (i) weight loss of product; (ii) unsightly pale, soft, wet appearance of meat on the supermarket shelves; (iii) unacceptable eating quality; (iv) low cook yields during processing of ham and bacon; (v) poor sliceability of ham and bacon; and (vi) unacceptable colour of two-toned ham and bacon.

Regardless of the electrical current applied, the application of electrical stunning for a long period (19s) resulted in a higher incidence of PSE compared to applying it for a shorter period (4s) (86–100%, 38–71% respectively) (Channon et al., 2003b). Furthermore, carbon dioxide stunning resulted in a lower incidence of PSE and lower blood splash than electrical stunning (Channon et al., 2002, 2003b). As described earlier, electrical stunning exacerbates the negative effects of the halothane gene (see above) on PSE incidence, as shown in Fig. 3.

3.8 Post-mortem technologies to improve meat quality

3.8.1 Ageing

The improvements in tenderness of pork during the first 1–2 days post-slaughter are rapid and then continue at a slower pace to plateau at around six days post-slaughter (Dransfield et al., 1980–81). Ageing of individually vacuum packaged pork loin, or of carcasses, for 4–7 days post-slaughter, improved sensory tenderness and shear force by 13–16% and 20–25%, respectively, compared to ageing for two days (Channon et al., 2001; Walker and Channon,

2003) (see Fig. 6). Extending the conditioning or ageing time for pork from 4 to 10–12 days post-slaughter has been shown to increase taste panel scores for tenderness, flavour intensity and overall liking scores and reduce the scores for intensity of abnormal flavours, compared with pork aged for only one day (Wood et al., 1996). Wood et al. (1996) reported that ageing pork for 10 days post-slaughter had a greater effect than both genotype (Duroc vs Large White) and feed level (high vs 0.8x high) in improving tenderness.

In an interaction with ageing, tenderness of pork may also be influenced by ultimate pH. Eikelenboom et al. (1996) reported that shear force and sensory tenderness at three and seven days post-slaughter could be related to ultimate pH. As shear force at seven days post-slaughter was more highly correlated with ultimate pH than at three days post-slaughter, ultimate pH may influence ageing rate as well as initial tenderness.

3.8.2 Aitch-bone hanging/tender stretching

Eating quality of pork can be influenced by the method used to suspend or hang carcasses pre-rigor. Hanging carcasses from the aitch bone has been shown to restrain or stretch the *M. longissimus, M. semimembranosus, M. biceps femoris* and *M. gluteus medius* muscles that are otherwise free to shorten on sides of carcasses hung from the Achilles tendon (Harris and Shorthose, 1988). As shown in Fig. 6, hanging pig carcasses by the aitch bone (tenderstretch) significantly improved tenderness of the *longissimus* muscle at two days post-slaughter, which persisted through to seven days post-slaughter. However, the adoption of aitch-bone hanging on a substantial scale requires innovative engineering approaches to minimise labour input, address occupational health and safety concerns, address changes in muscle shape and to minimise risks of bone breakages due to inaccurate splitting. Until there

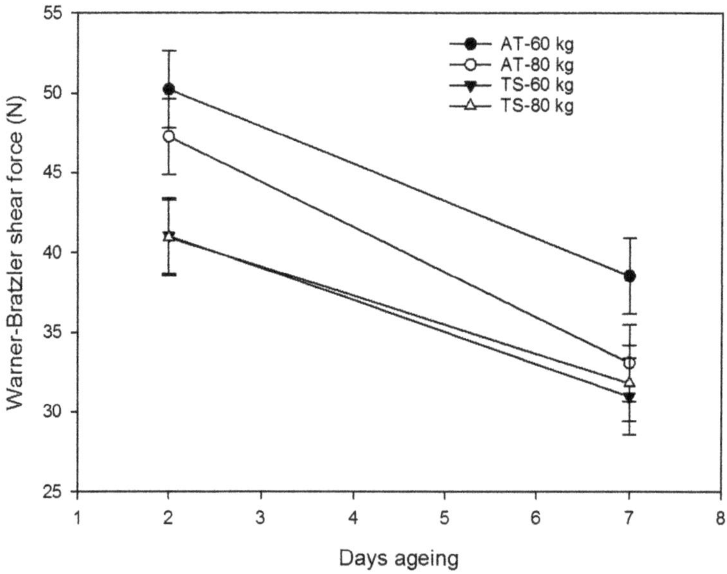

Figure 6 Effect of hanging (AT – Achilles tendon; TS – tenderstretch), ageing (2 or 7 days post-mortem) and hot carcass weight (60 vs 80 kg) on Warner–Bratzler shear force (N) of porcine *longissimus lumborum*. Derived from Channon et al. (2014).

are clear market signals in relation to eating quality and these issues are attended to, pork processors will remain resistant to commercial introduction of aitch-bone hanging on a routine basis. Interestingly, some progressive beef and sheep meat processors in Australia and New Zealand have adopted aitch-bone hanging in order to assure tenderness to their consumers.

3.8.3 Electrical stimulation

Electrical stimulation applied to carcasses post-slaughter results in an acceleration of rigor development, as a consequence of extensive membrane depolarisation and muscular contractions during stimulation (Aalhus et al., 1994). Electrically stimulated beef carcasses require shorter ageing time than unstimulated carcasses in order to reach an acceptable level of tenderness in the meat (Savell, 1979). Issues with drip loss and increased PSE incidence are reported in pig carcasses stimulated with constant voltage systems (Bowker et al., 2000; Taylor and Martoccia, 1995; Taylor et al., 1995b). Electrical stimulation using a constant current system with 150 mA for 30 sec at 2 min post-slaughter has not been adopted by Australian pork processors. This is despite Channon et al. (2003) showing that consumer scores for eating quality are improved when eelctrical stimulation is applied to pork carcasses, with no detrimental effects on drip loss, colour or PSE incidence. Anecdotal evidence has indicated issues with increased carcass shrinkage in pork carcasses undergoing electrical stimulation.

3.8.4 Enhancement of pork

The adoption of moisture infusion (enhancement) for pork meat has provided the pork industry in some countries, such as Australia and United States, with an alternative means of consistently producing high eating quality pork. Brines used for moisture-infused pork consist primarily of water and ingredients, and may include antimicrobial and antioxidant agents (e.g. lactates and citrates), phosphates and sodium chloride. Brines are generally added at 5–12% of the initial weight of the fresh pork. The flavour of moisture-infused pork can be influenced by these different brine ingredients and their levels of inclusion, even in neutrally flavoured brines. Technical issues with pork flavour, including excessive saltiness, metallic notes, soapiness and bitterness, can arise due to the form and levels of brine ingredients used and must be addressed to ensure consumer satisfaction is not affected. The injection of a 3 or 5% polyphosphate solution into pork was recommended by Sheard et al. (1999) for increased tenderness and juiciness of pork. The injection of 10% polyphosphate did not result in any further improvement in eating quality over and above that achieved with 3–5% polyphosphate. The positive effect of moisture infusion/enhancement on tenderness and juiciness is large (Channon et al., 2016b), particularly when compared with other factors including breed and feeding level, chilling rate, aitch-bone hanging (Channon et al., 2001), electrical stimulation (Channon et al., 2003b; Taylor et al., 1995a), ageing period (Bertram and Aaslyng, 2007; Channon et al., 2004; Wood et al., 1996) and cooking temperature (Bejerholm and Aaslyng, 2003; Moeller et al., 2010; Wood et al., 1995).

According to the FSANZ Food Standards Code (http://www.foodstandards.gov.au/code/pages/default.aspx), when brine ingredients are added to fresh pork, moisture-infused pork is considered to be a manufactured meat product in Australia, and must be labelled appropriately to differentiate it from non-infused pork at the retail level. In an Australian supply chain, the eating quality of branded moisture-infused pork loins from immunocastrated and female pigs were shown to have higher consumer scores for flavour, juiciness, tenderness and overall liking scores compared with non-infused loins (D'Souza et al., 2003a).

3.8.5 Cooking technique

The cooking temperature and technique/method are important determinants of final eating quality. For example, cooking pork loin (*longissimus*) and silverside (*semitendinosus*) at 70°C, rather than at 75°C, results in lower Warner–Bratzler shear force (WBSF) (more tender meat) (Channon et al., 2016a). This is most likely due to lower water lost during cooking and reduced denaturation of muscle proteins, such as actin, which denature at ~75°C (Purslow et al., 2016). Also, *sous vide* cooking in a vacuum sealed bag at 70°C for 12 h lowers WBSF in pork *longissimus* and *semitendinosus* when compared to roasting or grilling pork to an internal temperature of 70°C (29.4, 44.9, 38.6 N, respectively; averaged across both muscles) (Jin et al., 2016).

3.8.6 Novel and innovative technologies

Novel and innovative technologies for assisting in increasing the safety of meat products, as well as increasing tenderness, include pulsed electric field, high pressure processing (HPP; also called high hydrostatic pressure), ultrasound and shockwave (also called high hydrodynamic pressure). Of these technologies, the greatest improvements in tenderness have been achieved with HPP. Meta-analysis of data, across many studies and including beef, pork, chicken and lamb, has shown that compared to controls, HPP application to meat at 100–200 MPa results in a reduction in WBSF of 49 N when applied to post-rigor meat and 82 N when applied to pre-rigor meat (Warner et al., 2017). Of the pork studies, the application of HPP to cold-boned *longissimus thoracis et lumborum* (LTL) resulted in a reduction of WBSF of only 4–6 N, when applied at 200–300 MPa at 80°C (Hong et al., 2008). In addition, the application of HPP to pre-rigor (hot-boned) pork LTL and *semimembranosus* resulted in only small reductions in WBSF of 4–5 N (Souza et al., 2011). The smaller effects of HPP on pork tenderness, relative to those achieved for beef and lamb, are likely because the initial/control WBSF values for pork are low. The control values for WBSF for cold-boned pork were 10–30 N for two studies (Hong et al., 2008; Park et al., 2006) compared to an average of 98 N for controls across 11 studies on beef, pork and sheep meat (Warner et al., 2017). Similarly, for hot-boned (pre-rigor) pork the control muscles had a range of 17–24 N in WBSF (Souza et al., 2011), and the average WBSF across all four studies encompassing multiple species was 120 N (Warner et al., 2017). The one study on the application of explosive shockwaves to pork showed a reduction in WBSF of 2.8 N (Bowker et al., 2010), but the initial WBSF was only 47 N, and thus relatively tender before the application of explosive shockwaves. No studies were found which reported the application of ultrasound or pulsed electric field to pork carcasses or meat in order to achieve tenderisation (Warner et al., 2017).

4 Recommendations for overcoming challenges

4.1 Producing efficient lean carcasses with high meat quality

The reduction in meat quality, as a result of using metabolic modifiers, or entire males, can be grouped into (i) boar taint and associated flavour/odour problems, (ii) low IMF and associated problems with tenderness and juiciness and (iii) toughness (which is discussed in Section 4.3).

4.1.1 Boar taint

Genetics – Genetic markers for low boar taint are not yet available but will likely be available in the future (see Section 3.1.2). In their absence, and in the future, pork production systems which routinely monitor the levels of skatole and/or androstenone online in every carcass would enable genetic selection to be undertaken.

Improvac – Improvac®, a vaccine developed in Australia and manufactured by Zoetis Inc., is now being used to immunocastrate entire males in Australia to minimise flavour issues associated with the production of entire males, but this is not widespread. Worldwide interest in Improvac® use is increasing. In the EU, surgical castration of piglets can only be performed with prolonged analgesia and/or approved anaesthetic methods and it has been proposed that the procedure will be banned in the EU by 1 January 2018. The Improvac® vaccine contains a modified GnRH antigen in an aqueous adjuvant system that causes little tissue aggravation when injected. It requires the administration of two injections, one at ten weeks of age and the other from 4–5 weeks (Dunshea et al., 2001) to two weeks (Lealiifano et al., 2011) prior to slaughter. A delay in administration of the secondary injection of the vaccine to two weeks pre-slaughter presents economic advantages to producers by both maximising growth rate advantages associated with testicular hormone production of entire males and through minimising feed costs, as immunocastration can result in increased feed intake. Immunocastrated male pigs therefore have lower levels of skatole and androstenone in subcutaneous fat due to the inhibition of GnRH. Immunocastrated males are leaner and grow faster than surgical castrates and entire male pigs, when the second dose is administered at either 15 or 19 weeks of age for pigs slaughtered at 23 weeks of age or at either 18 or 22 weeks of age for pigs slaughtered at 26 weeks of age (Dunshea et al., 2001). This was proposed to be due to reduced sexual and aggression activities of immunocastrated male pigs. In general, immunologically castrated males produce pork of equivalent quality to females and surgically castrated males, and are superior to that from entire males (Allison et al., 2009). Whether the technology achieves global use will depend on the market acceptance (Fredriksen et al., 2011; Vanhonacker et al., 2009).

Other ways to reduce the problem of boar taint include:

- Infuse the pork with solutions containing compounds that may mask the smell/taste.
- Ensure feeding in the piggery is maintained to minimise any boar taint, for example, minimise feeding of protein of animal origin.
- Ensure the environment in the piggery is maintained to minimise any boar taint, for example, keep pigs and pens clean and free of faeces.

4.1.2 Low IMF

Where pig populations are producing very lean, muscular carcasses, the IMF content can be very low. The main problems with low IMF are the lack of juiciness and tenderness. There are several strategies that can be used alone, or in combination, to reduce the problem of low IMF pork. These are:

- Increase the content of Duroc in the pig population.
- Use moisture enhancement to increase the perceived juiciness and water-holding capacity (WHC) of the muscles.
- Develop and implement appropriate cooking procedures, particularly ensuring that overcooking is avoided. An example of an alternate cooking technique for consumers that can be beneficial is *sous vide* cooking.

4.2 Reducing the occurrence of aberrant quality due to low or high muscle pH (PSE and DFD)

Elimination of the HAL (*ryr1*) gene from the pig population was originally thought to dramatically reduce the frequency of PSE and DFD pork. Pigs carrying the HAL gene are certainly more susceptible to stress and will respond adversely to stressors around slaughter, such as electric goads during handling and stunning method, and exhibit a higher incidence of PSE (Channon et al., 2000). In fact, pigs which are not carrying the gene can exhibit high levels of PSE, particularly in response to electrical stunning at slaughter (Channon et al., 2000), use of electric goads pre-slaughter and negative pre-slaughter handling (Channon et al., 2000; D'Souza et al., 1998b,c). The use of dietary magnesium supplements for five days prior to slaughter has been successful in reducing the incidence of PSE in pig carcasses as has reducing the use of electric goads (D'Souza et al., 1998d).

In conclusion, strategies for reducing occurrence of PSE and DFD carcasses include:

- reducing the incidence of the halothane gene,
- using low stress handling,
- eliminate use of electric goads,
- feed magnesium for short period (3–5 days) pre-slaughter,
- consider using carbon dioxide stunning.

4.3 Meeting consumer demand for high quality

The main criteria for producing high-quality pork are to ensure that the visual, processing and sensory quality traits are considered. For visual and processing quality, attention needs to be paid to reducing the incidence of PSE and DFD carcasses, as described above. For sensory quality, the important attributes are tenderness, juiciness and flavour.

For juiciness:

- reduce the occurrence of PSE,
- if IMF % in muscles are low, use criteria above to address this.

For flavour:

- reduce and/or eliminate boar taint,
- ensure feeds which may affect the flavour of the meat/fat are monitored and either kept to low percentage of diet or eliminated, for example, fishmeal, fish oil or animal protein,
- consider feeding compounds which may enhance flavour, for example, chicory or hazelnuts,
- ageing may help with increasing flavour compounds.

For tenderness, each of the following can be used alone or in combination:

- Eliminate PSE.
- Consider ageing the meat for 5–7 days or longer in order to enhance proteolysis.
- Consider aitch-bone hanging or moisture infusion (enhancement).
- If the pH decline is slow, electrical stimulation of carcasses may assist.
- Innovative technologies such as shock wave and HPP may assist in improving tenderness, particularly if ready-to-eat meals for the supermarket shelves are the end product.

5 Case study: pork quality in Australia

Consumer decisions to purchase pork at the retail level can be influenced by its appearance, together with price. However, both consumption and re-purchasing frequency are largely dependent on whether the product satisfies the consumer's expectations on the basis of its eating quality. The ability to consistently provide fresh pork cuts that can meet consumer requirements for eating quality presents significant challenges for the pork industry, primarily due to the lack of a system for pork that enables it to be graded for eating quality. Such a system will need to take into account the plethora of pathway factors, from production to consumption, and their complex interactions that can influence tenderness, juiciness and/or flavour of pork.

As background, issues in Australia with variable pork quality of Australian pork loins purchased at the retail level were initially identified by Hofmeyr (1998), with 54% of pork samples found to be unacceptable to consumers, based on a WBSF value of 5 kg (49 N) being the maximum for consumer acceptability of lamb (Shorthose et al., 1986). This indicated to industry that efforts to improve pork eating quality consistency were needed and while objective measurements such as peak shear force are useful indicators of eating quality, sensory evaluations by consumers of pork, rather than trained taste panellists, were needed to ensure that the outcomes are reflective of consumer requirements for pork. The implementation of production and processing interventions, including aitch-bone hanging (rather than hanging carcasses from the Achilles tendon) and ageing period (from two days to seven days), were shown by Channon et al. (2001) to reduce the incidence of unsatisfactory pork from an eating quality perspective (based on a consumer having a negative re-purchase intention) from one in four (25%) to one in 20 (4%). The development of an eating quality pathway model has been a key priority of the Australian pork industry over the past decade, through the support of an extensive research and development programme utilising sensory evaluations of pork by consumers.

The first step was to quantify the size, effect and variability of key pathway factors from production to consumption (and the interactions between them) on the eating quality of pork. This was investigated by Channon et al. (2016b) using Monte Carlo simulation analyses, as an alternative approach to meta-analysis. This study involved the compilation of an extensive database comprising 328 research studies that reported effects of production, pre-slaughter, post-slaughter and/or cooking parameters on pork eating quality, namely tenderness, juiciness and flavour. For comparisons to be made, given the different factors investigated in different experimental designs, the approach used was to arbitrarily set variables for an Australian 'standard' pig (Table 1). Only studies that had included the variable for the 'standard' pig for each specific pathway parameter were included in the analysis for that parameter – noting that no studies involved all variables for the 'standard pig'. The proportional change of the mean data for each comparative variable compared with that for 'standard' pig variable for each study was then calculated. The correction factor for each 'standard' pig variable was therefore set at one. The normal distribution was used to characterise the uncertainty for the correction factors for each of the pathway parameters investigated.

Overall, for many of the pathway parameters investigated, only small shifts in sensory scores were found, with means of probability distributions being close to 1, as seen in Fig. 7 and 8. This analysis identified that positive improvements by more than 10% in means of probability distributions of correction factors were only obtained for a small number of parameters, including hanging method (tenderstretch/aitchbone vs Achilles) for tenderness and juiciness, ageing period (3–7 days vs 1–2 days), breed (White vs Hampshire) and sex

Table 1 Arbitrary variables of the Australian 'standard' pig set for each key pathway parameter

Pathway parameter	Variable (x)	Comparative variables
Gender	Female	Entire male, physical castrate, immunocastrated male
Genotype	White	≥50% Duroc, Pietrain, Hampshire, Berkshire
Halothane gene	Normal (NN)	Heterozygote carrier (Nn), homozygous reactor recessive (nn)
Housing	Indoor/conventional	Outdoor, Straw bedding
Plane of nutrition	Ad libitum	Restricted
Metabolic modifiers	None	Porcine somatotrophin (pST), ractopamine
Mixing	None	Mixed
Stunning method	CO_2	Electrical
Electrical stimulation	None	Applied
Ageing period	1–2 days	3–7 days, >7 days
Hanging method	Achilles	Aitchbone (tenderstretch)
Moisture infusion	No	Yes, moisture infusion
Final internal temperature	70–74°C	<70°C, >74°C
Intramuscular fat content	<1.6%	>1.6%
Ultimate pH	5.5–5.7	>5.7, <5.5

Source: From Channon et al. (2016a).

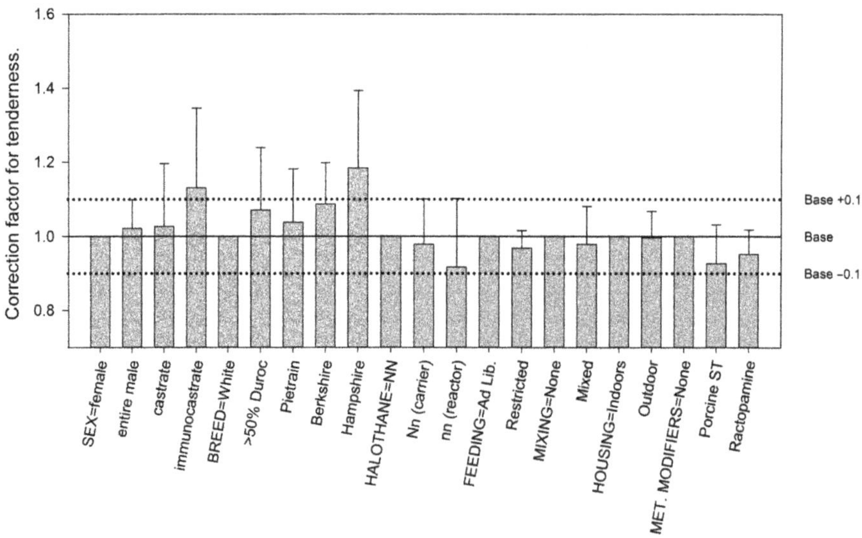

Figure 7 Effect of various on-farm factors on consumer tenderness score. The horizontal dotted black lines show base ±0.1. All terms on the x-axis are defined in Table 1, except MET. MODIFIERS = metabolic modifiers and Ad Lib. = Ad libitum. The base value is '1' for the 'standard' pig and is indicated by capitals on the x-axis. The correction factor for each comparative variable is shown after the 'standard' pig base parameter on the x-axis. Derived from Channon et al. (2016b).

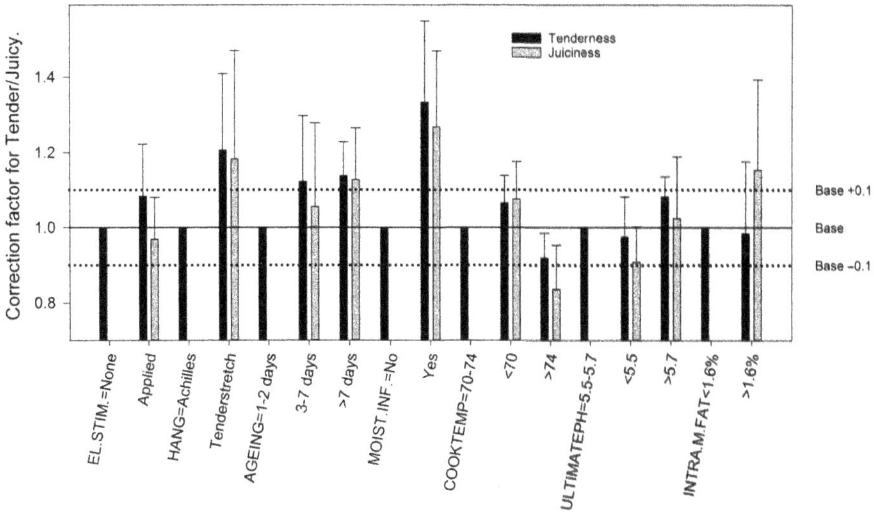

Figure 8 Effect of various post-mortem factors on consumer tenderness and juiciness scores. The horizontal dotted black lines show base ±0.1. All terms on the x-axis are defined in Table 1, except EL.STIM. = electrical stimulation, MOIST.INF. = moisture infusion and INTRA.M.FAT = intramuscular fat content. The base value is '1' for the 'standard' pig and is indicated by capitals on the x-axis. The correction factor for each comparative variable is shown after the 'standard' pig base parameter on the x-axis. Derived from Channon et al., (2016b).

(female vs immunocastrate) for tenderness, ageing period (>7 days vs 1–2 days) for tenderness and juiciness and moisture infusion/enhancement (moisture infusion vs no moisture infusion) for tenderness and juiciness (Figs. 7 and 8). These findings also indicate that moisture infusion is a more successful intervention in improving tenderness and juiciness compared with hanging method or ageing period. Notably, none of the pathway parameters investigated in the Monte Carlo analysis resulted in improvements in flavour scores comparable to the increases seen for tenderness or juiciness. The mean of the probability distribution of correction factors for consumer tenderness scores for pork from immunocastrated pigs was increased by 13% compared with females; however, this distribution only included data from three studies, indicating the need for more data. Interestingly, the mean of the probability distribution of correction factors for juiciness of pork cooked to a final endpoint temperature of >74°C was reduced by 16% compared with pork cooked to 70–74°C. Furthermore, the mean of the probability distribution of the correction factor for juiciness of pork with low ultimate pH of <5.5 was also reduced by about 10% compared with pork with a pH of 5.5–5.7.

In order to develop a non-prescriptive, cuts-based predictive model for use by industry to enable the consistent production of pork of high eating quality, significant data and knowledge gaps were identified. Deficiencies in the existing data and new studies which are needed to address these gaps include the following.

- Many multi-factorial studies did not include data for means or averages, of interacting terms, where interactions were not significant.
- Correction factors for multi-factorial interactions between different pathway parameters could not be determined due to lack of data.

- The majority of studies utilised only the loin muscle – primarily grilled, oven broiled or roasted as either a steak or chop, with variations in cut thickness.
- For a cuts-based model, data for additional primals and cuts cooked using a variety of cooking methods and to different cooking endpoint temperatures is needed.
- Additional eating quality data are needed for immunocastrated male pigs due to only a small number of studies suitable for inclusion in this analysis. This is a result of having to exclude studies with methodological problems or deficiencies.
- It is not clear whether the various factors which influence eating quality of pork have an additive effect, as suggested by Channon et al. (2016a) and Moore et al. (2015), or a multiplying effect.

Channon et al. (2016a) used the knowledge gaps stated above to design a multi-factorial experiment and then conducted it in commercial facilities. The aim was to determine the interacting effects of gender (entire male, physical castrate males; female), ageing period (2 days; 7 days), endpoint cooking temperature (70; 75°C), cut/muscle (leg, *M. biceps femoris*; shoulder *M. triceps brachii*; *M. supraspinatus*; loin, *LTL*) and cooking method (roast; stir fry; grill; not all cuts used for all cook methods) on consistency of pork eating quality. Pigs were slaughtered at 22 weeks of age at 90–105 kg liveweight and evaluated by consumers.

Across all cut types, pork from entire males scored lower for juiciness and flavour compared with physical castrates, with females intermediate. Boar taint was also found in 15% of entire males involved in this study. In addition, a higher proportion of consumers assessed pork from entire males as either unsatisfactory or below average (fail rate), compared with castrates and females (Fig. 9). Together, these results strongly suggest that the inclusion of entire males into an eating quality system would contribute significant product variability issues. This was confirmed by a study in the Australian pork industry research programme, where higher fail rates were obtained for pork loins from entire males compared with immunocastrated male pigs even though tenderness, juiciness and flavour scores were not influenced by gender (Moore et al., 2017).

Across all cuts in this study, there was no effect of ageing period (Fig. 9), even though it was previously identified that ageing of pork loin for >2 days can improve tenderness (Channon et al., 2016b). It is likely that the apparent lack of effect of ageing on consumer tenderness scores was caused by (1) taking samples at two days post-slaughter, rather than one day post-slaughter, as some ageing occurs between 1 and 2 days post-slaughter and/or (2) a prevalence of low ultimate pH in the silverside (*biceps femoris*) and loin (*LTL*); thus a rapid attainment of tenderness and lack of further ageing would occur, as previously observed in beef and sheep meat (Kim et al., 2014; Warner et al., 2014a,b).

Interestingly, differences in eating quality between the different cut types that were evaluated were greater in magnitude than any of the pathway interventions investigated (Fig. 9). Overall, lower fail rates were obtained for shoulder stir fry (5.6%) followed by shoulder roast (10.0%), loin stir fry (15.2%), loin roast (19.2%), silverside stir fry (21.5%), loin steak (30.2%) and finally silverside roast (36.0%) (see Fig. 9 for some of the data). This again may be related to the ultimate pH in the different muscles, as the loin and blade (*triceps brachii*) had an ultimate pH range of 5.41–5.48 and 5.61–5.68, respectively. Although these differences seem small, pork quality is well-known to improve as the ultimate pH increases above 5.5. These results identified the need for continuing focus on development of interventions for the loin and silverside in order to provide industry with feasible solutions to improve their eating quality performance.

Although endpoint temperature as a main effect did not improve eating quality scores, interestingly, cooking pork steaks to a 75°C endpoint temperature reduced both juiciness

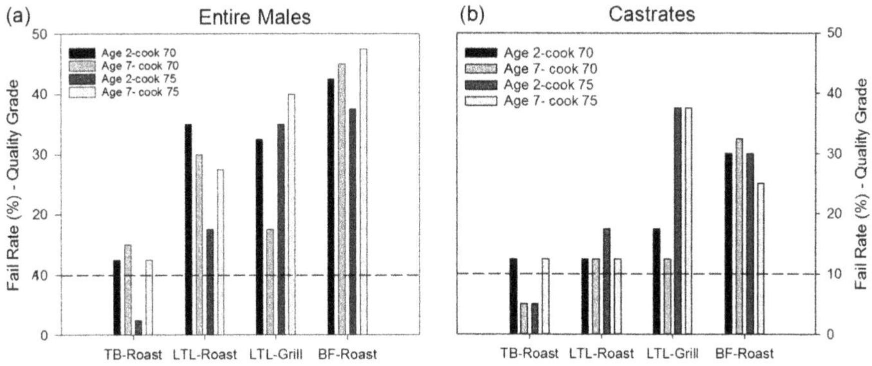

Figure 9 Effect of sex [(a) entire males; (b) castrates], ageing of meat (Age 2, 2 days post-slaughter; Age 7, 7 days post-slaughter), cooking temperature (cook 70, 70°C; cook 75, 75°C), muscle (TB, *triceps brachii*; LTL, *longissimus thoracis et lumborum*; BF, *biceps femoris*) and cook method (roast; grill) on fail rates (%) for quality grades. The dotted line at 10% represents the defined acceptability. Below 10%, less than 10% of samples within the treatment combination were given a score of 1= unsatisfactory (this was terrible, I did not enjoy it at all) or 2= below average (this was not nice, I did not enjoy it). Derived from Channon et al. (2016b).

and flavour scores compared with 70°C. This finding highlights that continuing efforts to improve consumer confidence and knowledge on how to cook pork loin steaks is required, in order to ensure that overcooking does not cause a decline in pork eating quality. During 2016, the Australian pork industry commenced the communication of the simple and innovative '6-2-2' cooking message to consumers suitable for a 2 cm thick steak – preheat the pan, 6 minutes on one side without turning, 2 minutes on the other side and then rest for 2 minutes before serving (www.pork.com.au). Such messages to consumers will assist with increasing the purchasing frequency of loin steaks as well as support the industry's endeavours to ensure that the high-quality fresh pork that is produced is able to satisfy consumer expectations once cooked.

Subsequent studies used commercial supply chains to investigate the effects of immunocastration of males, extending the ageing period to 7–14 days, tender stretching, electrical stimulation and moisture infusion (Channon et al., 2015a,b,c). Pork cuts produced from immunocastrated male pigs were consistently comparable or better than pork from female pigs, supporting the inclusion of immunocastrate males in an eating quality system, together with females. The effect of the other interventions on pork eating quality was inconsistent between supply chains. This demonstrates the need to develop various and new intervention options relevant for specific supply chains. Again there was a lack of increase in consumer tenderness scores and acceptability in response to 7–14 days ageing of vacuum-packaged pork cuts. Reasons for this continue to be explored.

In summary, the Monte Carlo analysis conducted on a large number of studies showed the important positive effects of immunocastration, the Hampshire breed, tender stretch, ageing moisture infusion and IMF on consumer scores for tenderness and juiciness. The positive effects of immunocastration and moisture infusion were confirmed in subsequent studies but the positive effects of ageing and of IMF were not confirmed for the Australian pig population, Furthermore, muscles with a higher ultimate pH (blade) were consistently better in eating quality than muscles with a lower ultimate pH (especially loin). Thus it is evident that changes in production and post-mortem practices that have been globally found to improve eating quality always need to be tested at a local level, to confirm the effects in the local pig/pork population.

6 Conclusion and future trends

Research on pig production and growth utilising genetics, nutritional management and use of metabolic modifiers to optimise carcass composition has achieved much in the last 10–20 years. The changes have resulted in an efficiently produced modern pig carcass which is very lean and has low IMF. The industry has identified that the meat from these carcasses can tend to be tough and lack juiciness and may have 'boar taint'.

Solutions for boar taint include immunocastration, physical castration, online monitoring of skatole in the carcass fat and infusion of solutions to mask the odour/taste. Physical castration will likely cease in many countries, due to animal welfare and consumer concerns. Further research required includes:

- Identification of the causative genes and development of genetic tests for boar taint, to enable genetic selection to occur.
- Development of rapid, cheap, accurate methods for online monitoring of androstenone, as well as skatole.
- Research on novel, natural compounds to mask the odour and taste of boar taint.

The increase in pork toughness associated with the use of metabolic modifiers and with production of lean carcasses is comparatively small. In addition, the genetic markers for tenderness which have been identified are only small in their effect and unlikely to assist in the drive for increased consistency for tenderness. Research and development that is required includes:

- Some processing interventions have been extensively researched and the adoption of these technologies should be encouraged (e.g. tender stretch and moisture infusion).
- Demonstration that various existing processing interventions can be used to increase tenderness and accelerate tenderisation post-mortem.
- The application of innovative and novel technologies, such as HPP, ultrasound, shockwave and pulsed electric field, for tenderisation have had little or no investigation on the application to pork. Thus the potential for these technologies to assist in providing consistently tender pork requires research.
- Considering the low WBSF often reported for pork in the existing literature, the importance of tenderness (relative to other quality traits) needs to be verified, using consumer sensory panels.

The decrease in juiciness associated with low IMF and low pH meat can be successfully overcome through moisture infusion but this sometimes results in bitter flavours and is not always acceptable to the consumer demanding 'fresh, natural' pork. In the future, research which would assist in addressing the issue includes:

- Development and commercialisation of genetic tests for IMF, to enable genetic selection to occur.
- Research on natural additives to the meat which may enhance the 'perception' of fat/juiciness.

The problem of low pH pork and rapid post-metabolism creates quality problems through effects on the visual appearance of the pork cuts on the retail shelves, the eating quality and the quality of processed products. To the credit of the pig industry, much research which has been conducted on the prevention of low pH, rapid post-mortem metabolism and PSE and DFD has been adopted, for example, CO_2 stunning, minimising the use of electric goads, reducing or eliminating the occurrence of the halothane reduction, feeding magnesium and so on.

Until the producer is rewarded for producing quality meat, there will be no incentive for the pig industry to implement systems for quality improvement. There has been excellent progress in developing a large database on the various on-farm and post-farm factors which influence the eating quality of pork. This research has identified the gaps and the future research needs include:

- Large multi-factorial studies on the modern pig which collect data on the interactions and use consumer panels to assess the eating quality.
- More data on the eating quality of cuts other than the loin, using a range of cooking methods and temperatures.
- Additional eating quality data is needed for immunocastrated male pigs.
- The eating quality data needs to be collected on a regular basis, recognising the temporal changes in the pig population and also in consumer requirements.
- The data need to be collected locally, as each pig population and consumer population differs. Ideally, procedures would be standardised globally.
- The global pig industry would greatly benefit if local data on eating quality could be collated globally, to enable efficiencies and rapid transfer and uptake of new knowledge. This is aspirational but the potential achievements are enormous.

7 Where to look for further information

Australasian Pig Science Association. http://www.apsa.asn.au/
Danish Meat Research Institute. http://www.dti.dk/services/
University of Illinois, Champaign-Urbana.

8 References

Aalhus, J. L., Best, D. R., Costello, F. and Schaefer, A. L. (1997). The effects of porcine somatotropin on muscle fibre morphology and meat quality of pigs of known stress susceptibility. *Meat Science*, 45(3), 283–95. doi:http://dx.doi.org/10.1016/S0309-1740(96)00111-8.

Aalhus, J. L., Jones, M. D. S., Lutz, S., Best, D. R. and Robertson, W. M. (1994). The efficacy of high and low voltage electrical stimulation under different chilling regimes. *Canadian Journal of Animal Science*, 74, 433–42.

Aaslyng, M. D., Oksama, M., Olsen, E. V., Bejerholm, C., Baltzer, M., Andersen, G., Bredie, W. L. P., Byrne, D. V. and Gabrielsen, G. (2007). The impact of sensory quality of pork on consumer preference. *Meat Science*, 76(1), 61–73.

Acton, J. and Dick, R. (1984). Protein-protein ineraction in processed meats. *Reciprocal Meat Conference Proceedings*, 37, 36–43.

Acton, J. and Dick, R. (1996). Functional roles of heat induced protein gelation in processed meat. In J. E. Kinsella and W. G. Soucie (Eds), *Food Proteins* (pp. 195–209). Illinois: The American Oil Chemists Society, Champaign. (Reprinted from: In File).

Adeola, O. and Ball, R. O. (1992). Hypothalamic neurotransmitter concentration and meat quality in stressed pigs offered excess dietary tryptohan and tyrosine. *Journal of Animal Science*, 70, 1888–94.

Allison, J., McKeith, F., Souza, C., Boler, D., Killefer, J. and Hennessy, D. (2009). *Impact of using vaccination with Improvac on the sensory perception of meat from finishing male pigs*. Paper presented at the Proceedings of 55th International Congress of Meat Science and Technology, Copenhagen, Denmark.

Aluwé, M., Bekaert, K. M., Tuyttens, F. A. M., Vanhaecke, L., De Smet, S., De Brabander, H. F., De Brabander, D. L. and Millet, S. (2011). Influence of soiling on boar taint in boars. *Meat Science*, 87(3), 175–9. doi:http://dx.doi.org/10.1016/j.meatsci.2010.10.005.

Asghar, A., Gray, J. I., Booren, A. M., Gomaa, E. A., Abouzied, M. M. and Miller, E. R. (1991). Effects of supranutritional dietary vitamin e levels on subcellular deposition of a-tocopherol in the muscle and on pork quality. *Journal of Science in Food and Agriculture*, 57, 31–41.

Asghar, A., Samejima, K., Yasui, T. and Henrickson, R. L. (1985). Functionality of muscle proteins in gelation mechanisms of structured meat products. *C R C Critical Reviews in Food Science and Nutrition*, 22(1), 27–106.

Babol, J. and Squires, E. J. (1995). Quality of meat from entire male pigs. *Food Research International*, 28(3), 201–12.

Barbut, S., Sosnicki, A. A., Lonergan, S. M., Knapp, T., Ciobanu, D. C., Gatcliffe, L. J., Huff-Lonergan, E. J. and Wilson, E. W. (2008). Progress in reducing the pale, soft and exudative (PSE) problem in pork and poultry meat. *Meat Science*, 79, 46–63.

Barton-Gade, P. A. (1997). The effect of pre-slaughter handling on meat quality in pigs. *Manipulating Pig Production*, 6, 100–15.

Beermann, D. H., Fishell, V. K., Roneker, K., Boyd, R. D., Armbruster, G. and Souza, L. (1990). Dose-response relationships between porcine somatotropin, muscle composition, muscle fiber characteristics and pork quality. *Journal of Animal Science*, 68(9), 2690–7. doi:10.2527/1990.6892690x.

Bejerholm, C. and Aaslyng, M. D. (2003). The influence of cooking technique and core temperature on results of a sensory analysis of pork – depending on the raw meat quality. *Food Quality and Preference*, 15, 19–30.

Bejerholm, C. and Barton-Gade, P. (1986). Effect of intramuscular fat level on the eating quality of pig meat (Vol. Manuscript No. 720E): Danish Meat Research Institute, Roskilde, Denmark.

Bertram, H. C. and Aaslyng, M. D. (2007). Pelvic suspension and fast post-mortem chilling: Effects on technological and sensory quality of pork – A combined NMR and sensory study. *Meat Science*, 76(3), 524–35.

Bonneau, M. (1998). Use of entire males for pig meat in the European Union. *Meat Science*, 49(Supplement 1), S257–72.

Bowker, B. C., Callahan, J. A. and Solomon, M. B. (2010). Effects of hydrodynamic pressure processing on the marination and meat quality of turkey breasts. *Poultry Science*, 89(8), 1744–9.

Bowker, B. C., Grant, A. L., Forrest, J. C. and Gerrard, D. E. (2000). Muscle metabolism and PSE pork. *Journal of Animal Science*, 79, 1–8.

Brewer, M. S., Jensen, J., Sosnicki, A. A., Fields, B., Wilson, E. and McKeith, F. K. (2002). The effect of pig genetics on palatability, color and physical characteristics of fresh pork loin chops. *Meat Science*, 61(3), 249–56.

Buckley, D. J. and Connolly, F. (1980). Influence of Alpha-Tocopherol (Vitamin E) on Storage Stability of Raw Pork and Bacon. *Journal of Food Protection*, 43(4), 265–7.

Bunter, K. L., Bennett, C., Luxford, B. G. and Graser, H. U. (2008). Sire breed comparisons for meat and eating quality traits in Australian pig populations. *Animal*, 2(8), 1168–77.

Cameron, N. and Enser, M. (1991). Fatty acid composition of lipid in Longissimus Dorsi muscle of Duroc and British Landrace pigs and its relationship with eating quality. *Meat Science*, 29, 295–307.

Cameron, N. D., Warriss, P. D., Porter, S. J. and Enser, M. B. (1990). Comparison of Duroc and British Landrace pigs for meat and eating quality. *Meat Science*, 27(3), 227–47.

Camou, J. P. and Sebranek, J. G. (1991). Gelation characteristics of muscle proteins from pale soft exudative pse pork. *Meat Science*, 30, 207–20.
Campbell, R. G. and Taverner, M. R. (1988). Genotype and sex effects on the relationship between energy intake and protein deposition in growing pigs. *Journal of Animal Science*, 66(3), 676–86.
Candek-Potokar, M., Zlender, B., Lefaucheur, L. and Bonneau, M. (1998). Effects of age and/or weight at slaughter on longissimus dorsi muscle: Biochemical traits and sensory quality in pigs. *Meat Science*, 48(3–4), 287–300.
Channon, H. A., Baud, S. R., Kerr, M. G. and Walker, P. J. (2003a). Effect of low voltage electrical stimulation of pig carcasses and ageing on sensory attributes of fresh pork. *Meat Science*, 65(4), 1315–24.
Channon, H. A., Baud, S. R. and Reynolds, J. (2001). Identifying pathways to ensure acceptable eating quality of pork (DV 163/1385). Final Report to Australian Pork Limited, Canberra ACT.
Channon, H. A., D'Souza, D. N. and Dunshea, F. R. (2015a). Aitchbone hanging or moisture infusion, but not ageing, influenced eating quality of pork cuts. *Animal Production Science*, 55(12), p 1474.
Channon, H. A., D'Souza, D. N. and Dunshea, F. R. (2015b). Pork eating quality was not improved by extended ageing for 14 days. *Animal Production Science*, 55(12), 1468.
Channon, H. A., D'Souza, D. N. and Dunshea, F. R. (2015c). Response to different pathway interventions to improve pork eating quality consistency. *Animal Production Science*, 55(12), 1473.
Channon, H. A., D'Souza, D. N. and Dunshea, F. R. (2016a). Developing a cuts-based system to improve consumer acceptability of pork: Impact of gender, ageing period, endpoint temperature and cooking method. *Meat Science*, 121, 216–27.
Channon, H. A., Hamilton, A. J., D'Souza, D. N. and Dunshea, F. R. (2016b). Estimating the impact of various pathway parameters on tenderness, flavour and juiciness of pork using Monte Carlo simulation methods. *Meat Science*, 116, 58–66.
Channon, H. A., Jongman, E. C., Warner, R. D. and Pengelly, A. M. (1997). The effect of carbon dioxide and elecrical stunning on meat quality and the welfare of pigs. DAV106P. A final report prepared for the Pig Rsearch and Development Corporation (pp. 1–115).
Channon, H. A., Kerr, M. G. and Walker, P. J. (2004). Effect of Duroc content, sex and ageing period on meat and eating quality attributes of pork loin. *Meat Science*, 66(4), 881–8.
Channon, H. A., Payne, A. and Warner, R. D. (2000). Halothane genotype, pre-slaughter handling and stunning method all influence pork quality. *Meat Science*, 56, 291–9.
Channon, H. A., Payne, A. M. and Warner, R. D. (2002). Comparison of CO2 stunning with manual electrical stunning (50 Hz) of pigs on carcass and meat quality. *Meat Science*, 60(1), 63–8.
Channon, H. A., Payne, A. M. and Warner, R. D. (2003b). Effect of stun duration and current level applied during head to back and head only electrical stunning of pigs on pork quality compared with pigs stunned with CO_2. *Meat Science*, 65, 1325–33.
Ciobanu, D. C., Bastiaansen, J. W. M., Lonergan, S. M., Thomsen, H., Dekkers, J. C. M., Plastow, G. S. and Rothschild, M. F. (2004). New alleles in calpastatin gene are associated with meat quality traits in pigs. *Journal of Animal Science*, 82, 2829–39.
Claus, R., Lösel, D., Lacorn, M., Mentschel, J. and Schenkel, H. (2003). Effects of butyrate on apoptosis in the pig colon and its consequences for skatole formation and tissue accumulation1. *Journal of Animal Science*, 81(1), 239–48.
Claus, R., Weiler, U. and Herzog, A. (1994). Physiological aspects of androstenone and skatole formation in the boar – A review with experimental data. *Meat Science*, 38(2), 289–305.
Crawford, S. M., Moeller, S. J., Zerby, H. N., Irvin, K. M., Kuber, P. S., Velleman, S. G. and Leeds, T. D. (2010). Effects of cooked temperature on pork tenderness and relationships among muscle physiology and pork quality traits in loins from Landrace and Berkshire swine. *Meat Science*, 84, 607–12.
D'Souza, D. N., Dunshea, F. R., Hewitt, R. J. E., Luxford, B. G., Meaney, D., Schwenke, F., Smits, R. J. and Van Barneveld, R. J. (2011). In R. J. Barneveld (Ed.), *Manipulating Pig Production XIII* (XIII ed., pp. 259). Werribee, Victoria, Australia: Australasian Pig Science Association (Inc.). (Reprinted from: Not in File).

D'Souza, D. N., Dunshea, F. R., Leury, B. J. and Warner, R. D. (1999a). Effect of mixing boars during lairage and pre-slaughter handling on pork quality. *Australian Journal of Agricultural Research*, 50, 109–13.

D'Souza, D. N., Dunshea, F. R., Warner, R. D. and Leury, B. J. (1998a). The Effect of Handling Pre-Slaughter and Carcass Processing Rate Post-Slaughter on Pork Quality. *Meat Science*, 50, 429–37.

D'Souza, D. N., McCollough, S., Brennan, S., Penn, R. and Mullan, B. P. (2003a). Benchmarking the eating quality of branded pork in Western Australia. In P. D. Cranwell (Ed.), *Manipulating Pig Production XI* (XIII ed., Vol. XI, pp. 25). Australasian Pig Science Association (Inc.). (Reprinted from: Not in File), Perth, Western Australia.

D'Souza, D. N., Pethick, D. W., Dunshea, F. R., Pluske, J. R. and Mullan, B. P. (2003b). Nutritional manipulation increases intramuscular fat levels in the *Longissimus* muscle of female finisher pigs. *Australian Journal of Agricultural Research*, 54, 745–9.

D'Souza, D. N., Pethick, D. W., Dunshea, F. R., Pluske, J. R. and Mullan, B. P. (2008). Reducing the lysine to energy content in the grower growth phase diet increases intramuscular fat and improves the eating quality of the longissimus thoracis muscle of gilts. *Australian Journal of Experimental Agriculture*, 48, 1105–9.

D'Souza, D. N., Warner, R. D., Dunshea, F. R. and Leury, B. J. (1998c). Effect of on-farm and pre-slaughter handling of pigs on meat quality. *Australian Journal of Agricultural Research*, 49, 1021–5.

D'Souza, D. N., Warner, R. D., Dunshea, F. R. and Leury, B. J. (1999b). Comparison of different dietary magnesium supplements on pork quality. *Meat Science*, 51, 221–5.

D'Souza, D. N., Warner, R. D., Leury, B. J. and Dunshea, F. R. (1998d). The effect of dietary magnesium aspartate supplementation on pork quality. *Journal of Animal Science*, 76, 104–9.

D'Souza, D. N., Warner, R. D., Leury, B. J. and Dunshea, F. R. (2000). The influence of dietary magnesium supplement type, and supplementation dose and duration, on pork quality and the incidence of PSE pork. *Australian Journal of Agricultural Research*, 51(2), 185–90.

Deng, Y., Rosenvold, K., Karlsson, A. H., Horn, P., Hedegaard, J. and Andersen, H. J. (2002). Relationship between thermal denaturation of procine muscle proteins and water-holding capacity. *Journal of Food Science*, 67(5), 1642–7.

DeVol, D. L., McKeith, F. K., Bechtel, P. J., Novakofski, J., Shanks, R. D. and Carr, T. R. (1988). Variation in composition and palatability traits and relationships between muscle characteristics and palatability in a random sample of pork carcasses. *Journal of Animal Science*, 66(2), 385–95.

Dikeman, M. E. (2007). Effects of metabolic modifiers on carcass traits and meat quality. *Meat Science*, 77, 121–35.

Dransfield, E., Jones, R. and MacFie, H. (1980–81). Tenderising in *M. Longissimus dorsi* of beef, veal, rabbit, lamb and pork. *Meat Science* 5, 139–47.

Duijvesteijn, N., Knol, E. F., Merks, J. W., Crooijmans, R. P., Groenen, M. A., Bovenhuis, H. and Harlizius, B. (2010). A genome-wide association study on androstenone levels in pigs reveals a cluster of candidate genes on chromosome 6. [journal article]. *BMC Genetics*, 11(1), 42.

Dunshea, F. R., Colantoni, C., Howard, K., McCauley, I., Jackson, P., Long, K. A., Lopaticki, S., Nugent, E. A., Simons, J. A., Walker, J. and Hennessy, D. P. (2001). Vaccination of boars with a GnRH vaccine (Improvac) eliminates boar taint and increases growth performance. *Journal of Animal Science*, 79(10), 2524–35.

Dunshea, F. R., D'Souza, D. N., Pethick, D. W., Harper, G. S. and Warner, R. D. (2005). Effects of dietary factors and other metabolic modifiers on quality and nutritional value of meat. *Meat Science*, 71, 8–38.

Dunshea, F. R., D'Souza, D. N. and Channon, H. A. (2016). Metabolic modifiers as performance-enhancing technologies for livestock production. *Animal Frontiers*, 6(4), 6–14.

Edwards, S. A., Wood, J. D., Moncrieff, C. B. and Porter, S. J. (1992). Comparison of the Duroc and Large White as terminal sire breeds and their effect on pigmeat quality. *Animal Production*, 54, 289–97.

Eikelenboom, G., Hoving-Bolink, A. H. and van der Wal, P. G. (1996). The eating quality of pork I. The influence of ultimate pH. *Fleischwirtschaft*, 76, 392–3.

Essén-Gustavsson, B., Karlsson, A., Lundström, K. and Enfält, A. C. (1994). Intramuscular fat and muscle fibre lipid contents in halothane-gene-free pigs fed high or low protein diets and its relation to meat quality. *Meat Science*, 38(2), 269–77.

Estrade, M., Vignon, X. and Monin, G. (1993). Effect of the RN- gene on ultrastructure and protein fractions in pig muscle. *Meat Science*, 35(3), 313–19.

Fernandez, X., Monin, G., Talmant, A., Mourot, J. and Lebret, B. (1999). Influence of intramuscular fat content on the quality of pig meat – 2. Consumer acceptability of m. longissimus lumborum. *Meat Science*, 53, 67–72.

Fjelkner-Modig, S. and Persson, J. (1986). Carcass properties as related to sensory properties of pork. *Journal of Animal Science*, 63(1), 102–13.

Fox, J. B. (1987). The pigments of meat. In J. F. Price and B. S. Schweigert (Eds), *The Science of Meat and Meat Products* (3rd ed., pp. 1193–216). Food and Nutrition Press, Connecticut, USA.

Fredriksen, B., Johnsen, A. M. S. and Skuterud, E. (2011). Consumer attitudes towards castration of piglets and alternatives to surgical castration. *Research in Veterinary Science*, 90(2), 352–7.

Fuji, J., Otsu, K., Zorzato, F., De Leon, S., Khanna, V. K., Weiler, J. E., O'Brien, P. J. and MacLennan, D. H. (1991). Identification of a mutation in porcine ryandine receptor associated with malignant hypothermia. *Science*, 253, 448–51.

Gower, D. B. (1972). 16-Unsaturated C 19 steroids. A review of their chemistry, biochemistry and possible physiological role. *Journal of Steroid Biochemistry*, 3(1), 45–103.

Grindflek, E., Meuwissen, T. H. E., Aasmundstad, T., Hamland, H., Hansen, M. H. S., Nome, T., Kent, M., Torjesen, P. and Lien, S. (2011). Revealing genetic relationships between compounds affecting boar taint and reproduction in pigs. *Journal of Animal Science*, 89, 680–92.

Große-Brinkhaus, C., Storck, L. C., Frieden, L., Neuhoff, C., Schellander, K., Looft, C. and Tholen, E. (2015). Genome-wide association analyses for boar taint components and testicular traits revealed regions having pleiotropic effects. *BMC Genetics*, 16, 36.

Hagdrup, C. (2009). The Danish experience, from 27 years with skatole analysis of samples from entire male pigs. In H. J.E. (Ed.), *Report of the EAAP/ALCASDE Boar Taint Detection Workshop, Bologna, 28 October* 2009 (pp. 34–7).

Hansen, L. L., Larsen, A. E. and Hansen-Moller, J. (1995). Influence of Keeping Pigs Heavily Fouled with Faeces plus Urine on Skatole and Indole Concentration (Boar Taint) in Subcutaneous Fat. *Acta Agriculturae Scandinavica, Section A – Animal Science*, 45(3), 178–85.

Hansen, L. L., Larsen, A. E., Jensen, B. B., Hansen-Møller, J. and Barton-Gade, P. (1994). Influence of stocking rate and faeces deposition in the pen at different temperatures on skatole concentration (boar taint) in subcutaneous fat. *Animal Science*, 59(1), 99–110.

Hansen, L. L., Mejer, H., Thamsborg, S. M., Byrne, D. V., Roepstorff, A., Karlsson, A. H., Hansen-Møller, J., Jensen, M. T. and Tuomola, M. (2006). Influence of chicory roots (Cichorium intybus L) on boar taint in entire male and female pigs. *Animal Science*, 82(3), 359–68.

Hansson, K. E., Lundstrom, K., Fjelkner-Modig, S. and Persson, J. (1980). The importance of androstenone and skatole for boar taint. *Swedish Journal of Agricultural Research*, 10, 167–73.

Harris, P. V. and Shorthose, W. R. (1988). Meat Texture. In R. A. Lawrie (Ed.), *Developments in Meat Science* – 4 (pp. 245–96). Elsevier: London.

Harrison, G. G. (1979). Porcine malignant hypothermia. *International Anaesthesiology Clinics*, 17, 25–61.

Haugen, J. E., Brunius, C. and Zamaratskaia, G. (2012). Review of analytical methods to measure boar taint compounds in porcine adipose tissue: The need for harmonised methods. *Meat Science*, 90(1), 9–19.

Hemsworth, P. H. and Barnett, J. L. (1991). The effects of aversively handling pigs, either individually or in groups, on their behaviour, growth and corticosteroids. *Applied Animal Behaviour Science*, 30(1–2), 61–72.

Hemsworth, P. H., Barnett, J. L. and Hansen, C. (1986). The influence of handling by humans on the behaviour, reproduction and corticosteroids of male and female pigs. *Applied Animal Behaviour Science*, 15(4), 303–14.

Hemsworth, P. H., Barnett, J. L. and Hansen, C. (1987). The influence of inconsistent handling by humans on the behaviour, growth and corticosteroids of young pigs. *Applied Animal Behaviour Science*, 17(3–4), 245–52.

Hennessy, D. P., Dunshea, F. R., Jackson, P., Long, K., Lopaticki, S., McCauley, I. and Sali, L. (1997). Boar taint in Australian and New Zealand pigs. In P. D. Cramwell (Ed.), *Manipulating Pig Production Vi* (XIII ed., p. 144). Australasian Pig Science Association (Inc.). (Reprinted from: Not in File) Werribee, Victoria, Australia.

Hermesch, S. (1997). Genetic influences on pork quality. In P. D. Cranwell (Ed.), *Manipulating Pig Production VI* (Vol. VI, pp. 82–90). Australasian Pig Science Association, Werribee, Australia

Hertzman, C., Goransson, A. and Ruderus, H. (1988). Influence of fishmeal, rape-seed, and rape-seed meal in feed on the fatty acid composition and storage stability of porcine body fat. *Meat Science*, 23, 37–53.

Hocquette, J. F., Gondret, F., Ba, za, E., M,dale, F., Jurie, C. and Pethick, D. W. (2010). Intramuscular fat content in meat-producing animals: Development genetic and nutritional control and identification of putative markers. *Animal*, 4(2), 303–19.

Hofmeyr, C. D. (1998). Pork quality measured in Melbourne retail outlets. *Milnes Pork Journal*, 20, 24–5.

Honeyman, M. S., Pirog, R. S., Huber, G. H., Lammers, P. J. and Hermann, J. R. (2006). The United States pork niche market phenomenon1. *Journal of Animal Science*, 84(8), 2269–75. doi:10.2527/jas.2005-680.

Hong, G.-P., Shim, K.-B., Choi, M.-J. and Min, S.-G. (2008). Effects of thermal processing combined with high pressure on the characteristics of cooked pork. *Korean Journal for Food Science of Animal Resources*, 28(4), 415–21.

Hughes, D. (2008). *Managing Today, Planning for Tomorrow*. Skamania Lodge, WA: *PNW Agribusiness Executive Seminar*.

Hughes, J. M., Oiseth, S., Purslow, P. P. and Warner, R. D. (2014). A structural approach to understanding interactions between colour, water-holding capacity and tenderness. *Meat Science*, 98, 520–532.

Janss, L. L. G., Van-Arendonk, J. A. M. and Brascamp, E. W. (1997). Bayesian statistical analyses for presence of single genes affecting meat quality traits in a crossed pig population. *Genetics*, 145, 395–408.

Jensen, M. T. and Hansen, L. L. (2006). Feeding with chicory roots reduces the amount of odorous compounds in colon and rectal contents of pigs. *Animal Science*, 82(3), 369–76.

Jeremiah, L. E., Gibson, J. P., Gibson, L. L., Ball, R. O., Aker, C. and Fortin, A. (1999). The influence of breed, gender, and PSS (Halothane) genotype on meat quality, cooking loss, and palatability of pork. *Food Research International*, 32(1), 59–71.

Jin, S., Cottrell, J. J., Padayachee, A., Channon, H. A., Warner, R. D. and Dunshea, F. R. (2016). Sous-vide cooking and aging can reduce the shear force and hardness of pork. *Advances in Pork Production*, 27, Abstract #25.

Jonsdottir, R., Valdimarsdottir, T., Baldursdottir, B. and Thorkelsson, G. (2003). Influence of low fat fishmeal on fatty acid composition and sensory quality of pork. *Journal of Muscle Foods*, 14(1), 51–65.

Joo, S. T., Kim, G. D., Hwang, Y. H. and Ryu, Y. C. (2013). Control of fresh meat quality through manipulation of muscle fiber characteristics. *Meat Science*, 95(4), 828–36.

Judge, M., Aberle, E., Forrest, J., Hedrick, H. B. and Merkel, R. (1989). *Principles of Meat Science* (2 ed.), Kendall/Hunt Publishing Co. Iowa, USA

Karlsson, A., Enfält, A. C., Essén-Gustavsson, B., Lundström, K., Rydhmer, L. and Stern, S. (1993). Muscle histochemical and biochemical properties in relation to meat quality during selection for increased lean tissue growth rate in pigs. *Journal of animal science*, 71(4), 930–8.

Kauffman, R. G. and Marsh, B. B. (1987). Quality characteristics of muscle as food. In J. F. Price and B. S. Schweigert (Eds), *The Science of Meat and Meat Products* (pp. 349–69). Westport, CT: Food and Nutrition Press

Kim, Y. H. B., Warner, R. D. and Rosenvold, K. (2014). Influence of high pre-rigor temperature and fast pH fall on muscle proteins and meat quality: A review. *Animal Production Science*, 54, 375–95.

Kuo, C. C. and Chu, C. Y. (2003). Quality characteristics of Chinese sausages made from PSE pork. *Meat Science*, 64(4), 441–9.

Lanari, M. C., Schaefer, D. M. and Scheller, K. K. (1995). Dietary vitamin E supplementation and discoloration of pork bone and muscle following modified atmosphere packaging. *Meat Science*, 41(3), 237–50. doi:10.1016/0309-1740(95)00006-7.

Lane, G. A. and Fraser, K. (1999). A comparison of phenol and indole flavour compounds in fat, and of phenols in urine of cattle fed pasture or grain. *New Zealand Journal of Agricultural Research*, 42(3), 289–96.

Lealiifano, A. K., Pluske, J. R., Nicholls, R. R., Dunshea, F. R., Campbell, R. G., Hennessy, D. P., Miller, D. W., Hansen, C. F. and Mullan, B. P. (2011). Reducing the length of time between harvest and the secondary gonadotropin-releasing factor immunization improves growth performance and clears boar taint compounds in male finishing pigs. *Journal of Animal Science*, 89, 2782–92.

Lee, S. H., Choe, J. H., Choi, Y. M., Jung, K. C., Rhee, M. S., Hong, K. C., Lee, S. K., Ryu, Y. C. and Kim, B. C. (2012). The influence of pork quality traits and muscle fiber characteristics on the eating quality of pork from various breeds *Meat Science*, 90(2), 284–91.

Leong, J., Morel, P. C. H., Purchas, R. W. and Wilkinson, B. H. P. (2011). Effects of dietary components including garlic on concentrations of skatole and indole in subcutaneous fat of female pigs. *Meat Science*, 88(1), 45–50.

Ling, J. L. W. (2011). *Studies into factors responsible for the acceptability of pork on the Singaporean market*. PhD, Massey University, Palmerston North, New Zealand.

Lo, L., McLaren, D., McKeith, F., Fernando, R. and Novakofski, J. (1992a). Genetic analyses of growth, real time ultrasound, carcase, and pork quality traits in Duroc and Landrace pigs: II. Heritabilities and Correlations. *Journal of Animal Science*, 70, 2387–96.

Lo, L. L., McLaren, D. G., McKeith, F. K., Fernando, R. L. and Novakofski, J. (1992b). Genetic analyses of growth, real-time ultrasound, carcass, and pork quality traits in Duroc and Landrace pigs: I. Breed effects. *Journal of Animal Science*, 70(8), 2373–86.

Lundstrom, K., Malmfors, B., Malmfors, G., Stern, S., Patersson, H., Mortensen, A. B. and Sorensen, S. E. (1988). Skatole, androstenone and taint in boars fed two different diets. *Livestock Production Science*, 18, 55–67.

Lyford, C., Thompson, J., Polkinghorne, R., Miller, M., Nishimura, T., Neath, K., Allen, P. and Belasco, E. (2010). Is willingness to pay (WTP) for beef quality grades affected by consumer demographics and meat consumption preferences? *Australian Agribusiness Review*, 20(January 2010), 1–17.

Martel, J., Minveille, F. and Poste, L. M. (1988). Effects of crossbreeding and sex on carcass composition, cooking properties and sensory characteristics of pork. *Journal of Animal Science*, 66, 41–6.

McGloughlin, P., Allen, P., Tarrant, P. V., Joseph, R. L., Lynch, P. B. and Hanrahan, T. J. (1988). Growth and carcass quality of crossbred pigs sired by Duroc, Landrace and Large White boars. *Livestock Production Science*, 18, 275–88.

Meyers, S. N. and Beever, J. E. (2008). Investigating the genetic basis of pork tenderness: Genomic analysis of porcine CAST. *Animal Genetics*, 39, 531–43.

Meyers, S. N., Rodriguez-Zas, S. L. and Beever, J. E. (2007). Fine-mapping of a QTL influencing pork tenderness on porcine chromosome 2. *BMC Genetics*, 8, p 69.

Milan, D., Jeon, J. T., Looft, C., Amarger, V., Robic, A. and Thelander, M. (2000). A mutation in *PRKAG3* associated with excess glycogen content in pig skeletal muscle. *Science*, 288, 1248–51.

Miller, M. F., Carr, M. A., Ramsey, C. B., Crockett, K. L. and Hoover, L. C. (2001). Consumer thresholds for establishing the value of beef tenderness. *Journal of Animal Science*, 79(12), 3062–8.

Moeller, S. J., Miller, R. K., Edwards, K. K., Zerby, H. N., Logan, K. E., Aldredge, T. L., Stahl, C. A., Boggess, M. and Box-Steffensmeier, J. M. (2010). Consumer perceptions of pork eating quality as affected by pork quality attributes and end-point cooked temperature. *Meat Science*, 84(1), 14–22.

Monahan, F. J., Gray, J. I., Booren, A. M., Miller, E. R., Buckley, D. J., Morrissey, P. A. and Gomaa, E. A. (1992). Influence of dietary treatment on lipid and cholesterol oxidation in pork. *Journal of Agricultural and Food Chemistry*, 40(8), 1310–15.

Moore, K. L., Mullan, B. P. and Dunshea, F. R. (2017). Boar taint, meat quality and fail rate in entire male pigs and male pigs immunized against gonadotrophin releasing factor as related to body weight and feeding regime. *Meat Science*, 125, 95–101.

Moore, K. L., Mullan, B. P., Kim, J. C., Trezona, M. and Dunshea, F. R. (2015). Immunisation against gonadotrophin releasing factor reduces pork eating quality fail rates. *Animal Production Science*, 55(12), 1469–.

Naveau, J., Sellier, P. and Roy, P. L. (1996). Evidence for a new major gene influencing meat quality in pigs. *Genetics Research Cambridge*, 55, 33–40.

Newton, K. G. and Gill, C. O. (1981). The microbiology of DFD fresh meat: A review. *Meat Science*, 5, 223–32.

Offer, G. and Knight, P. (1988). The structural basis of water holding in meat. Part 2: Drip losses. In R. A. Lawrie (Ed.), *Developments in Meat Science* – 4 (pp. 173–244). Elsevier Applied Science, London and New York.

Oksbjerg, N., Petersen, J. S., Sórensen, M. T., Henckel, P., Agergaard, N., Bejerholm, C. and Erlandsen, E. (1995). The influence of porcine growth hormone on muscle fibre characteristics, metabolic potential and meat quality. *Meat Science*, 39(3), 375–85.

Ono, Y., Solomon, M. B., Evock-Clover, C. M., Steele, N. C. and Maruyama, K. (1995). Effects of porcine somatotropin administration on porcine muscles located within different regions of the body. *Journal of Animal Science*, 73(8), 2282–8. (doi:10.2527/1995.7382282x).

Park, S., Ryu, H., Hong, G. and Min, S. (2006). Physical properties of frozen pork thawed by high pressure assisted thawing process. *Food Science and Technology International*, 12(4), 347–52.

Parr, T., Mareko, M. H. D., Ryan, K. J. P., Hemmings, K. M., Brown, D. M. and Brameld, J. M. (2016). The impact of growth promoters on muscle growth and the potential consequences for meat quality. *Meat Science*, 120, 93–9.

Pethick, D. W., Pleasants, A. B., Gee, A. M., Hopkins, D. L. and Ross, I. R. (2006). *Eating quality of commercial meat cuts from Australian lambs and sheep*. Paper presented at the Proceedings of the New Zealand Society of Animal Production, Volume 66.

Primary Industries Standing Committee (2008). Model Code of Practive for the Welfare of Animals: Pigs. PISC Report No. 92. CSIRO Publishing, Collingwood, Australia.

Purchas, R. W., Smith, W. C. and Pearson, G. (1990). A comparison of the Duroc, Hampshire, Landrace and Large White as terminal sire breeds of crossbred pigs slaughtered at 85 kg liveweight. 2. Meat quality. *New Zealand Journal of Agricultural Research*, 33, 97–104.

Purslow, P. P., Oiseth, S., Hughes, J. and Warner, R. D. (2016). The structural basis of cooking loss in beef: Variations with temperature and ageing. *Food Research International*, 89(1), 731–48.

Rothschild, M. F., Bidanel, J. P. and Ciobanu, C. (2004). Genome analysis of QTL for muscle tissue development and meat quality. In M. F. W. Te Pas, M. E. Everts and H. P. Haagsman (Eds), *Muscle Development of Livestock Animals. Physiology, Genetics and Meat Quality* CABI Publishing. (Reprinted from: Not in File)Wallingford, Oxfordshire, U.K., pp. 247–66.

Rowe, S. J., Karacaören, B., de Koning, D.-J., Lukic, B., Hastings-Clark, N., Velander, I., Haley, C. S. and Archibald, A. L. (2014). Analysis of the genetics of boar taint reveals both single SNPs and regional effects. [journal article]. *BMC Genomics*, 15(1), 424.

Ryu, Y. C., Choi, Y. M., Lee, S. H., Shin, H. G., Choe, J. H., Kim, J. M., Hong, K. C. and Kim, B. C. (2008). Comparing the histochemical characteristics and meat quality traits of different pig breeds. *Meat Science*, 80(2), 363–9.

Salmi, B., Trefan, L., Bloom-Hansen, J., Bidanel, J. P., Doeschl-Wilson, A. B. and Larzul, C. (2010). Meta-analysis of the effect of the halothane gene on 6 variables of pig meat quality and on carcass leanness1. *Journal of Animal Science*, 88(9), 2841–55.

Sanchez, M. P., Iannuccelli, N., Basso, B., Bidanel, J. P., Billon, Y., Gandemer, G., Gilbert, H., Larzul, C., Legault, C., Riquet, J., Milan, D. and Le Roy, P. (2007). Identification of QTL with effects on intramuscular fat content and fatty acid composition in a Duroc x Large White cross. *BMC Genetics*, 8, 55.

Savell, J. W. (1979). *Industry acceptance of electrical stimulation*. Paper presented at the Proceedings of the 3rd Reciprocal Meat Conference.

Sayre, R. N., Briskey, E. J. and Hoekstra, W. G. (1963). Comparison of muscle characteristics and post-mortem glycolysis in three breeds of swine. *Journal of Animal Science*, 22, 1012–20.

Sellier, P. (1998). Genetics of Meat and Carcass Traits. In M. F. Rothschild and A. Ruvsinsky (Eds), *Genetics of the pig* (pp. 463–510). CABI Publishing (Reprinted from: In File). Wallingford, UK

Sheard, P. R., Nute, G. R., Richardson, R. I., Perry, A. and Taylor, A. A. (1999). Injection of water and polyphosphate into pork to improve juiciness and tenderness after cooking. *Meat Science*, 51(4), 371–6.

Shorthose, W. R., Powell, V. H. and Harris, P. V. (1986). Influence of electrical stimulation, cooling rates and aging on shear force values of chilled lamb. *Journal of Food Science*, 51, 889–92.

Solomon, M. B., Campbell, R. G. and Steele, N. C. (1990). Effect of sex and exogenous porcine somatotropin on longissimus muscle fiber characteristics of growing pigs. *Journal of Animal Science*, 68(4), 1176–81.

Solomon, M. B., Campbell, R. G., Steele, N. C. and Caperna, T. J. (1991). Effects of exogenous porcine somatotropin administration between 30 and 60 kilograms on longissimus muscle fiber morphology and meat tenderness of pigs grown to 90 kilograms. *Journal of Animal Science*, 69(2), 641–5.

Sosnicki, A. A. and Newman, S. (2010). The support of meat value chains by genetic technologies. *Meat Science*, 86, 129–37.

Souza, C. M., Boler, D. D., Clark, D. L., Kutzler, L. W., Holmer, S. F., Summerfield, J. W., Cannon, J. E., Smit, N. R., McKeith, F. K. and Killefer, J. (2011). The effects of high pressure processing on pork quality, palatability, and further processed products. *Meat Science*, 87, 419–27.

Squires, E. J. (1999). Genetics of boar taint: Implications for the future use of intact males *Annual Conference, November* 16–19, 1999. National Swine Improvement Federation (Reprinted from: Not in File).Des Moines, Iowapp. 1–5.

Squires, E. J. and Schenkel, F. S. (2010). Managing boar taint: Focus on genetic markers *London Swine Conference – Focus on the Future March* 31 – *April* 1. (Reprinted from: Not in File), pp. 99–102.

Stearns, T. M., Beever, J. E., Southey, B. R., Ellis, M., McKeith, F. K. and Rodriguez-Zas, S. L. (2005). Evaluation of approaches to detect quantitative trait loci for growth, carcass, and meat quality on swine chromosomes 2, 6, 13, and 18. I. Univariate outbred F2 and sib-pair analyses1. *Journal of Animal Science*, 83(7), 1481–93.

Stoller, G. M., Zerby, H. N., Moeller, S. J., Baas, T. J., Johnson, C. and Watkins, L. E. (2003). The effect of feeding ractopamine (Paylean) on muscle quality and sensory characteristics in three diverse genetic lines of swine12. *Journal of Animal Science*, 81(6), 1508–16.

Taylor, A. A. and Martoccia, L. (1995). The effect of low voltage and high voltageelectrical stimulation on pork quality. *Meat Science*, 39(3), 319–26.

Taylor, A. A., Nute, G. R. and Warkup, C. C. (1995a). The effect of chilling, electrical stimulation and conditioning on pork eating quality. *Meat Science*, 39(3), 339–47.

Taylor, A. A., Perry, A. M. and Warkup, C. C. (1995b). Improving pork quality by electrical stimulation or pelvic suspension of carcasses. *Meat Science*, 39(3), 327–37.

Thomsen, R., Edwards, S. A., Jensen, B. B., Rousing, T. and Sørensen, J. T. (2015). Effect of faecal soiling on skatole and androstenone occurrence in organic entire male pigs. *Animal*, 9(9), 1587–96.

Thornton, J. W., Alsmeyer, R. H. and Davey, R. J. (1968). Genetic, diet and pig age effects on pork tenderness and palatability. *Journal of Animal Science*, 27, 1229–32.

Touraille, C., Monin, G. and Legault, C. (1989). Eating quality of meat from European x Chinese crossbred pigs. *Meat Science*, 25, 177–86.

Trefan, L., Bünger, L., Bloom-Hansen, J., Rooke, J. A., Salmi, B., Larzul, C., Terlouw, C. and Doeschl-Wilson, A. (2011). Meta-analysis of the effects of dietary vitamin E supplementation on [alpha]-tocopherol concentration and lipid oxidation in pork. *Meat Science*, 87, 305–14.

Trefan, L., Doeschl-Wilson, A., Rooke, J. A., Terlouw, C. and Bünger, L. (2013). Meta-analysis of effects of gender in combination with carcass weight and breed on pork quality1. *Journal of Animal Science*, 91(3), 1480–92.

Tuomola, M., Vahva, M. and Kallio, H. (1996). High-Performance Liquid Chromatography Determination of Skatole and Indole Levels in Pig Serum, Subcutaneous Fat, and Submaxillary Salivary Glands. *Journal of Agricultural and Food Chemistry*, 44(5), 1265–70.

Vanhonacker, F., Verbeke, W. and Tuyttens, F. A. M. (2009). Belgian consumers' attitude towards surgical castration and immunocastration of piglets. *Animal Welfare*, 18(4), 371–80.

Walker, P. J. and Channon, H. A. (2003). *Pork eating quality varies between butchers*. Paper presented at the Manipulating Pig Production, Fremantle, Western Australia.

Warner, R. D. (1994). *Physical properties of porcine musculature in relation to post mortem biochemical changes in muscle proteins*. PhD thesis, University of Wisconsin-Madison, USA.

Warner, R. D., Eldridge, G. A., Hofmeyr, C. D. and Barnett, J. L. (1998). The effect of dietary tryptophan on pig behaviour and meat quality – preliminary results. *Animal Production in Australia*, 22, 325.

Warner, R. D., Kauffman, R. G. and Greaser, M. L. (1997). Muscle protein changes post mortem in relation to pork quality traits. *Meat Science*, 45(3), 339–52.

Warner, R. D., Kerr, M., Kim, Y. H. B. and Geesink, G. (2014a). Pre-rigor carcass stretching counteracts the negative effects of high rigor temperature on tenderness and water-holding capacity –using lamb muscles as a model. *Animal Production Science*, 54, 494–503.

Warner, R. D., Thompson, J. M., Polkinghorne, R., Gutzke, D. and Kearney, G. A. (2014b). A consumer sensory study of the influence of rigor temperature on eating quality and ageing potential of beef striploin and rump. *Animal Production Science*, 54, 396–406.

Warner, R., McDonnell, C., Bekhit, A. A., Claus, J., Vaskoska, R., Sikes, A. L., Dunshea, F. R. and Ha, M. (2017). Systematic review of emerging technologies for tenderising meat. *Meat Science*, 132, 72–89.

Weikard, R., Rehfeldt, C. and Ender, K. (1992). Changes in muscle structure and protein metabolism of pigs in response to porcine somatotrophin (pST). *Archiv Fur Tierzucht*, 35, 273–84.

Whipple, G., Hunt, M. C., Klemm, R. D., Kropf, D. H., Goodband, R. D. and Schricker, B. R. (1992). Effects of porcine somatotropin and supplemental lysine on porcine muscle histochemistry1. *Journal of Muscle Foods*, 3(3), 217–27.

Wood, J. D., Brown, S. N., Nute, G. R., Whittington, F. M., Perry, A. M., Johnson, S. P. and Enser, M. (1996). Effects of breed, feed level and conditioning time on the tenderness of pork. *Meat Science*, 44, 105–12.

Wood, J. D., Nute, G. R., Fursey, G. A. J. and Cuthbertson, A. (1995). The effect of cooking conditions on the eating quality of pork. *Meat Science*, 40(2), 127–35.

Woods, A. L., Armstrong, T. A., Anderson, D. B., Elam, T. E. and Sutton, A. L. P. (2011). CASE STUDY: Environmental benefits of ractopamine use in United States finisher swine1,2. *Professional Animal Scientist*, 27(5), 492–9.

Yang, X. Q., Liu, H., Guo, L. J., Xu, Y. and Liu, D. (2007). The mutation site analysis on CAPN1 gene of Wild boar, Min pig and Yorkshire. *Yi Chuan*, 29(581–6).

Ying, J. L. W. (2011). *Studies into factors responsible for the acceptability of pork on the Singaporean market*. Doctor of Philosophy, Massey University, Palmerston North.

Zadinová, K., Stupka, R., Stratil, A., Vehovský, K., Urbanová, D. and Čítek, J. (2016). Boar taint – the effects of selected candidate genes associated with androstenone and skatole levels – a review. [Article]. *Animal Science Papers & Reports*, 34(2), 107–27.

Zamaratskaia, G., Babol, J., Andersson, H. K., Andersson, K. and Lundström, K. (2005). Effect of live weight and dietary supplement of raw potato starch on the levels of skatole, androstenone, testosterone and oestrone sulphate in entire male pigs. *Livestock Production Science*, 93(3), 235–43.

Zamaratskaia, G., Lou, Y., Peacock, J., Rydhmer, L., Andersson, H. K., Juneja, R. K., Chen, G., Lundström, K. and Squires, E. J. (2008). Effect of polymorphism in the porcine cytochrome b5 (*CYB5A*) gene on androstenone and skatole concentrations and sexual development in Swedish pig populations. *Animal*, 2, 190–6.

Chapter 6

Factors affecting pork flavour

Mingyang Huang and Yu Wang, University of Florida, USA; and Chi-Tang Ho, Rutgers University, USA

1 Introduction

Pork and pork-related products are very popular among consumers due to their high nutritional value and appealing sensory qualities. Generally speaking, uncooked meats have a bloody taste with very little flavour[1], and raw pig meat also tends to have a 'sour' smell. However, once cooked, pork has a desirable flavour dependent on a variety of factors that include not only the methods used for processing and cooking, but also any additives, such as spices, used during processing or cooking. The 'meaty' flavour of pork results from interactions of flavour precursors with the appropriate amount of heat[2]. Two major pathways are involved in the formation of meat volatile flavour compounds during the cooking process. These include the Maillard reaction that occurs when reducing sugars and amino acids interact at high temperatures, resulting in water-soluble components, and the oxidative degradation of lipids[3]. These water-soluble pork flavour precursors consist of a large number of organic compounds, including free sugars, sugar phosphates, sugar amines, free amino acids, peptides, nucleotides, nucleosides, nucleic acids, glycogen, amines and other nitrogenous compounds, such as thiamine[1, 4]. The fatty aromas generated by lipid degradation determine the differences between the flavour profiles of cooked meats from different species[4].

There have been extensive investigations into the flavour constituents of pork. However, the literature focusing on key aroma compounds in pork-related products and factors affecting flavour constituents during heating is limited. Therefore, the purpose of this chapter is to provide an integrated overview of the essential flavour constituents in pork products and the presumptive factors affecting pork flavour. Despite large differences

http://dx.doi.org/10.19103/AS.2017.0030.06

among animal species, genetics and methods used in processing and cooking, this chapter will attempt to focus on consensus odorants that have been studied by at least two independent groups and emphasize the heat-induced formation of pork odorants' pathways. Conversely, this chapter will not address any influences of an animal's pre-slaughter, post-mortem, age or nutritional status on aroma characteristics.

2 Essential aroma compounds and processing effects

2.1 Aroma compounds essential to pork flavour

Odorants generated when pork is cooked have been well studied, identifying over 300 volatile compounds[5]. However, there is no single compound identified in the pork aroma profile described as uniquely meaty or that accounts for cooked 'meat' flavour. Therefore, the major attributes for overall pork flavour have been studied using aroma extraction dilution analysis (AEDA)[6], with results indicating that it is the combination of major attributes that contributes to overall pork flavour[2]. These major aroma compounds are mainly formed through the Maillard reaction and lipid oxidation.

In addition, it has been well established that meat flavour is associated with both water-soluble and lipid-soluble fractions and that water-soluble precursors such as carbohydrates and amino acids play a major role in the formation of volatile compounds during cooking[2]. As previously mentioned, the Maillard reaction and lipid oxidation contribute to the formation of the thermally induced characteristic flavour of cooked meat[4]. Through a series of complex pathways, both of these reactions are responsible

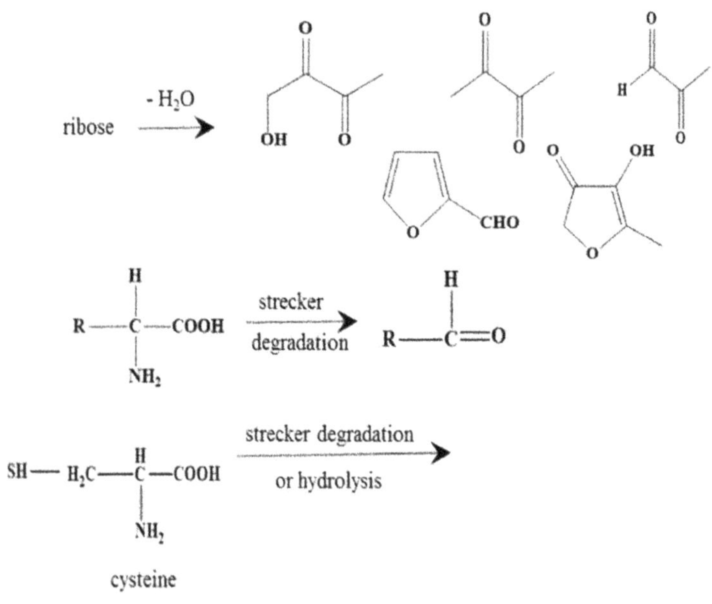

Figure 1 Some important flavour intermediates in the Maillard reaction in meat[4].

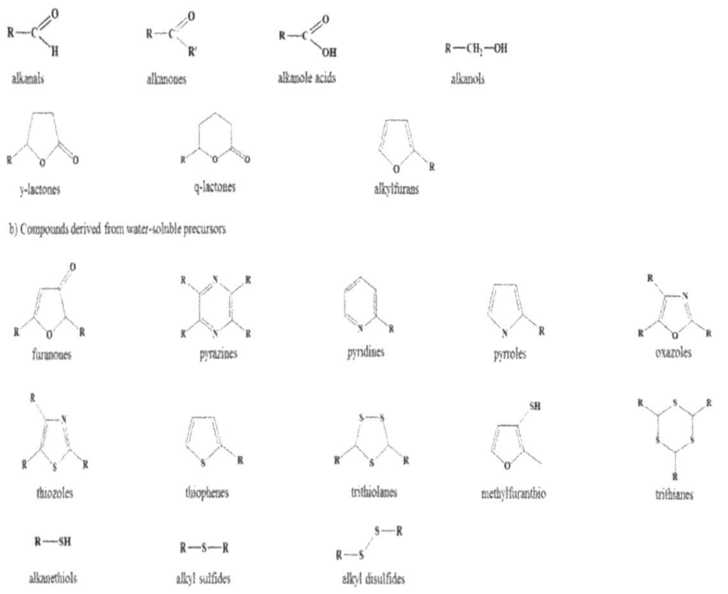

Figure 2 Some classes of volatile compounds produced during the cooking of meat[4].

for the formation of a large group of flavour constituents that account for the overall flavour profile of cooked meat. The intermediates formed during the Maillard reaction may also react with compounds formed during lipid oxidation. Some important flavour intermediates in the Maillard reaction in meat are shown in Fig. 1. This gives rise to additional aroma compounds and adds to the complexity of the overall aroma notes[4]. For example, lipid-derived saturated and unsaturated aldehydes can react with Maillard reaction intermediates to form alkylpyrazines, alkylthiazoles, alkylpyridines and so on. Pathways responsible for thiazole formation in cooked meat involve aldehydes, ammonia, α-hydroxyketones and hydrogen sulphide. It is also possible that aldehydes formed by lipid oxidation provide a reaction source resulting in long-chain 2-alkylthiazoles.

Heterocyclic compounds especially those containing O-, N- and S- such as furans, furanones, thiazoles, thiazolines, thiophenes, oxazoles and pyrazines make a significant contribution to the overall pleasing aromas of pork[1], with only a small portion of other constituents possessing any odour activity[7]. A list of major volatiles identified from several studies is presented in Fig. 2.

Volatiles formed by the Maillard reaction

Early stages of the Maillard reaction involve interactions of primary amino acids with a carbonyl group of reducing sugars, resulting in glycosylamine formation. Subsequently, Amadori rearrangement, an alkali catalysed isomerization reaction occurs. Degradation of Amadori products results in deoxyosones, reactive α-dicarbonyl compounds responsible for the formation of secondary products such as furanones, pyridines and pyrroles. It is these secondary products derived from 3-deoxyosone, including pyridines, pyrroles and formylpyrroles, which are the important characteristic aroma compounds found in cooked meat[1, 4, 5, 7].

Subsequent stages occur between intermediates formed during the Maillard reaction and other reactive components such as aldehydes, amino acids, ammonia, hydrogen sulphide and amines. The Strecker reaction, degradation of amino acids by dicarbonyl compounds, is one of the most important associated reactions and occurs under harsh conditions such as high temperature or high pressure. The dicarbonyl is converted to an aminoalcohol or an α-aminoketone, while the amino acid undergoes decarboxylation and deamination resulting in an aldehyde. The Strecker reaction produces ammonia, hydrogen sulphide and acetaldehyde, if the amino acid is cysteine. This provides a rich source of intermediates for further reactions. Additionally, interaction of two aminoketones can yield pyrazine derivatives, which are strong aroma compounds in cooked meat[1, 3–6].

Heterocyclic compounds, such as thiazoles, oxazoles, pyridine, pyrazines and furans produced in the Maillard reaction, are generally associated with roasted, meaty, boiled and savoury flavours[8–10]. More specifically, 6,7-dihydro-5(*H*)-cyclopentapyrazines and pyrrolopyrazines are two unique volatiles of cooked meats as they have not been reported in any other food[11]. Furthermore, the flavours of thiazoles have been reported as bread-like, meaty, nutty, mocha and slightly malty, whereas pyrazines are usually described as sweet, corn-like, having a bitter note and pungent. It has been reported that furans, thiophenes and related disulphides are responsible for the meat-like aromas of cooked meat[11, 12]. These compounds also have exceptionally low odour threshold values, which make them very important characteristic aromas of meat. The heterocyclic compounds, particularly those containing O, furans and furanones are responsible for a fruity, nutty and caramel-like odour, which make extremely important contributions to the overall odour of heated meat[1, 8]. Important reviews of the formation pathways for furans in meat flavour have been published by Shahidi et al., and will be discussed in the bacon section[5]. By using gas chromatography–olfactometry (GC–O), the compounds containing 2-methyl-3-furanyl groups such as 2-mercapto-3-furanthiol, bis(2-methyl-3-furanyl) disulphide, 2-[(2-methyl-3-furanyl)dithio]-3-pentanone and 2-methyl-3-(methyldithio) furan possessed meaty characteristics, whereas 2-methylfuranyl groups 3-[(2-furanylmethyl) dithio]-2-pentanone and 3-[(2-furanylmethyl) dithio]-2-butanone all indicate nutty, burnt and roasted characteristics[4]. The route for the formation of 2-methyl-3-furanthiol, bis(2-methyl-3-furanyl) disulphide and 2-methyl-3-(methylthio)-furan from ribose phosphate is shown in Fig. 3.

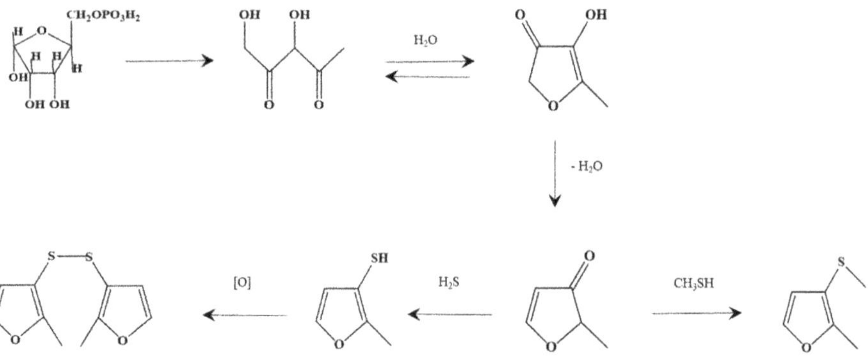

Figure 3 Route for the formation of 2-methyl-3-furanthiol, bis(2-methyl-3-furanyl) disulphide and 2-methyl-3-(methylthio)-furan from ribose phosphate[142].

Lipid-derived volatiles in cooked pork

The major classes of lipid-derived volatiles found in cooked pork consist of aldehydes, ketones, lactones, alcohols, esters, short-chain fatty acids and aliphatic hydrocarbons[2]. The total amount of these newly formed lipid-derived compounds is quite large, and correlates with molecule size, substitution type and degree of saturation. Generally, lipid oxidation during cooking is the most important associated reaction, and gives rise to compounds that include carbonyl compounds and some oxygenated heterocyclic compounds including alkylfurans and lactones[4]. The overall mechanism of lipid oxidation consists of three parts: initiation, which stands for the formation of free radicals; propagation, which refers to the free radical chain reactions; and termination, which presents the formation of final non-radical products. Once hydroperoxides form in the propagation stage, they are capable of breaking through dismutation and generate important volatile compounds such as aldehydes, alcohols, ketones and hydrocarbons[33, 34].

The most quantitatively dominant volatiles found in cooked meat are, without a doubt, those that are lipid-derived, as the triacylglycerols and phospholipids found in meat readily provide a large source of lipids. However, these lipid-derived compounds are not aroma significant due to their comparatively high odour threshold values, especially when compared to key aroma heterocyclic compounds with low odour threshold values. The basic aroma profiles of meat from different species have been reported to be quite similar. However, when comparing these meat aroma profiles, it is the lipid-derived aromas that make the greatest contribution to the differences found among the species[2]. In particular, aldehydes derived from lipid degradation in cooked pork are believed to be the most characteristic flavour. When comparing lamb or beef, to chicken or pork, the latter two have higher amounts of unsaturated fatty acids in their triacylglycerols, thus yielding more unsaturated aldehydes. This, in turn, may influence characteristic aromas for each of the specific species[31, 32]. These aldehyde aromas are reported as fatty and green[4].

2.2 Processing effects on pork flavour

With the increase in number of identified meat flavour compounds comes a greater understanding of how meat processing alters the flavour of cooked pork. There are numerous processing methods that include cooking, smoking, curing, canning, deboning, grinding, mincing, irradiation and freeze-drying. Add to those the multitude of additives, including spices, applied during processing, and the result is a varied composition of not only the cooked meat's flavour profile, but also the overall aroma of the final product[13–26]. For example, the degradation of methyl mercaptan while canning, which requires longer heating times, causes accumulation of hydrogen sulphide. This leads to the formation of a stronger cooked meat flavour[7]. Additionally, cooking plays a vital role in determining final meat flavour. In particular, methods of applying heat, such as frying, boiling, roasting, retort heating, pressure-cooking and so on, plus the intensity of the heat treatment applied to pork and its related products induces a varying series of chemical changes to occur during the process[1, 5].

Not only do the smoking and curing processes contribute directly to meat flavour, but the addition of curing agents and smoking materials also add flavour notes to meat products. It is believed that the combination of salt and nitrite found in the curing treatment applied to meat produces a characteristic flavour[9, 10, 18, 28]. Chemical changes that take place during the smoking process introduce a smoke flavour to cooked meat, whereas curing ingredients usually add a certain degree of salty flavour to the meat, and is dependent upon the amount applied[7]. The pleasing aroma of cured and smoked sausage coincides with the decrease in volatile reducing compounds such as aldehydes and carbonyls that occurs during the smoking process. The accumulation of non-volatile free amino acids, creatine, fatty acids and hypoxanthine also leads to the desirable taste of curing products, and typically increases during the cooking process[7]. Prior to cooking, meats to be cured or smoked undergo numerous reactions in the various stages of preparation. Thus, curing, drying and ageing may each account for different chemical changes attributed to flavour development[27].

Grinding, mincing and deboning lead to the disruption of the meat muscle membrane system, and subsequently cause the acceleration of lipid oxidation in cooked meat[29]. Freeze-drying raw meat results in little change of the final flavour of cooked meat. It has been demonstrated that the process of freeze-drying meat results in cooked meat that has less flavour when compared to freshly cooked meat[27]. This might be the result of freeze-drying causing the loss of flavour precursors such as the oxidation of lipids and carbonyls and/or the removal of the major flavour-producing reactants necessary for the browning reactions to occur[27]. Irradiation is a meat sterilization processing method that causes a very characteristic, newly developed odour called irradiated odour (IO). In particular, methional appears to be an important factor accounting for off-flavours[29]. However, there are other components involved in a series of chemical reactions that influence the final meat flavour. The formation of new compounds, condensation and degradation of the carbon chains, and alteration of lipid components are also responsible for the addition of any new flavour to the original basic cooked meat odour, thus producing the off-flavour[27]. It has been observed that irradiation with a higher temperature (5°C) releases more IO than irradiation at a frozen state (−35°C)[30]. IO development in meat also appears to be affected by the presence of oxygen during the irradiation process. Conversely, the addition of certain spices or additives such as thyme, sage, pepper, turmeric, mace and celery have been shown to reduce IOs in meats[30].

3 Bacon

As previously stated, the aroma profile of a specific meat product is not only related to its composition, but also to the methods used to process the meat. There have been numerous investigations leading to the identification of pork aroma, especially the aroma compounds comprising the most popular processed pork product among consumers – bacon. To date, a total of 135 compounds have been identified from fried bacon, including pyrazines, furans, thiazoles, oxazoles, oxazolines, pyrroles, pyridines, hydrocarbons, ketones, alcohols, ethers, esters, acids, phenols and other miscellaneous compounds[38]. Meat flavours generated during processing are due to complex interactions of precursors that include the Maillard reaction; pyrolysis of amino acids and peptides; lipid oxidation; and sugar, thiamine and ribonucleotides degradation. The major aroma compounds found in roasted pork have been characterized by the oxygenous benzene derivatives such as eugenol, estragole and ethyl cinnamate, which are derived from both spices and the large quantities of lipid-degraded aldehydes[59]. The spices added during meat processing typically possess woody and herbaceous odours that may modify overall meat flavour.

Nitrate is a well-known key contributor producing the characteristic cured meat aroma in all cured meat products. This distinctive and pleasing flavour results from interactions between nitrate and the meat. Cured meat flavour is the result of appreciable amounts of pyridines developing from the applied nitrites[62]. Additionally, it has been reported that application of ionizing radiation yields a specific irradiation flavour to all meat products that decreases overall customer acceptance as the irradiation dose and/or temperature increases[20]. Similarly, the higher the dose and temperature of irradiation applied to meat, the higher the amount of volatile compounds that can be identified or isolated. Numerous studies have reported that there is no single volatile compound responsible for bacon's pleasing aroma. Conversely, this desirable odour is due to a combination of various aroma-active compounds all contributing to the overall flavour of bacon.

3.1 Heterocyclic aroma compounds

Pyrazines

The roasted nut-like aroma of pyrazines is associated with the volatiles found in a wide variety of foods. The suggested mechanism for pyrazine formation has been proposed by Flament et al. and Hodege and Osman[51, 52]. Flament et al. described the α-dicarbonyl (Strecker) compounds as major reactant sources, giving rise to the carbon skeletons of pyrazines. These carbon skeletons further interact with amino acids to form α-aminoketones that dehydrate and eventually oxidized to pyrazines during heating[51, 52].

Butyl and pentyl pyrazines are two major alkyl-substituted heterocyclic compounds identified in cooked meat[2, 4, 56]. The possible pathway to form these compounds might be through the reaction of hexanal or pentanal with a dihydropyrazine, which are formed during Maillard reaction by the condensation of two aminoketone molecules[2, 4, 56]. The possible pathway for the formation of alkylpyrazines is shown in Fig. 4. A total of 22 pyrazines have been reported as volatile flavour constituents of pan-fried bacon[38]. In particular, 2-ethyl-3-methylpyrazine, 2-ethyl-6-vinylpyrazine, 2-acetyl-3-methylpyrazine,

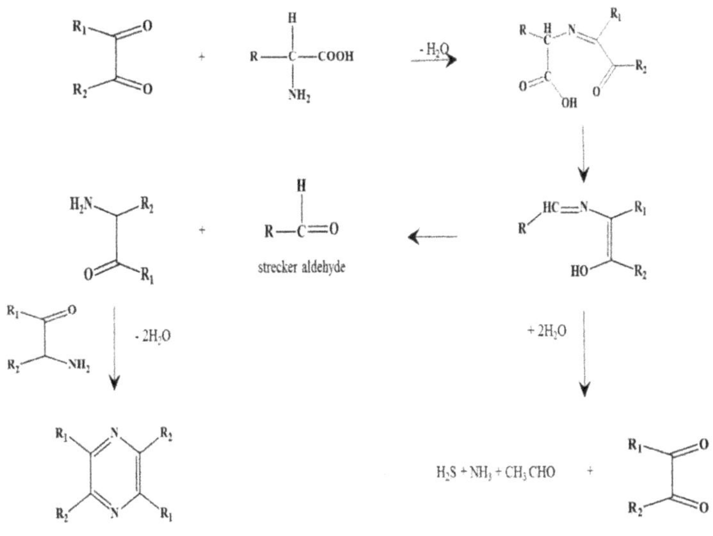

Figure 4 Strecker degradation of amino acids, showing a route for the formation of alkylpyrazines, and the formation of ammonia, hydrogen sulphide and acetaldehyde from cysteine[4].

2-ethyl-3,5,6-trimethylpyrazine and 2-isobutyl-3-methylpyrazine were not identified before these recent studies. Two important factors influencing the formation of specific pyrazines are temperature and pH[1]. Any processing methods affecting the temperature and pH of the reaction environment may influence the overall aroma profile of heated bacon through the disruption of pyrazine formation. The proposed pathways to form pyrazines are shown in Fig. 5 and 6.

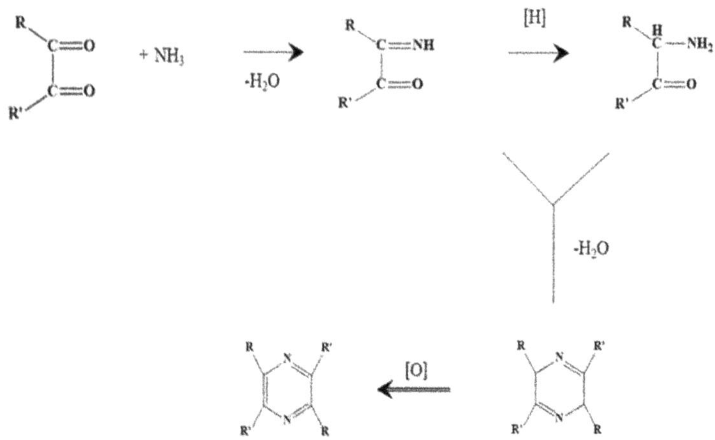

Figure 5 Interaction of Maillard reaction products to form pyrazines[149].

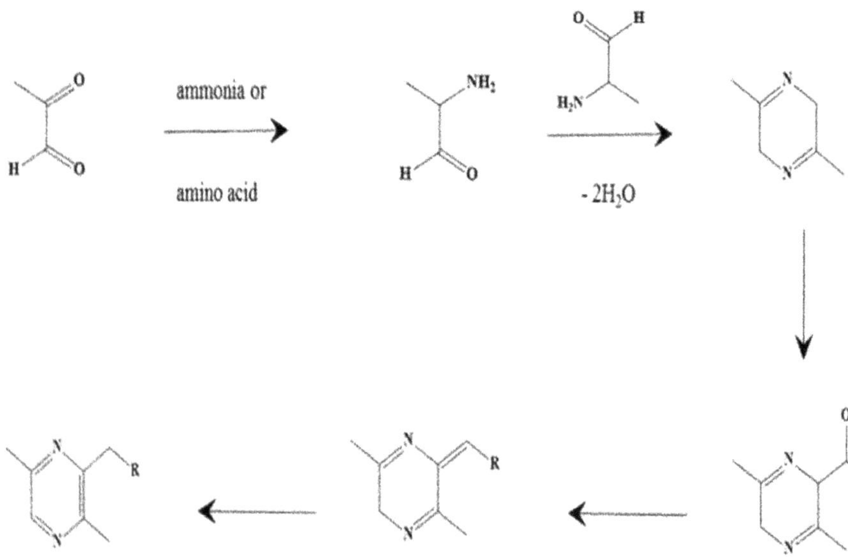

Figure 6 Route to alkyldimethylpyrazines from the interaction of lipid-derived aldehydes with the Maillard reaction[4, 143].

Thiazoles and thiazolines

The possible formation of thiazoles in pork results from a combination of carbohydrates or carbonyls with sulphur-containing amino acids[38]. In particular, these compounds are formed by combining acetaldehyde, ammonia or hydrogen sulphide with pentane-2,3-dione, butane-2,3-dione (diacetyl) and pyruvaldehyde[37]. Kato et al. identified thiazoles, stating that they are formed through interactions between cysteine and pyruvaldehyde or glucose under temperatures of 160°C[49]. A similar system was employed to investigate the formation of thiazoles during heating at 125°C for 24 hours[50]. Due to the oxidation–reduction of thiazolines induced while cooking, thiazoles and the corresponding thiazolines frequently coexisted[43]. Three different thiazoles, 2-isopropyl-4,5-dimethylthiazole, 2-butyl-4,5-dimethylthiazole and 2,5-dimethyl-4-ethylthiazole, have been identified in fried bacon at 350°C for 5 min. It has been reported that alkylthiazoles process a nutty, vegetable-like, green aroma[35]. The related pathways to form thiazoles and thiazolines are shown in Fig. 7 and 8.

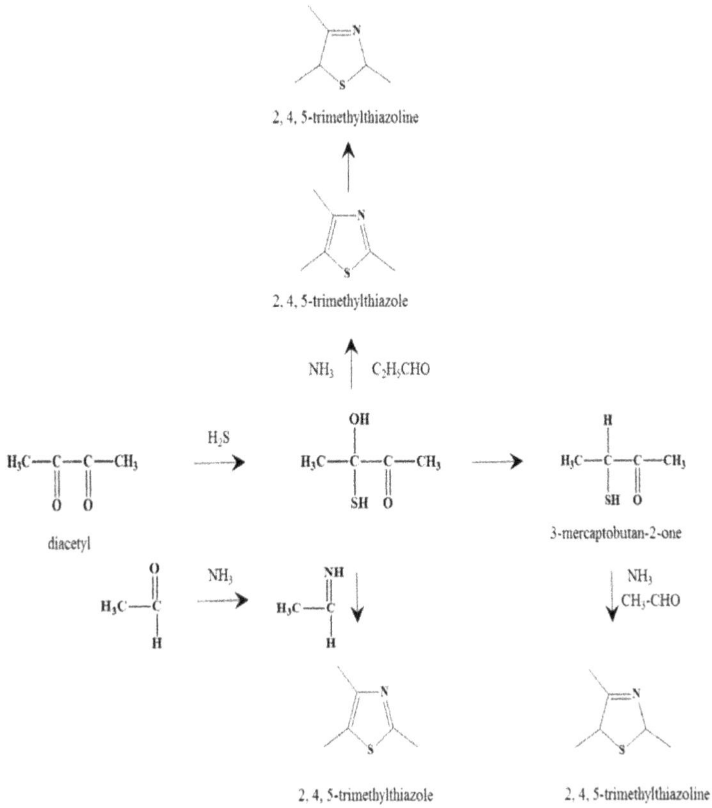

Figure 7 Reaction pathways for formation of thiazoles and thiazolines by heating Maillard degradation products of meat precursors[147].

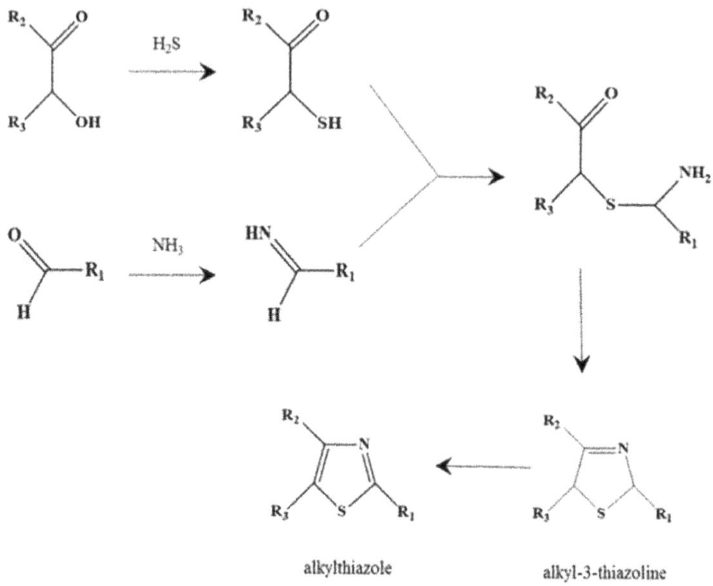

Figure 8 Route for the formation of thiazolines and thiazoles in the Maillard reaction from the reaction of hydroxyketones, aldehydes with ammonia and hydrogen sulphide[4, 144].

In a study of identified volatiles found in roasted pork, it was reported that despite proportions of benzothiazole and sulphured 2-acetylthiazole accounting for only 0.09% of total volatiles, these two compounds contribute significantly to the overall flavour of the meat due to low thresholds and distinguishable odours[59].

Oxazoles and oxazolines

Oxazoles have been identified in numerous food sources, but only 2-methylbenzoxazole and 2,4,5-trimethyloxazole are identified as volatiles of heated meat. The formation of 4,5-dimethyloxazole in the flavour profile of fried bacon has been associated with the reaction mixture of diacetyl and L-cysteine[36]. Oxazolines are rarely found in foods other than meat. The 2,4-dimethyl-3-oxazoline identified in fried bacon flavour is described as vegetable-like and nutty[35, 37]. The proposed pathway to form oxazoles and oxazolines is shown in Fig. 9.

Furans

As one of the most important heterocyclic compounds in the volatile profile of cooked meat, furans contribute a caramel-like, nutty and sweet aroma that accounts for the overall aroma impression of heated meat and of the smoky odours of smoked meat[8, 38]. The possible mechanisms of furan formation might involve thermal degradation of carbohydrates, the Maillard reaction, thermal oxidation of lipids and the decomposition of ascorbic acid or its derivatives. Different precursors leading to parent furan upon thermal treatment is shown in Fig. 10. Furan formation mechanisms were elucidated using C13-labelled ascorbic acid[69]. The authors indicated two possible pathways associated with furan formation, one

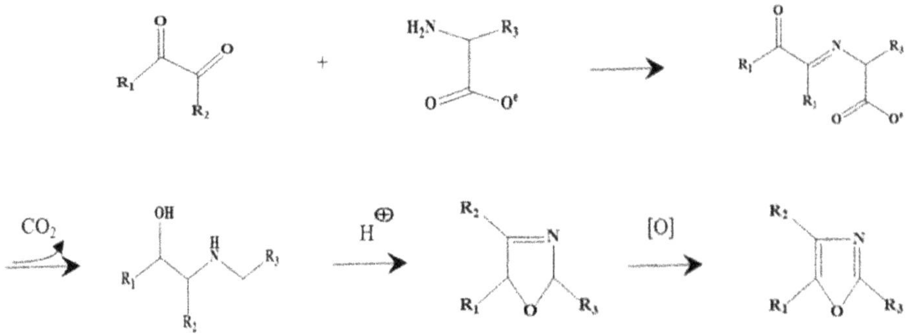

Figure 9 Proposed pathway of oxazoles and oxazolines formation[148].

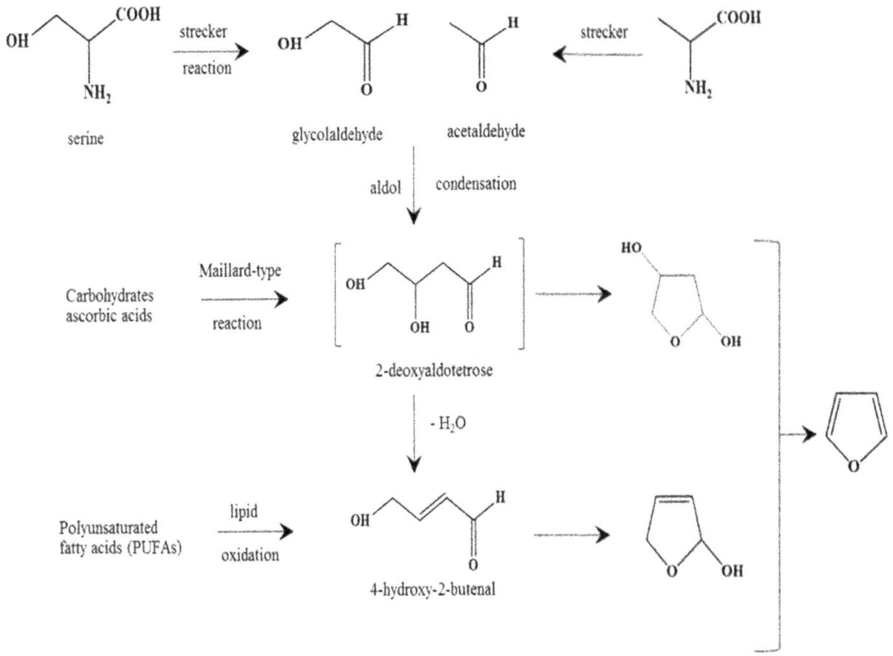

Figure 10 Different precursors leading to parent furan upon thermal treatment[67].

that forms from an intact C4 skeleton and the other that forms through a recombination of fragments. Possible reaction sources include amino acids, carbohydrates, polyunsaturated lipids and ascorbic acid[68–73]. Cysteine and serine are capable of forming glycolaldehyde and acetaldehyde and subsequently undergo aldol condensation followed by dehydration and cyclization to generate furans. Upon heating, linoleic and linolenic acids can generate furans[68]. The correlated triacylglycerols of linolenic and linoleic acids have also been found to actively form furans through lipid oxidation. Specifically, as shown in Fig. 11, furans can be generated from 4-hydroxy-2-butenal through cyclization and dehydration reactions[72].

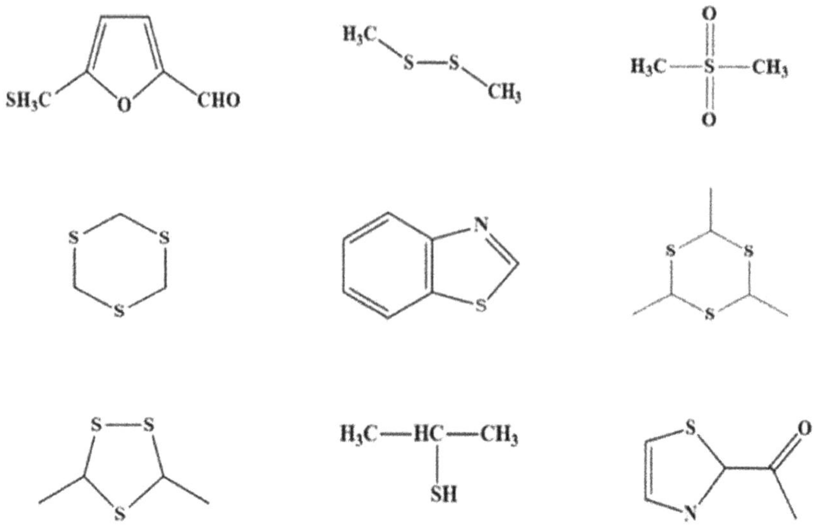

Figure 11 Sulphur compounds in meat volatiles[5].

Several model studies have revealed that under pyrolytic conditions (250–350°C for 20 seconds), furans are prominently generated by ascorbic acid derivatives. This was confirmed by Seaman and Becalski who employed a pressure-cooking condition (118°C for 30 minutes)[67, 68]. Key intermediates responsible for producing furans include acetaldehyde, aldotetrose derivatives, 4-hydoxy-2-butenal and glycolaldehyde[67]. These authors also discussed the important roles of polyunsaturated fatty acids (PUFAs), carotenoids and triacylglycerols as precursors of furan formation during cooking and processing[68].

The use of vegetable oil results in a grassy and beany flavour in cooked meat, presumably due to the furan-related compound, 2-pentylfuran[39, 40]. In addition, the aroma compound, ethyl 2-furoate, has been described as buttery, burnt and vanilla-like in other investigations[35, 41]. Among all aroma compounds synthesized for structure confirmation, 2,4,5-trimethyl-3(2*H*)-furanone is significant as it is associated with the caramel odour of cooked bacon[46].

Pyridines

There are four pyridines reported as volatile flavour compounds of pan-fried bacon. Studies into these indicate 2-methylpyridine is characterized as hazelnut-like and astringent in aroma[35], whereas the formation of 2-pentylpyridine in typical meat species likely forms through a reaction between ammonia and 2,4-decadienal[4, 48, 56].

3.2 Phenols

A significant portion of the smoked flavour of cooked meat has been well attributed to phenolic compounds. The smoking processing method provides the major aroma components of wood smoke vapour to smoked bacon flavour compounds, including guaiacol, phenol and 4-methylguaiacol[42, 43]. The pathway for the formation of phenolic compounds in wood smoke is through the pyrolysis of lignin in the wood used during the smoking process[38]. Similar results were published in a study of volatile constituents of bacon fried at 350°C and using smoked sawdust instead of liquid smoke[35]. In an investigation comparing phenolic components of laboratory-smoked pork belly with commercial beef sausage, temperature and time mimicked that of commercial preparation for bacon. Results from an investigation comparing phenolic components of laboratory-smoked pork belly with commercial beef sausage, where temperature and time mimicked that of commercial preparation for bacon, revealed syringol was the major flavour component of smoked pork belly samples, whereas phenol was only detected in minute quantities[23]. During a subsequent recovery experiment, researchers recovered 4-methylguaiacol and guaiacol from the pork samples after exposure to smoke at identical time and temperature conditions. This phenomenon suggests that other pathways may be involved as the mechanism involved more than simple retention and deposition of these compounds.

Another investigation into volatile constituents of the phenolic fraction from cooked bacon using a steam-jacketed kettle with rapid refluxing at temperatures ranging from 20 to 175°C revealed similar results. A total of four 2,6-dimethylphenols instead of a single 3,5-dimethylphenol were isolated in this study, a difference that may result from the smoke used in the bacon process[43, 46]. Four fractions including phenolic, basic, acidic and neutral were obtained using chemical separation by ether extraction[46]. Results indicated that the phenolic fraction appeared to be the part that processes the characteristic aroma for bacon.

A total of 12 phenols identified from different fractions of the aroma concentrates of smoked-cured bacon using the nitrogen purge-and-steam distillation method with GC and GC–MS have been reported[53]. Phenols account for 30% of the aroma constituents identified in this study, and were judged as tangy in flavour. Among the 12 phenols detected, 2,3,5-trimethoxytoluene, *o*-tertbutylphenol, 2,2′-methylenebis(butylhydroxylto luene) and 3,4-dimethylphenol had not been reported by Toth and Potthast, who studied chemical aspects of smoked meat products using liquid smoke preparations as opposed to raw firewood smoke preparations[54]. These phenol derivatives contribute to the flavour of liquid smoke with aromas described as smoky, burnt and pungent[57, 58].

3.3 Hydrocarbons, aldehydes, alcohols, lactones and ketones

Ketone, alcohol, lactone and lower fatty acid formation are all related to lipid oxidation that occurs upon heating during the cooking of meat. If free radicals are present, lipid autoxidation might occur around 60°C, whereas most lipid degradation occurs under higher temperatures in the 200–300°C range[2]. It was also reported that undesirable acrid and bitter compounds may form through pyrolysis around 600°C.

Most carbonyl compounds, alcohols and hydrocarbons are not major characteristic aroma components in bacon flavour, with the exception of the Strecker aldehydes such as 3-methylbutanal[35]. One product of the Maillard reaction is 2-hydroxy-3-2-cyclopenten-1-one, which is also known as cyclotene. This compound had been unreported in heated meat until its identification in an investigation into the flavour profile of fried bacon[44]. Detected cyclotene and its related compounds may be synthesized from sugar during the Maillard reaction that occurs by heating[47]. Lustre and Issenberg attributed the flavour-enhancing function of cyclotene to bacon flavour[45]. This investigation of the phenolic fraction of pork belly smoked with hardwood sawdust as the smoking material identified cyclotene, maltol, vanillin and its related organic compound – acetovanillone[23]. The smoking apparatus used in this study was similar to those used in the commercial market and gave rise to a pleasant, heavily smoky odour. A total of eight lactone compounds (γ-valerolactone, γ-hexalactone, γ-nonalactone, δ-octalactone, δ-nonalactone, 2,4-methyl-2-butenolide, δ-hexalactone and δ-heptalactone) were identified in a study of volatile constituents of cooked bacon[46]. The isolated lactones, derived from lipids during heating, were all odorous compounds that may provide either desirable or undesirable notes to the overall flavour profile[48].

Volatile flavour constituents of cured roasted pork from mini-pigs were studied using solid-phase microextraction (SPME) and solvent extraction (SDE) combined with GC–MS. Among all the aroma compounds identified using SPME, the aldehydes fraction including saturated aldehydes, unsaturated aldehydes and benzene-derived aldehydes appeared to be predominant. Amounts of alcohols and ketones were comparatively lower than other identified volatiles, only totalling 0.25% and 2.04% of all volatiles, respectively[59]. Similarly, from a qualification point of view, using the SDE method, aldehydes were the largest proportion, totalling 38.13% of all volatiles. Notably, 2, 4-decadienal isomers and hexanal were the major compounds. The majority of aldehydes were greater than five carbon atoms such as nonanal, (*E*)-2-nonenal, hexanal, (*E*)-2-hexenal and so on, and were produced through lipid oxidation[59, 60]. Additionally, the identified phenylacetaldehyde and methyl-branched aldehyde (3-methylbutanal) were both derived from amino acid degradation. It has also been suggested that these aldehydes should be key contributors to overall pork flavour since they possessed low odour thresholds and distinctive odours. When compared to a group of 19 short-chain (C3–C9) aliphatic alcohols identified from volatile compounds of fried bacon and pork loin, the group of aliphatic alcohols was much smaller[59, 61].

Irradiated meats have a well-documented associated change in flavour chemicals. Among all identified trace volatiles, hydrocarbons, alcohols and carbonyl compounds are predominant. They are typically formed from lipids. In particular, hydrocarbons are the most abundant radiolytic compounds isolated from meat, and the fatty acid composition of lipids accounts for their distribution[20]. The detailed mechanism of hydrocarbon formation involves rupturing triacylglycerol bonds adjacent to or near the bond linkages[62, 63]. It is known that higher amounts of oxygen exposure during the irradiation process results in a greater number of identified carbonyl compounds[64–66].

4 Sausage

It is well established that the main sources of aroma compounds of dry fermented sausages include lipid autoxidation; addition of spices; and microbial metabolism of proteins, lipids and carbohydrates[75]. The mixture of odour-active constituents in appropriate amounts yields the typical aroma of dry fermented sausage and is reported to consist of 55 aroma-active compounds[85]. The type and amount of spices, such as garlic or pepper, added during sausage processing and factors of the fabrication process, such as fermentation stage and curing agents, also contribute to the overall aroma profile of sausage[84, 85]. Several odorants are known to be affected by the addition of nitrite or nitrate[86]. Compounds that originate from lipid autoxidation, amino acid degradation and microbial metabolism appeared to be in high concentration in nitrate-added dry fermented sausages. Conversely, in nitrite-added sausage, compounds derived from lipid β-oxidation were more concentrated[86–89].

Schmidt and Berger examined the aroma compounds of different dry fermented sausages, such as salami, derived from several origins using various extraction methods including high vacuum distillation, molecular distillation and vacuum steam distillation[74].

Some sulphur compounds in meat volatiles are shown in Fig. 11. A variety of aroma compounds were identified, including alcohols, aldehydes, ketones, terpenes, phenols and so on. Conversely, sulphur compounds and esters accounted for minor constituents contributing less than 1% of the total volatiles[74]. A German salami brand processes a strong overall sour and buttery aroma, with some limited fruity and spicy odours as well. Eugenol and diallyl disulphide appear as the most active odorants with a dilution factor of 1:10000 among a total of 40–50 odorants. They are followed by linalool, acetic acid, methylallyl sulphide, 3-methylbutanoic acid and diallyl sulphide. The addition of spices, such as garlic and nutmeg, during the ripening procedure yields aroma-active compounds such as sulphur compounds and eugenol, respectively[74].

4.1 Aldehydes

The aldehydes identified from Italian dried sausages by Sunesen et al. include hexanal, nonanal, heptanal, pentanal, benzeneacetaldehyde and 2- and 3-methylbutanal. Hexanal was the most abundant during the entire ripening period, and was described as green odour[75, 78]. This distribution and identification of aldehydes agrees with findings reported by Meynier et al.[76]. Notably, decanal was not present in mince, then appeared and rose significantly after 28 days of processing[75]. Masson indicated that microbial-mediated conversions led to the formation of benzeneacetaldehyde, 2- and 3-methylbutanal with the precursors phenylalanine, isoleucine and leucine, respectively[82]. Furthermore, the study by Sunesen et al. isolated benzeneacetaldehydes, whereas earlier studies by Stahnke et al. described characteristics of three other sausages investigated as hyacinth in aroma[83].

Aldehydes ranged between 7.72% and 13.98% of the total peak area[91]. Most of them, including octanal, 2-octenal, 2-decenal, 2,4-decadienal, hexanal, heptanal and 2-heptenal were concluded to be derived from lipid oxidation and certain off-flavours appeared to be associated to these aldehydes[78, 94, 95]. The aldehyde 2-phenylacetaldehyde, derived from the catabolism of phenylalanine, may be an indicator of proteolysis, and imparts a hawthorn and harsh odour note[94]. Benzaldehyde is associated with an almond flavour that is isolated in all sausages and attributes a characteristic pork flavour note[96].

A study of early lipid oxidation in smoked, pork sausages with spices found hexanal to be one of the major components contributing to rancid odour during storage[90]. Compared to samples stored in vacuum packages, concentrations of hexanal were much higher in samples stored in air. It was reported that the decomposition of hydroperoxides, derived from the autoxidation of n-6 fatty acids, might be the major contributor to hexanal formation[90]. Linoleic acid accounts for the most abundant PUFAs in sausages, reasonably leading to high hexanal concentrations identified in the air-stored samples[90].

The major aroma compounds attributing the typical dry-cured aroma in fermented French sausages prepared with curing salts such as nitrite and nitrate have been well discussed[84]. Higher detection frequency (DF) values of nonanal and 2-heptenal were exhibited in nitrite-added sausages, as compared to nitrate-added sausages with higher DF values for 3-methylbutanal and phenylacetaldehyde[84]. Major odorants essential for the aroma of dry, fermented sausage in different stages of the curing process have been examined, as has the influence of curing agents on the generation of these odorants[86]. Eleven aldehydes were found to be dominant for odour-active compounds analysed using multiple headspace SPME (HS-SPME). The compounds with the highest oil odour-activity values (OAVs) such as octanal, 2-methylbutanal and 3-methylbutanal, which also increase during the drying process, were the major contributors to the aroma of dry fermented sausages. Aldehydes such as pentanal, hexanal, nonanal and propanal were also important odorants as they were generated towards the end of the process[86].

4.2 Acids

From a quantitative point of view of all the identified compounds of a typical Spanish dry fermented sausage using a simultaneous distillation-extraction (SDE) system, acids were predominant, accounting for at least 60% of the total peak area[91]. Compared to medium- and long-chain fatty acids, the strong cheesy odour and lower threshold values of short-chain fatty acids were responsible for greater contributions to flavour development[91]. The degradation of lipids and hydrolysis of triacylglycerols and phospholipids has led to the formation of medium- and long-chain fatty acids in dry fermented sausages[92]. Microbial metabolism-derived acids including 2-methylbutanoic acid, 3-methylbutanoic acid and 2-methylpropanoic acid impart a sweet note to the overall aroma[93]. The conversion of these three acids into fruity esters may also be responsible for the overall aroma of sausages[78]. These acids include 2-methylpropanoic acid and 3-methylbutanoic acid and are both important contributors to the aroma of dry fermented sausages identified at the end of the multiple HS-SPME process[86].

4.3 Alcohols, ketones, esters and phenolic compounds

It has been discussed that lipid oxidation possibly generates alcohols from a corresponding aldehyde with enzyme dehydrogenase[79]. Among all the alcohols isolated, 1-octen-3-ol was found in the highest quantity during the entire ripening period, and imparts a mushroom note. 2-heptanol and 1-propanol are typically not present during the early stages of ripening, but gradually increase after 4 days of processing[75]. Propanal may be the source for producing propanol through carbohydrate metabolism[80]. 1-octen-3-ol was isolated from a typical Spanish dry fermented sausage as well. This afforded a mushroom odour, plus had a very low odour threshold[91]. A study analysing early lipid oxidation in smoked pork sausages with spices reported that 1-penten-3-ol originated from autoxidation of n-3 fatty acids as one of the major components contributing to rancid odour during storage[90].

It has been reported that nitrite-added sausages exhibit higher DF values for 1-hexanol and ethanol than the rest of identified alcohols[84]. The identified alcohols in the dry fermented sausages, such as 1-octen-3-ol and 3-methyl-1-butanol, do not increase in concentration throughout the drying process as do most of the other compounds[86]. 4-terpineol, which is derived from a typical Spanish dry fermented sausage and possesses a terpenic structure, was also identified. Additional alcohols identified include geraniol, linalool, α-terpineol and phenylethylalcohol, and are described as rose, floral, peach and warm rose-honey, respectively[91].

Ketones such as 2-heptanone, 2-pentanone, 2-nonanone, 2-octanone and so on were also identified. Among these, 2-heptanone appeared to be the most abundant ketone during the entire ripening period[75]. Similarly, 2-heptanone, isolated from a typical Spanish dry fermented sausage, affords a spicy, blue cheese odour[91]. The microbial β-oxidation of saturated fatty acids is responsible for producing methyl ketones, subsequently followed by the reaction of β-keto acid decarboxylation[81]. The most plausible pathway for 4-heptanone formation is fungal β-oxidation of fatty acid, not lipid autoxidation, as the compound only appeared after 39 days of processing[75].

The odorant ester ethyl 2-methyl butanoate is the major contributor from the beginning of the process, whereas ethyl 3-methyl butanoate and ethyl hexanoate also appear to be important odorants as they are generated at the end of the process[86]. Particularly, ethyl 2-methyl butanoate consistently increased in concentration during the drying process. Numerous ethyl and methyl esters were found in five different brands of dry fermented sausages. These were derived from esterification of acids and alcohols accounting for 5% of the total peak area[91, 101].

As previously stated in the bacon section, Lustre and Issenberg compared the phenolic components of laboratory-smoked pork belly with commercial sausage. They reported that *trans*-isoeugenol appeared to have the highest concentration, followed by 2,6-dimethoxy-4-methylphenol and phenol in smoked sausages[23]. The smoking process increases phenolic compounds isolated from sausages due to the pyrolysis of lignin[102]. Phenylalanine and benzaldehyde with certain bacterial involvement are responsible for producing phenols[103]. Among all identified phenolic compound, 4-methylphenol and guaiacol were the most predominantly detected[91], and are described as pungent and smoky odours[102].

4.4 Spice-derived compounds

In a study investigating the evolution of volatiles during the ripening of dried sausages, quantitatively dominant terpenes were found mainly as derived from added peppers[75]. The concentration of all the isolated pepper compounds peaked during the middle of the processing method[75]. Similar findings reportedly have shown major constituents of terpenes as D-limonene, α-pinene, α-phellandrene, β-phellandrene, β-myrcene and so on, with concentrations peaking at day 18 of the entire ripening process. Among all the volatiles identified by Schmidt and Berger, terpenes were the most dominant, followed by ketones and aliphatic alcohols. The isolated odorants such as thujene, sabinene, 3-carene, y-terpinene, α-pinene, α-phellandrene and terpinolene were all related to the use of pepper in sausages preparation[91, 97]. Some of these compounds impart a fruity, fresh and floral odour note to the overall aroma profile. However, it has been reported that terpenes may result from added pepper during fermenting and ripening, and are not major contributors to salami aroma. Notably, only linalool contributed greatly to the overall odour[74]. Some non-terpenic compounds including eugenol (honey-like, spicy), safrole and myristicine (nutmeg, spicy) have been found as well, and are thought to be associated with the application of pepper in sausage preparation[99].

Some typical garlic compounds were identified using dynamic headspace sampling methods with GC–MS[75]. It has been well established that added spices contribute to the overall odour of different species of sausages, garlic being one of the most frequently added seasonings. The total concentration of garlic compounds increases during ripening, particularly diallyl disulphide which rises as an individual garlic compound towards the end of ripening. Besides diallyl disulphide, allyl methylsulphide was also identified as quantitatively dominant among all the isolated garlic compounds[75]. Isolated sulphur compounds such as diallyl trisulphide, diallyl disulphide, methyl allyl disulphide and 1-propene result from the use of garlic as an ingredient[93, 100]. Some thiols, sulphides and disulphides found in the volatiles of cooked meat are shown in Fig. 12. Limonene, cubebene and β-caryophyllene from paprika all used in preparation for a typical Spanish sausage have also been isolated[98]. Some non-terpenic compounds such as methylsalicylate and tetramethylpyrazine have also been detected in paprika[98].

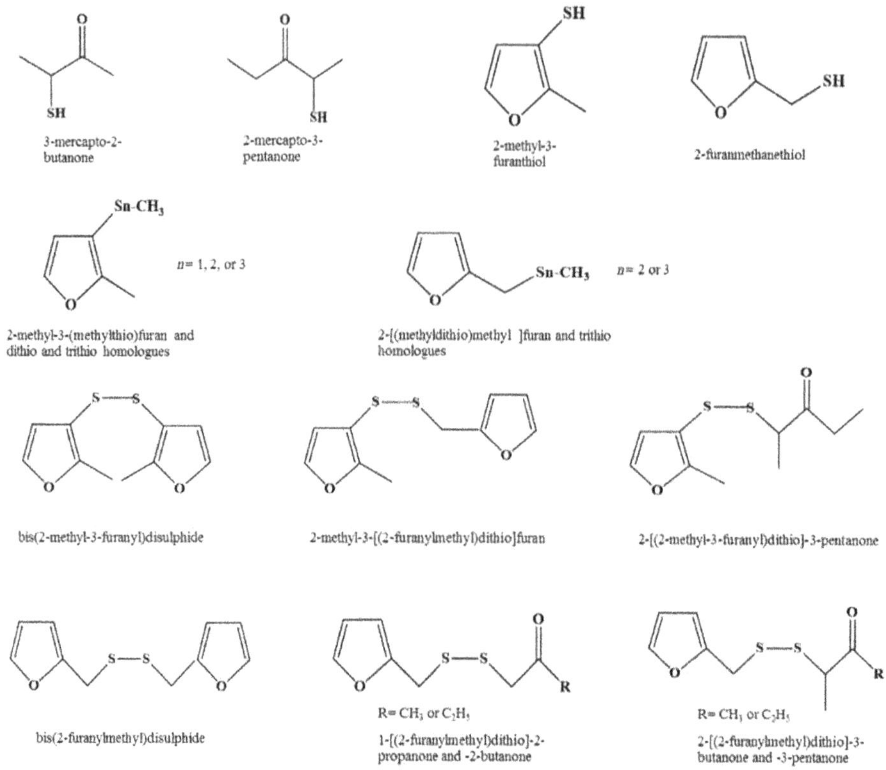

Figure 12 Some thiols, sulphides and disulphides found in the volatiles of cooked meat [11, 12, 145–147].

5 Ham

It is well known that ham is one of the most popular pork-related products worldwide. Over the past decades, numerous investigations have been conducted on the aroma components of various species of ham. These identified odorants are typically generated from varying sources including lipids, amino acids and carbohydrates[26]. Studies have shown that the main aroma-active constituents are composed of aldehydes, alcohols, ketones, lactones and hydrocarbons in different species of ham including Iberian ham, country-style ham, Italian- and Parma-type ham, and so on[27, 106–108, 140]. Additionally, it has been discussed that intense lipolysis and lipid oxidation are involved in generating the numerous volatile compounds that contribute to some characteristic aroma notes of meat products such as aged and rancid ham[26].

Phospholipases and lipases, two enzymes required for lipolysis, lead to the formation of free fatty acids in fresh meat during processing[115]. These lipases remain active during a large part of dry-cured ham processing, even though their activity decreases[116, 117]. These neutral and basic lipases are very active during the early stage of meat processing. Conversely, acid lipase is less active during the entire dry-curing process of ham[117]. These neutral lipases have been confirmed as the main enzymes responsible for lipolysis in adipose tissue triacylglycerols because these enzymes remain active for more than one year during the dry-curing process of ham[115, 118]. The amount of free fatty acid during the dry-curing process of ham is much higher than those present in fresh meat products[117, 119].

Lipid oxidation contributes to the formation of a variety of typical meat product aromas, and is also the main contributor to the deterioration of meat during processing[5, 120–122]. The overall mechanism of fatty acid oxidation has been well established. In meats, autoxidation is the major process for lipid oxidation and consists of three stages that include initiation, propagation and termination, as mentioned in Section 1[123, 124]. Among various factors affecting aroma extracted from meat products, the structure of fatty acids is the most important as it is related to the proportion and the number of hydroperoxide isomers[123]. Moreover, the medium conditions such as pH, temperature, ion presence and so on also influence reactions[124, 125]. A large group of odorants, especially those with low odour thresholds, such as C5–C8 unsaturated ketones, pentenyl or pentyl furan, and C3–C10 aldehydes significantly impact the overall aroma of final products. The previously mentioned odorants are described as green, cucumber, fruity, oily, deep-fried and mushroom[125]. In particular, hexanal exhibits a greenish aroma, whereas 2,4-decadienal, *t*-2-heptenal, nonanal and 2-pentylfuran process rancid, oily and deep-fried odours[26]. A low initial pH and high salt content exacerbate the lipid oxidation process and eventually increase correlated oxidation products in dry-cured ham[115, 126]. There are some factors that affect quality differences of volatile compounds obtained from lipid oxidation during the curing process. These include temperature processing length, lipid content, amount of salt or other agents applied, and the degradation of carbohydrates and amino acids[127–130].

A large variety of factors affecting meat aroma profiles have been well documented. For example, as mentioned in the introduction, the IO derived from irradiation processing can be reduced through application of spices such as thyme, sage, onion and so on[29]. Due to off-odour formation by free radical oxidation, antioxidants have been applied to slow down this process[30]. When comparing irradiation strengths and times, high doses for shorter times produce more off-odours than low doses for longer times[134]. Therefore, the undesirable chemical changes in food during irradiation processing can be reduced by lowering the temperature[135].

It has been well documented that nitrite is one of the most popular agents applied during the curing process of meat. The amount of pickling nitrite, typically within a range of 100 to 1500 mg/L, is capable of producing a desirable ham flavour[136, 138]. In a study examining the effects of temperature, time and curing ingredients during the curing process of ham, Kemp et al. reported that both curing treatment and storage conditions influence the flavour of dry-cured ham[137, 138]. Comparatively, pickle-cured ham only contains sodium chloride and samples with nitrite levels as low as 50 mg/kg yield a characteristic cured meat flavour[133]. Nitrite's potent antioxidant effect may prevent the formation of an unpleasant warmed-over flavour during storage[104, 105]. In addition, off-odour formation in ham has been significantly reduced through the slowing of lipid oxidation by nitrites during 14 days of aerobic storage[132].

5.1 Aldehydes

Significant differences have been reported in *n*-hexanal and *n*-pentanal concentrations from a study of volatile fractions comparing heating-cured and uncured ham. The authors indicated that hexanal and valeraldehyde were present in appreciable quantities in uncured ham but were detected in minute amounts in cured ham[14]. At an internal temperature of 70°C, hexanal and pentanal were not formed during heating-cured hams, whereas substantial quantities of these compounds appeared in uncured hams[14]. Although aldehydes including propionaldehyde, acetaldehyde and butyraldehyde exist in larger quantities in uncured ham as opposed to cured ham, differences in their content were less pronounced. These authors suggested the oxidative cleavage of unsaturated fatty acid residues such as linoleate might lead to the formation of valeraldehyde, hexanal and butyraldehyde. They also concluded that curing with nitrite and cyanide might interfere with the oxidation of unsaturated lipids through deactivation of related hematin catalysts. However, the application of nitrite did not affect the conversion of amino acids to the corresponding aldehydes[14].

Several studies indicate the positive aroma notes from dry-cured ham, cured ham and aged odours are associated with branched aldehydes that are produced by the degradation of amino acids. Pathways to form branched aldehydes such as 2- and 3-methylbutanal are shown in Fig. 13 and 14. In a study of volatile components of non-smoked dry-cured ham, the largest proportion of the total volatiles detected was aldehydes, with their total relative area equal to 26.23% in which aliphatic aldehydes accounted for 25.02%[13].

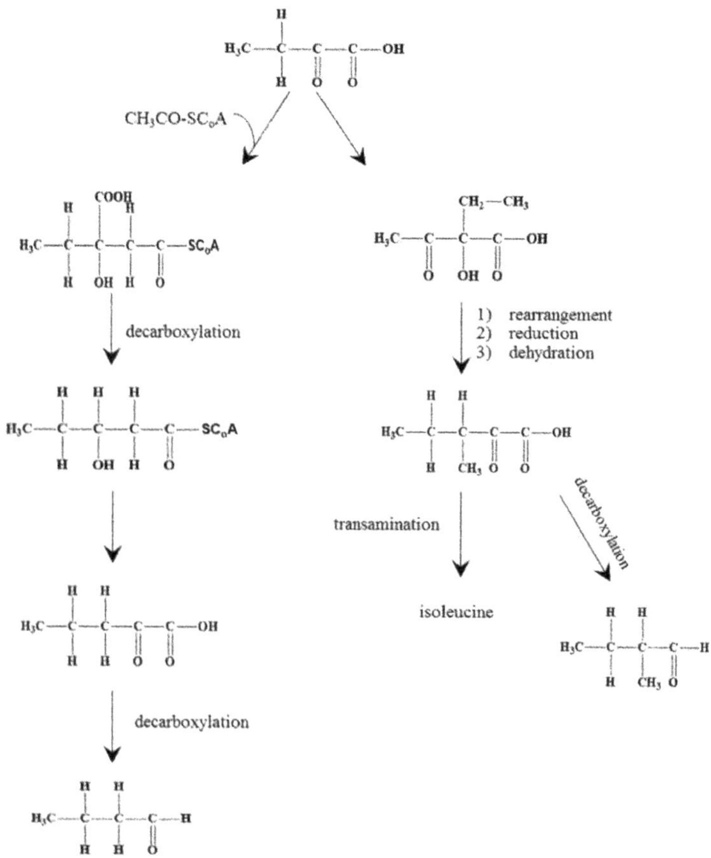

Figure 13 Pathway to form 2-methylbutanal[60].

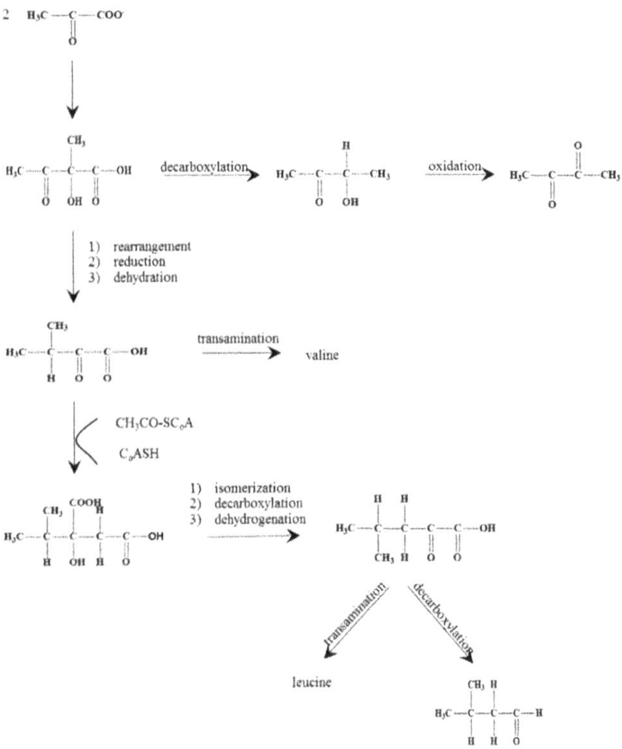

Figure 14 Pathway to form 3-methylbutanal[60].

Subsequent olfactory tests of the same samples revealed that certain aldehydes such as nonanal, 2-nonenal and hexanal contribute fat, oil and tallow aromas, which may also be major contributors for rancid aroma of dry-cured ham. The possible pathways that were associated with the formation of these aldehydes may be lipid oxidation, sugar catabolism, meat proteolysis and lipolysis[112]. In particular, lipid oxidation is responsible for the alkenals and alkanals with more than six carbon atoms such as octanal, nonanal, decanal, octadecanal and 9-octadecenal[13]. The inferior homologous series of these detected aldehydes are due to autoxidative mechanisms of the free fatty acids[109, 113]. Among all the detected aldehydes, 3-methylbutanal processes sweaty, pungent and rancid aromas, while benzeneacetaldehyde exhibits hawthorn and harsh odours.

Among all the aldehydes identified in an investigation into aroma components in American country ham, decanal and (*E*)-2-nonenal were two of the major odorants found with FD factors ≥125 using a dynamic headspace dilution analysis (DHDA) method. However, the neutral/basic fraction was dominated by hexanal, 3-methylbutanal and phenylacetaldehyde using a direct solvent extraction–solvent-assisted flavour evaporation (DSE–SAFE) method. All these detected aldehydes process olfactive characteristics: decanal (green, orange), (*E*)-2-nonenal (stale, hay), hexanal (cut-grass, green), 3-methylbutanal (dark chocolate) and phenylacetaldehyde (rosy)[140].

An investigation into the volatile compounds of Italian-type dry-cured ham included two phases: salting, drying and ripening as the first phase, and ripening and post-ripening as the second phase, and identified a total of 21 aldehydes including acetaldehyde, hexanal, octanal, nonanal and decanal. The ham was subjected to salting, drying and the first ripening stage where predominant quantities of hexanal and pentanal were detected, revealing lipid autoxidation occurring at this stage of the curing process, as these compounds are secondary end products derived from lipid autoxidation[139].

5.2 Ketones

1-octen-3-one and 1-nonen-3-one are predominant with FD factors three 125 among all aroma compounds identified using DHDA in the three samples of American country ham[140]. Specifically, the OAV of 1-octen-3-one is high due to its low threshold, and contributes a mushroom-like odour to the overall note of American country ham[140]. Similar results were found in this same study using the DSE–SAFE method with GC–O, as 1-octen-3-one was the dominant ketone identified and afforded a mushroom odour[140].

In a study that investigated the lipid changes during meat processing, Gandemer concluded that the C5 and C10 containing methylketones were the main ketones resulting from lipid oxidation. Some related key aroma notes included 2-dodecanone (fatty), 2-heptanone (blue cheese), 2-decanone and 2-undecanone (fruity)[5]. An additional study of volatile components of non-smoked dry-cured ham reported the main ketones as *trans*-geranylacetone and methyl ketones such as 3-methylbutan-2-one. It has been stated that the β-oxidation of saturated fatty acids and/or the decarboxylation of β-keto acid are the two pathways producing methyl ketones[114]. Among identified ketones, 3-hydroxybutan-2-one was described as a strong smell of butter, octan-2-one processed herbaceous and green odours, heptan-2-one had blue cheese and spicy aromas, and *trans*-geranylacetone exhibited a parsley odour. The 3-hydroxy-2-bytanone has been detected in an investigation into lipid changes in dry-cured meat products, and was derived from the oxidation of hydroxyl fatty acids, describing as a buttery aroma[114].

5.3 Lactones, carboxylic acids, alcohols and hydrocarbons

Two γ-lactones, γ-hexalactone and γ-nonalactone have been reported as products of γ-hydroxyacids cyclization and dehydration. The lactone γ-nonalactone processes a smell of coconut, licorice and musk. Similar observations have shown that γ-nonalactone is predominant in the neutral/basic fraction of American country ham using a DSE–SAFE method with GC–O[140].

It has also been reported that free carboxylic acids represent 6.18% of the total relative peak area including hexanoic acid, octanoic acid, nonanoic acid and pentadecanoic acid. These may be produced through hydrolysis of phospholipids and triacylglycerols. Short-chain aromatic acids are the major components in acidic fraction of American country ham using an AEDA method with GC–O. All these detected compounds have olfactive characteristics: hexanoic acid (sour), pentanoic acid (rancid), acetic acid (vinegar), butanoic acid (cheesy), phenylacetic acid (rosy) and 2-methylpropanoic acid (cheesy, faecal)[140].

The aroma compounds, acetic acid and *n*-pentanol, were identified from a group of hams that underwent 25 days of salting, then 90 days of drying and finally 90 days to first ripening. This observation was attributed to these compounds being end products derived from the autoxidation of lipids occurring at this stage of processing[139]. In addition to *n*-pentanol, more alcohols identified in this study include 3-methylbutanol,

1-octen-3-ol and hexanol[13]. Among the alcohols with strong odours, 1-octen-3-ol perceives as mushroom; phenylethanol describes as rose; octan-1-ol as fatty, waxy and sharp odours; and farnesol exhibits a musk-like aroma. Lipid oxidation is responsible for the formation of saturated and unsaturated alcohols (C4–C8), as discussed in a study of lipid changes seen in dry-cured meat products. The main contributors to the overall aroma of dry-cured meat are unsaturated alcohols such as 1-penten-3-ol, which describes as a grass odour, and 1-octen-3-ol with a mushroom aroma note[5, 26].

All identified aroma compounds have been grouped on the basis of their chemical families such as aldehydes, alcohols, ketones, lactones, carboxylic acid and hydrocarbons, as previously mentioned in each section. The possible mechanism for these identified odorants may be due to the catabolism of the main constituent parts found in ham, including lipids, glucids and parotids, during the curing process of non-smoked ham[108]. Results reveal that the identified number of hydrocarbons appears to be the most abundant. These include eight *n*-alkanes, six methyl-branched alkanes and five aromatics such as toluene, ethylbenzene and *n*-xylene. The autoxidation of the lipids might be related to the *n*-alkanes formation[109]. On the other hand, the branched alkanes might be associated with oxidation of branched fatty acids present in animal tissues[110, 111].

6 Conclusion

6.1 Bacon

Heterocyclic aroma compounds including pyrazines, thiazoles, thiazolines, oxazoles, oxazolines, furans and pyridines accounted for the major aroma-active groups in bacon[38]. As the most commonly used curing agent, the application of nitrite for developing cured meat flavour leads to appreciable amount of pyridines, which imparts a roasted nut-like aroma[51, 52, 62]. Furan was one of the most important heterocyclic compounds in the volatile profile of cooked meat, which imparts a caramel-like, nutty, smoke and sweet aroma[8, 38]. Despite the minor constituents of benzothiazole and sulphured 2-acetylthiazole in roasted pork, they could significantly contribute to the overall meat flavour due to their low thresholds and distinguishable odours[59]. Oxazolines have scarcely been found in any foods other than meat. 2,4-dimethyl-3-oxazoline was identified in fried bacon affording a vegetable-like, nutty aroma[35, 37]. Only two oxazoles including 2-methylbenzoxazole and 2,4,5-trimethyloxazole were identified in the volatiles of heated meat[36].

In addition to heterocyclic odorants, some other aroma compounds contributed to the aroma profile of cooked bacon as well. Phenolic compounds such as guaiacol, phenol and 4-methylguaiacol were major smoked bacon flavour constituents, which were described as smoky, burnt and pungent[42, 43, 57, 58]. One Maillard reaction product, 2-hydroxy-3-2-cyclopenten-1-one, also known as cyclotene, was summarized to enhance the flavour of bacon[41, 45, 47]. Among all the identified aldehydes, those with more than five carbon atoms were in the large level using SDE method, such as nonanal, (*E*)-2-nonenal, hexanal, (*E*)-2-hexenal and so on[59, 60].

6.2 Sausage

Spice-derived aroma compounds contributed the most to the overall aroma profile of sausages. Terpenes such as D-limonene, α-pinene, α-phellandrene, linalool, β-phellandrene, β-myrcene thujene, sabinene, 3-carene, y-terpinene and terpinolene mainly derived from pepper-added sausages were quantitatively dominant[75]. Specifically, linalool contributed greatly to the overall odour[74]. Some non-terpenic compounds such as eugenol (honey-like, spicy), safrole and myristicin (nutmeg, spicy) were reported to be associated with the application of pepper in the preparation stage of sausages[99]. Some garlic-derived sulphur compounds such as diallyl disulphide, diallyl trisulphide and allyl methylsulphide turned out to be quantitatively dominant among all the isolated garlic compounds[75, 93, 100].

Besides spice-derived aroma compounds, aldehydes such as octanal, 2-methylbutanal and 3-methylbutanal with the highest OAVs were major contributors to the aroma of dry fermented sausages[86]. Particularly, benzaldehyde was associated with an almond aroma, which was isolated in all sausages and attributed to a characteristic flavour note in pork[96]. Hexanal was found to be the most abundant during the whole ripening period in Italian dried sausages[78, 79]. Despite minor constituents contributing less than 1% of the total volatiles, some sulphur compounds and acids isolated from sausages appeared to be relatively active odorants such as diallyl disulphide, acetic acid, methylallyl sulphide, 3-methylbutanoic acid and diallyl sulphide[74]. The microbial metabolism-derived acids such as 2-methylbutanoic acid and 2-methylpropanoic acid were two important contributors to the aroma of dry fermented sausages, imparting a sweet note to the overall aroma[93]. The 1-octen-3-ol isolated from a typical Spanish dry fermented sausage was reported that it afforded a mushroom odour and possessed a very low odour threshold[91]. Among ketone groups identified in typical Spanish dry fermented sausage, 2-heptanone appeared to be the most abundant, which afforded a spicy, blue cheese odour[91].

6.3 Ham

The main aroma-active constituents in different species of hams were composed of aldehydes, alcohols, ketones, lactones and hydrocarbons through intense lipid lipolysis and lipid oxidation[26, 27, 106–108, 140]. Particularly, those with low odour thresholds, such as C5–C8 unsaturated ketones, pentenyl or pentyl furan, and C3–C10 aldehydes revealed significant impact on the overall aroma profile, mainly affording green, cucumber, fruity, oily, deep-fried and mushroom odour notes[125].

Despite high constituents of aroma-active aldehydes in hams, differences in identification and concentration of aldehydes were reported for various species of hams. For example, hexanal and valeraldehyde were shown to be present in appreciable quantities in uncured ham but were with minute amounts in cured ham[14]. Aldehydes such as nonanal, 2-nonenal and hexanal were identified as the highest proportion of the total volatiles contributing fat, oil and tallow smell to a non-smoked dry-cured ham[112]. Decanal and (*E*)-2-nonenal were two major odorants found with FD factors n 125 in an American country ham using DHDA method[140]. In an investigation on the volatile compounds in Italian-type dry-cured ham, hexanal and pentanal were predominant in quantities after hams were subjected to salting, drying and the first ripening stage[139].

1-octen-3-one was the dominant ketone identified among all the aroma compounds in American country hams, contributing a mushroom-like odour to the overall note with high OAV due to its low threshold[140]. The odorant γ-nonalactone was reported as predominant lactone in American country ham, which processed a smell of coconut, licorice and musk[140]. Even though relative limited numbers and amount of free carboxylic acids were identified in American county ham, all of the detected compounds yielded olfactive characteristics: hexanoic acid (sour), pentanoic acid (rancid), acetic acid (vinegar), butanoic acid (cheesy), phenylacetic acid (rosy) and 2-methylpropanoic acid (cheesy, faecal)[140]. Among all the identified alcohols such as l-pentanol, 3-methylbutanol, 1-octen-3-ol and hexanol, the main contributors to the overall aroma of dry-cured meat were 1-penten-3-ol (grass) and 1-octen-3-ol (mushroom)[5, 13, 26].

In summary, as discussed above, there is no single compound identified in the pork aroma profile described as uniquely meaty or that accounts for cooked 'meat' flavour[6]. Upon heating and various processing methods, different types of pork-related products yield characteristic aroma[13–26]. The major attributes for overall pork flavour of major related products have been discussed in this book chapter. All previous findings indicate that it is the combination of major attributes that contributes to overall flavour of individual pork-related product[2]. Two major chemical reactions involved in these aroma-active compounds formation are Maillard reaction and lipid oxidation[2, 4]. Additionally, among numerous processing methods such as cooking, smoking, curing, canning, deboning, grinding, mincing, irradiation and freeze-drying, cooking plays a vital role in determining final meat flavour[7, 27, 29, 30]. In particular, methods of applying heat, such as frying, boiling, roasting, retort heating and pressure-cooking induce a varying series of chemical changes that yield various odorants[1, 5]. Besides various processing methods, the addition of spices, additives or agents employed during processing such as garlic, thyme, sage, pepper, nitrite, nitrate and so on lead to the formation of new aroma compounds as well[1, 5, 9, 10, 18, 28].

7 Where to look for further information

There are a couple of reviews written by Mottram[4, 32] on meat flavour that give a general account of the types of flavour compounds in meat including pork. The review of Bailey[1] discusses in detail the effect of the Maillard reaction on meat flavour development. The review of Shahidi et al[5] gives more details information on the flavor compounds and their formation in processed meat products.

8 References

1 Bailey, M. E. (1994). The Maillard reaction and meat flavor development. In *Flavor of Meat and Meat Products, Chapman and Hall, London, UK*, ed. F. Shahidi, pp. 153–73.
2 Wasserman, A. E. (1972). Thermally produced flavor components in the aroma of meat and poultry. *J. Agric. Food Chem.* 20, 737.
3 Watkins, P. J.1, Frank, D., Singh, T. K., Young, O. A. and Warner, R. D. (2013). Sheepmeat flavor and the effect of different feeding systems: A review. *J. Agric. Food Chem.* 61(15): 3561–794.
4 Mottram, D. (1998). Flavour formation in meat and meat products: a review. *Food Chem.* 62, 415–24.
5 Shahidi, F., Rubin, L. J. and D'Souza, L. A. (1986). Meat flavor volatiles: A review of the composition, techniques of analysis, and sensory evaluation. *Crit. Rev. Food Sci. Nutr.* 24, 141–243.
6 Ramarathnam, N., Rubin, L. J. and Diosady, L. L. (1993b). Studies on meat flavor. 4 fractionation, characterization, and quantitation of volatiles from uncured and cured beef and chicken. *J. Agric. Food Chem.* 41, 939–45.

7 Landmann, W. A. and Batzer, O. F. (1966). Influence of processing procedures on the chemistry of meat flavors. *J. Agric. Food Chem.* 14, 210–14.
8 Shibamoto, T. (1980). Heterocyclic compounds in cooked meat. *J. Agric. Food Chem.* 27, 237–43.
9 Pittet, A. O. and Hruza D. E. (1974). Comparative study of flavor properties of thiazole derivatives. *J. Agric. Food Chem.* 22, 264–9.
10 Chang, S. S. and Peterson, R. J. (1977). The basis of quality in muscle foods. *J. Food Sci.* 42, 296–305.
11 Farmer, L. J. and Patterson, R. L. S. (1991). Compounds contributing to meat flavor. *Food Chem.* 40, 201–5.
12 Madruga, M. S. and Mottram D. S. (1995). The effect of pH on the formation of Maillard-derived aroma volatiles using a cooked meat system. *J. Sci. Food Agric.* 68, 305–10.
13 Berdague, J. L., Denoyer, C., SanLuc, J., Quere, L. and Semon, E. (1991) Volatile components of dry-cured ham. *J. Agric. Food Chem.* 39(7), 1257–61.
14 Cross, C. K. and Ziegler, P. (1965). A Comparison of the volatile fractions from cured and uncured meat. *J. Food Sci.* 30, 610–14.
15 Mottram, D. S., Edwards, R. A. and MacFie, H. J. H. (1982). A comparison of the flavour volatiles from cooked beef and pork meat systems. *J. Sci. Food Agric.* 33, 934–44.
16 Fiddler, W., Gates, R. A., Pensabene, J. W., Phillips, J. G. and Wierbicki, E. (1981). Investigation on nitrosamines in irradiation-sterilized bacon. *J. Agric. Food Chem.* 29(3), 551–4.
17 Olsen, E.1, Vogt, G., Veberg, A., Ekeberg, D. and Nilsson, A. (2005). Analysis of early lipid oxidation in smoked, commuted pork or poultry sausages with spices. *J. Agric. Food Chem.* 53(19), 7448–57.
18 John, W. P., Gates, R. A., Jenkins, R. K. and Fiddler, W. (1987). Effect of cesium-137 radiation on the formation of N-nitrosopyrrolidine in bacon. *J. Agric. Food Chem.* 35(2), 192–4.
19 Ramarathnam, N., Rubin, L. J. and Diosady, L. L. (1991). Studies on meat flavor. 1. Qualitative and quantitative differences in uncured and cured pork. *J. Agric. Food Chem.* 39, 344–50.
20 Merritt, C., Jr. and Angelini, P. (1975). Chemical changes associated with flavor in irradiated meat. *J. Agric. Food Chem.* 23, 1037.
21 Xie, J., Sun, B., Zheng, F. and Wang, S. (2008). Volatile flavor constituents in roasted pork of mini-pig. *Food Chem.* 109, 506–4.
22 Ramarathnam, N., Rubin, L. J. and Diosady, L. L. (1993). Studies on meat flavor. 3. A novel method for trapping volatile components from uncured and cured pork. *J. Agric. Food Chem.* 41, 933–8.
23 Lustre, A. O. and Issenberg, P. (1970). Phenolic components of smoked meat products. *J. Agric. Food Chem.* 18, 1056–60.
24 Byun, M. W. Lee, J. W., Jo, C. and Yook, H. S. (2001). Quality properties of sausage made with gamma-irradiated natural pork and lamb casing. *Meat Sci.* 59(3), 223–8.
25 Brewer, M. S. (2009). Irradiation effects on meat flavor: a review. *Meat Sci.* 81, 1–14.
26 Gandemer, G. (2002). Lipids in muscles and adipose tissues, changes during processing and sensory properties of meat products. *Meat Sci.* 62, 309.
27 Ockerman, H. W., Blumer, T. N. and Craig, H. B. (1964). Volatile chemical compounds in dry-cured hams. *J. Food Sci.* 29, 123–9. 73.
28 Brooks, J., Haines, R. B., Moran, T. and Pace, J. (1940). The function of nitrate, nitrite and bacteria in the curing of bacon and hams. Department of Science and Industrial Research, Food Investigative Special Report, No. 49.
29 Raharjo, S. (1996). Review of flavor differences in meat and meat products. *Indonesian Food and Nutrition Progress* Vol. 3, No. 2 30.
30 Huber, W., Brasch, A. and Waly, A. (1953). Effects of processing conditions and organoleptic changes in food stuffs sterilized with high intensity electrons. *National Agricultural Library (AGRIS)* 7(3), 109–15.
31 Noleau, I. and Toulemonde, B. (1987). Volatile components of roast chicken fat. *Lebensm. Wiss. Technol.* 20, 37–41.

32 Mottram, D. S. (1991). Meat. In *Volatile Compounds in Foods and Beverages*, ed. H. Maarse. Marcel Dekker, New York, pp. 107–77.
33 Frankel, E. N. (1980). Lipid oxidation. *Prog. Lipid Res.* 19, 1–22.
34 Grosch, W. (1982). Lipid degradation products and flavours. In *Food Flavours*, eds I. D. Morton and A. J. MacLeod. Elsevier, Amsterdam, pp. 325–98.
35 Ho, C. T., Ken, N., Lee, Q. and Zhang, J. (1983). Isolation and Identification of volatile flavor compounds in fried bacon. *J. Agric. Food Chem.* 31(2), 336–42.
36 Ho, C.-T. and Hartman, G. J. (1982). *J. Agric. Food Chem.* 30, 793.
37 Mussinan, C. J., Wilson, R. A., Katz, I., Hruza, A. and Vock, M. H. (1976). *ACS Symp. Ser.* 26, 133.
38 Kim, K., Kurata, T. and Fujimaki, M. (1974). *Agric. Biol. Chem.* 38, 53.
39 Ho, C. T., Smagula, M. S. and Chang, S. S. (1978). *J. Am. Oil Chem. Soc.* 55, 233.
40 Smouse and Chang, (1967). A systematic characterization of the reversion flavor of soybean oil, *J. Amer. Oil Chem. Soc.* 44, 509–14.
41 Maga, J. A. (1979). CRC *Crit. Rev. Food Sci Nutr.* 10, 355.
42 Kornreich, M. R. and Issenberg, P. J. (1972). *Agric. Food Chem.* 20, 110943.
43 Knowles, M. E., Gilbert, J. and McWeeny, D. J. (1975). *J. Sci. Food Agric.* 26, 189.
44 Lustre, A. O. and Issenberg, P. (1969). *J. Agric. Food Chem.* 17, 1387.
45 Lustre, A. O. and Issenberg, P. (1970). *J. Agric. Food Chem.* 18, 1056.
46 Shu, C.-K. and Mookherjee, B. D. (1985). Volatile components of the phenolic fraction of cooked bacon. *J. Agric. Food Chem.* 33, 1107–9.
47 Vernin, G. and Parkanyi, C. (1982). *Heterocyclic Flavoring and Aroma Compounds*, Wiley: Chichester, p. 151.
48 Watanabe, K. and Sato, Y. (1969). *Agric. Biol. Chem.* 33, 242.
49 Kato, S., Kurato, T. and Fujimaki, M. (1973). *Agric. Biol. Chem.* 37, 539.
50 Mulders, E. J. Z. (1973). LebensmUnters-Forsch, 152, 193.
51 Flament, I., Kohler, M. and Aschiero, R. (1976). *Helv. Chim. Acta*, 59, 2308.
52 Hodge, J. E. and Osman, E. M. (1976). Principles of food science. In *Part I, Food Chemistry*, ed. Fennema, O. R. Marcel Dekker, Inc.: New York, pp. 41–138.
53 Yu, A-N. and Sun, B-G. (2005). Flavor substances of Chinese traditional smoke-cured bacon. *Food Chem.* 89, 227–3.
54 Toth, L. and Potthast, K. (1984). Chemical aspects of the meat smoking of meat and meat products. *Adv. Food Res.* 29, 87–158.
55 Mottram, D. S. (1985). The effect of cooking conditions on the formation of volatile heterocyclic compounds in pork. *J. Sci. Food Agric.* 36, 377–82.
56 Tang, J., Jin, Q. Z., Shen, G. H., Ho, C. T. and Chang, S. S. (1983). Isolation and identification of volatile compounds from fired chicken. *J. Agric. Food Chem.* 31, 1287–92.ll
57 Fujimaki, M., Kim, K. and Kurata, T. (1974). Analysis and comparison of flavor constituents in aqueous smoke condensates from various woods. *Agric. Biol. Chem.* 38 (1), 45–52.
58 Baltes, W. and Sochtig, I. (1979). Low molecular weight ingredients of smoke flavour preparations. *Zeitschrift fur Lebensmittel-Untersuchung und-Forschung* 169, 9–16.
59 Xie, J., Sun, B., Zheng, F. and Wang, S. (2008). Volatile flavor constituents in roasted pork of mini-pig. *Food Chem.* 109, 506–14.
60 Belitz, H.-D. and Grosch, W. (1987). *Lipids in Food Chemistry*, Berlin: Springer Verlag, pp. 128–200.
61 Timon, M. L., Carrapiso, A. I., Jurado, A. and Lagemaat, J. V. (2004). A study of the aroma of fried bacon and fried pork loin. *J. Sci. Food Agric.* 84, 825–31.
62 Dubravcic, M. F. and Nawar, W. W. (1968). Radiolysis of lipids: mode of cleavage in simple triglycerides. *J. Am. Oil Chem. Soc.* 45, 656.
63 LeTellier, P. R. and Nawar, W. W. (1972). 2-Alkylcyclobutanones from the radiolysis of triglycerides. *J. Agric. Food Chem.* 20, 129.
64 Batzer, O. J., Scribney, M., Doty, D. M. and Scweigert, B. S. (1957). Irradiation of Food. *J. Agric. Food Chem.* 5, 700.

65 Monty, J. J., Tappel, A. L. and Groninger, H. S. (1961). Carbonyl Compounds of Irradiated Meats *J. Agric. Food Chem.* 9, 55.
66 Wick, E. L., Murray, E., Mizutani, J. and Koshika, M. (1967). In: Josephson, E.S., Peterson, M. S. (Eds.), Radiation Preservation of Foods 65. *Advanced Chemistry Series*, Washington, DC, pp. 12–25.
67 Perez, L. C. and Yaylayan, V. A. (2004). Origin and mechanistic pathways of formation of the parent furan – A food toxicant. *J. Agric. Food Chem.* 52, 6830–6.
68 Becalski, A. and Seaman, S. (2005). Furan precursors in food: A model study and development of a simple headspace method for determination of furan. *J. AOAC Int*, 88, 102–6.
69 Limacher, A., Kerler, J., Conde-Petit, B. and Blank, I. (2007). Formation of furan and methylfuran from ascorbic acid in model systems and food. *Food Addit. Contam.* 24(S1), 122–35.
70 Yaylayan, V. (2003). Recent advances in the chemistry of Strecker degradation and Amadori rearrangement: Implications to aroma and color formation. *Food Sci. Technol. Res.* 9. 1–6.
71 Yaylayan, V. and Wnorowski, A. (2001). The role of L-serine and L-threonine in the generation of sugar-specific reactive intermediates during Maillard reaction, in Food Flavors and Chemistry-Advances of the New Millennium. *Royal Soc. Chem.*, Cambridge, UK, pp. 313–17.
72 Sayre, L. M., Arora, P. K., Iyer, R. S. and Salomon, R. G. (1993). Pyrrole formation from 4-hydroxynonenal and primary amines. *Chem. Res. Toxicol.* 6, 19–22.
73 Liao, M. L. and Seib, P. A. (1987). Selected reactions of L-ascorbic acid related to foods. *Food Technol.* 41(11), 104–7, 111.
74 Schmidt, S. and Berger, R. G. (1998). Aroma Compounds in Fermented Sausages of Different Origins. *National Agricultural Library (AGRIS)*, 31, 559–67.
75 Sunsen, L. O., Dorigoni, V., Zanardi, E. and Stahnke, L. (2001). Volatile compounds released during ripening in Italian dried sausage. *Meat Sci.* 58, 93–7.
76 Meynier, A., Novelli, E., Chizzolini, R., Zanardi, E. and Gandemer, G. (1998). Volatile compounds of commercial Milano salami. *Meat Sci.* 51, 175–83.
77 Viallon, C., BerdagueÂ J. L., Montel, M. C., Talon, R., Martin, J. F., Kondjoyan, N. and Denoyer, C. (1996). The effect of stage of ripening and packaging on volatile content and flavor of dry sausages. *Food Res. Int.* 29, 667–74.
78 Stahnke, L. H. (1994). Aroma components from dried sausages fermented with Staphylococcus xylosus. *Meat Sci.* 38, 39–53.
79 Shahidi, F. 1998. Flavor of muscle foods: an overview. In *Flavor of Meat*, Meat Products, and Seafoods. Ed. F. Shahidi. Blackie Academic and Professional, London, pp. 1–4.
80 Halvarson, H. (1973). Formation of lactic acid, volatile fatty acids and neutral, volatile mocarbonyl compounds in Swedish fermented sausage. *J. Food Sci.* 38, 310–12.
81 Okumura, J. and Kinsella, J. E. (1985). Methyl ketone formation by Penicillium camemberti in model systems. *J. Dairy Sci.* 68, 11–15.
82 Masson, F., Hinrichsen, L., Talon, R. and Montel, M. C. (1999). Factors influencing leucine catabolism by a strain of Staphylococcus carnosus. *Int. J. Food Microbiol.* 49, 173–8.
83 Stahnke, L. H., Sunesen, L. O. and Smedt, A. D. (1999). Sensory characteristics of European, dried, fermented sausages and the correlation to volatile profile. Proceedings of the 13th Forum for Applied Biotechnology, 22–23 September, Gent, Belgium, Med. Fac. Landbouw. Univ. Gent, 64/5b, 559–66.
84 Marco, A., Navarro, J. L. and Flores, M. (2007). Quantitation of selected odor-active constituents in dry fermented sausages prepared with different curing salts. *J. Agric. Food Chem.* 55, 3058–65.
85 Marco, A., Navarro, J. L. and Flores, M. (2007). Quantification of selected odor-active constituents in dry fermented sausages prepared with different curing salts. *J. Agric. Food Chem.* 55, 3058–65.
86 Olivares, A., Navarro, J. L. and Flores, M. (2009). Establishment of the contribution of volatile compounds to the aroma of fermented sausages at different stages of processing and storage. *Food Chem.* 115(4, 15), 1464–72.

87 Flores, M. and Hernández, D. (2007). Optimization of multiple headspace solid-phase microextraction for the quantification of volatile compounds in dry fermented sausages. *J. Agric. Food Chem.* 55, 8688–95.
88 Olesen, P. T., Meyer, A. S. and Stahnke, L. H. (2004). Generation of flavour compounds in fermented sausages-the influence of curing ingredients, Staphylococcus starter culture and ripening time. *Meat Sci.* 66, 675–87.
89 Marco, A., Navarro, J. L. and Flores, M. (2006). The influence or nitrite and nitrate on microbial, chemical and sensory parameters of show dry fermented sausage. *Meat Sci.* 73, 660–73.
90 Olsen, E., Vogt, G., Veberg, A., Ekeberg, D. and Nilsson, A. (2005). Analysis of early lipid oxidation in smoked, comminuted pork or poultry sausages with spices. *J. Agric. Food Chem.* 53(19), 7448–57.
91 Ansorena, D., Gimeno, O., AstiasaraÂn, I. and Bello, J. (2001). Analysis of volatile compounds by GC±MS of a dry fermented sausage: chorizo de Pamplona. *Food Res. Int.* 34, 67–75.
92 Girard, J. P. and Bucharles, C. (1991). La acidicacioÂ n. In *TecnologõÂa de la carne y los productos caÂrnicos*, ed. J. P. Girard, Zaragoza, Spain: Acribia, S. A.
93 Mateo, J. and Zumalacarregui, J. M. (1996b). Volatile compounds in Chorizo and their changes during ripening. *Meat Sci.* 44(4), 255–73.
94 BerdagueÂ, J. L., Monteil, P., Montel, M. C. and Talon, R. (1993). Effects of starter cultures on the formation of flavour compounds in dry sausage. *Meat Sci.* 35, 275–87.
95 MacLeod, G. (1994). The flavour of beef. In *Flavour of Meat and Meat Products*, ed. F. Shahidi, Blackie Academic and Professional, London, pp. 4–37
96 Shahidi, F. (1994). *Flavour of Meat and Meat Products*. Blackie Academic and Professional, London.
97 Ekundayo, O., Laakso, I., Adegbola, R. M., Oguntimein, B., Sofowora, A. and Hiltunen, R. (1988). Essential oil constituents of Ashanti pepper (Piper guineense) fruits (Berries). *J. Agric. Food Chem.* 36, 880–2.
98 Guadayol, J. M., Caixach, J., Cabañas, J. and Rivera, J. (1997). Extraction, separation and identication of volatile organic compounds from paprika oleoresin (Spanish type). *J. Agric. Food Chem.* 45, 1868–72.
99 Russel, and Jennings (1969). Constituents of black pepper. Some oxigenated compounds. *J. Agric. Food Chem.* 17, 1107–12.
100 Kuo, M. C., Chien, M. and Ho, C. T. (1990). Novel polysulfides identified in the volatile components from Welsh Onions (*Allium fistulosum* L. var. maichuon) and Scallions (*Allium fistulosum* L. var. caespitosum). *J. Agric. Food Chem.* 38, 1378–81.
101 Shahidi, F., Rubin, L. J. and D'Souza, L. A. (1986). Meat flavour volatiles: a review of the composition, technique of analysis and sensory evaluation. *CRC Crit. Rev.n Food Sci. Nutr.* 24, 219–27.
102 Hollenbeck, C. M. (1994). Contribution of smoke flavourings to processed meats. In *Flavour of Meat and Meat Products*, ed. F. Shahidi, Blackie Academic and Professional, London.
103 GenomeNet. (1998, December). WWW Server: www.genome.ad.jp
104 Pearson, A. M., Love, J. D. and Shorland, F. B. (1977). Warmed-over flavour in meat, poultry and fish. *Adv. Food Res.* 23, 1–74.
105 Fooladi, M. H., Pearson, A. M., Coleman, T. H., and Merkel, R. A. (1979). The Role of nitrite in preventing development of warmed-over flavour. *Food Chem.* 4, 283–92.
106 Lillard, D. A. and Ayres, J. C. (1969). Flavor compounds in country cured hams. *Food Technol.* 23, 251–4.
107 Giolitti, G., Cantoni, C. A., Bianchi, M. A., and Renon, P. (1971). Microbiology and chemical changes in raw hams of Italian type. *J. Appl. Bacteriol.* 34, 51–61.
108 Garcia, C., Berdague, J. L., Antequera, T., Lopez-Bote, C., Cordoba, J. J. and Ventanas, J. (1991). Volatile components of dry cured Iberian hams. *Food Chem.* In press.
109 Loury, M. (1972). Possible mechanisms of autoxidative rancidity. *Lipids.* 7, 671–5.
110 Van Straten, S. (1977). *Volatile Compounds in Food*, 4th ed.; Civotno: Zeist, The Netherlands.

111 Rembold, H., Wallner, P., Nite, S., Kollmannsberger, H. and Drawert, F. (1989). Volatile components of chickpea (Cicer arietinum L.) seed. *J. Agric. Food Chem.* 37, 659–62.
112 Hertz, K. O. and Chang, S. S. (1970). Meat flavour, *Adv. Food Res.* 18, 2–83.
113 Paquette, G., KKupranycz, D. B., and Van de Voort, F. R. (1985). The mechanism of lipid autoxidation. I: Primary oxidation products. *Can. Inst. Food Sci. Technol.* 2, 112–18.
114 Lehninger, A. L. (1981). *Flammarion Medecine-Science.* Paris, 1088pp.
115 Toldra, F. and Flores, M. (1998). The role of muscle proteases and lipases in flavor development during the processing of dry-cured ham. *Crit. Rev. Food Sci.* 38(4), 331–52.
116 Hernandez, P., Navarro, J. L. and Toldra, F. (1999). Lipolytic and oxidative changes in two Spanish pork loin products: dry-cured loin and pickled-cured loin. *Meat Sci.* 51, 123–8.
117 Motilva, M. J., Toldra, F., Nieto, P. and Flores, J. (1993). Muscle lipolysis phenomena in the processing of dry-cured ham. *Food Chem.*, 48, 121–5.
118 Motilva, M. J., Toldra, F., Aristoy, M. C. and Flores, J. (1993). Subcutaneous adipose tissue lipolysis in the processing of dry-cured ham. *J. Food Biochem.* 16, 323–35.
119 Coutron-Gambotti, C. and Gandemer, G. (1999). Lipolysis and oxidation in subcutaneous adipose tissue during dry-cured ham processing. *Food Chem.* 64, 95–101.
120 Asghar, A., Gray, J. I., Buckley, D. J., Pearson, A. M. and Booren, A. M. (1988). Perspectives on warmed-over flavor. *Food Technol.* 42, 102–8.
121 Gray, J. I., Gomaa, E. A. and Buckley, D. J. (1996). Oxidative quality and shelf life of meats. *Meat Sci.* 43, S111–23.
122 Morrissey, P. A., Sheehy, P. J. A., Galvin, K., Kerry, J. P. and Buckley, D. J. (1998). Lipid stability in meat and meat products. *Meat Sci.* 49, S73–86.
123 Frankel, E. N. (1982). Volatile lipid oxidation products. *Prog. Lipid Res.* 22, 133.
124 Frankel, E. N. (1985). Chemistry of autoxidation: Mechanism, products and flavor significance. In *Flavour Chemistry of Fats and Oils*, eds. D. B. Min and T. Smouse, American Oil Chemist Society Press, Champaign, IL, 99. 1–37.
125 Grosch, W. (1987). Reactions of hydroperoxide products of low molecular weight. In *Autoxidation of Unsaturated Lipids*, ed. H. W. W. S. Chan, Academic Press, London, UK, pp. 95–139.
126 Buscailhon, S., Berdague, J. L., Gandemer, G., Touraille, C. and Monin, G. (1994). Effects of initial pH on compositional changes and sensory traints of French dry-cured hams. *J. Muscle Foods* 5, 257–70.
127 Buscailhon, S., Berdague, J. L. and Monin, G. (1993). Time-related changes in volatile compounds of lean tissue during processing of French dry-cured ham. *J. Sci. Food Agric.* 63, 69–75.
128 Dirinck, P., Van Opstaele, F. and Vandendriessche, F. (1997). Flavour differences between northern and southern European cured hams. *Food Chem.* 59(4), 511–21.
129 Garcia, C., Berdague, J. L., Antequera, T., Lopez-Bote, C., Cordoba, J. J. and Ventanas, J. (1991). Volatile components of dry cured Iberian ham. *Food Chem.* 1, 23–32.
130 Ruiz, J., Ventanas, J., Cava, R., Andres, A. and Garcia, C. (1999). Volatile compounds of dry0cured Iberian ham as affected by the length of the curing process. *Meat Sci.* 52, 19–27.
131 Buscailhon, S., Berdague, J. L., Bousset, J., Cornet, M., Gandemer, G., Touraille, C. and Monin, G. (1994). Relations between compositional traits and sensory qualities of French dry-cured ham. *Meat Sci.* 37, 229–43.
132 MacDonald, B. and Gray, J. I. (1978). The role of nitrite in cured meat flavor. *Am. Chem Sec. Abator*, 176.
133 MacDonald, B., Gray, J. I., Kakuda, Y. and Lee, M. L. (1980). Role of nitrite in cured meat flavor: Chemical analysis. *J. Food Sci.* 45(4), 889–92.
134 Merrit, C. Jr., Angelini, P., Wierbicki, E. and Shutes, G. W. (1975). Chemical changes associated with flavor irradiated meat. *J. Agric. Food Chem.* 23, 1037.
135 Josephson, E. S. (1981). Food irradiation and sterilization. *Radiat. Phys. Chem.* 18(1–2), 223–39.
136 Barnett, H. W., Nordin, H. B., Bird, H. D., and Rubin, L. J. (1964). A study of factors affecting the flavor of cured ham. 11th Meeting European Meat Research Worker, Belgrade, 1501.

137 Kemp, J. D., Langlois, B. E., Fox, J. D. and Varney, W. Y. (1975), Effects of curing ingredients and holding times and temperatures on organoleptic and microbiological properties of dry-cured sliced ham. *J. Food Sci.* 40, 634–6.
138 J. I. Gary and Pearson, A. M. (1984). Cured meat flavour. *Adv. Food Res.* 29.
139 Belitz, H.-D. and Grosch, W. (1986). *Food Chemistry*; Springer and Verlag, Berlin, pp. 155–68, 255–304.
140 Song, H.1 and Cadwallader, K. R. (2008). Aroma components of American country ham. *J. Food Sci.* 73(1), C29–35.
141 Takken, H. J., van der Linde, L. M., Devalois, D. J., van Dort, H. M. and Boelens, M. (1976). Phenolic, Sulfur and nitrogen compounds in food flavors. In eds G. Charalambous and I. Katz, ACS Symposium Series 26, Washington, DC, pp. 114–21.
142 van den Ouweland, G. A. M. and Peer, H. G. (1972). Mercap-tofurane and mercaptothiophene derivatives. GB Patent 1,283,912.
143 Ho, C. T., Carlin, J. T. and Huang, T. C. (1987). Flavour development in deep-fat fried foods. In *Flavour Science and Technology*, eds M. Martens, G. A. Dalen and H. Russwurm. Wiley, Chichester, pp. 35–42.
144 Vernin, G. and Parkanyi, C. (1982). Mechanisms of formation of heterocyclic compounds in Maillard and pyrolysis reactions. In *Chemistry of Heterocyclic Compounds in Flavours and Aromas*, ed. G. Vemin, Ellis Horwood, Chichester, pp. 151–207.
145 MacLeod, G. and Ames, J. M. (1986). 2-Methyl-3-(methylthio)furan: a meaty character impact aroma compound identified from cooked beef. *Chem. Ind. (London)*, 175–6.
146 Gasser, U. and Grosch, W. (1988). Identification of volatile flavour compounds with high aroma values from cooked beef. *Z. Lebensm. Unters. Forsch.* 186, 489–94.
147 Madruga, M. S. (1994). Studies on some factors affecting meat flavour formation. PhD Thesis, The University of Reading, UK.
148 Rizzi, G. P. (1969). The formation of tetramethylpyrazine and 2-isopropyl-4,5-dimethyl-3-oxazoline in the Strecker degradation of dl-valine with 2,3-butanedione. *J. Org. Chem.* 34, 2002–4.
149 Waller, G. R. and Feather, M. S. (1983). The Maillard reaction in foods and nutrition, ACS Symposium Series 215, *Am. Chem. Soc*: Washington, 174–80.

Chapter 7

Factors affecting the colour and texture of pig meat

Xin Sun and Eric Berg, North Dakota State University, USA

1 Introduction

When purchasing meat, colour and texture are the most important characteristics that influence the consumer's decision (Mancini and Hunt, 2005). Colour, as detected by the eye, describes the wavelength of light radiation. There are three routes by which the light wavelength will react upon contacting an object: 1) certain wavelengths may pass through, 2) some are absorbed and 3) some are reflected (Hui et al., 2001). When wavelengths representing blue and green pass through or are absorbed by an object, the object will show more of red colour to the human eye. This is the case for what we perceive as the colour of lean meat, and this is the first factor influencing the consumer's purchase decision in the retail store. According to Hui et al. (2001), the colour of lean tissue is an important indicator of meat quality to consumers. It is critically important for pork processors and retailers to understand how muscle and (or) meat pigments work and how both ante- and post-mortem conditions can impact the perception of colour.

Hammond (1932) determined that the bundles of muscle fibres (fascicle) (Fig. 1) were most associated with meat texture. The muscle fascicle is a collection of muscle cells (myofibres) held together by a membrane of connective tissue called the perimysium. In meat, muscle fascicles vary in size, length and thickness depending on the anatomical

http://dx.doi.org/10.19103/AS.2017.0030.07

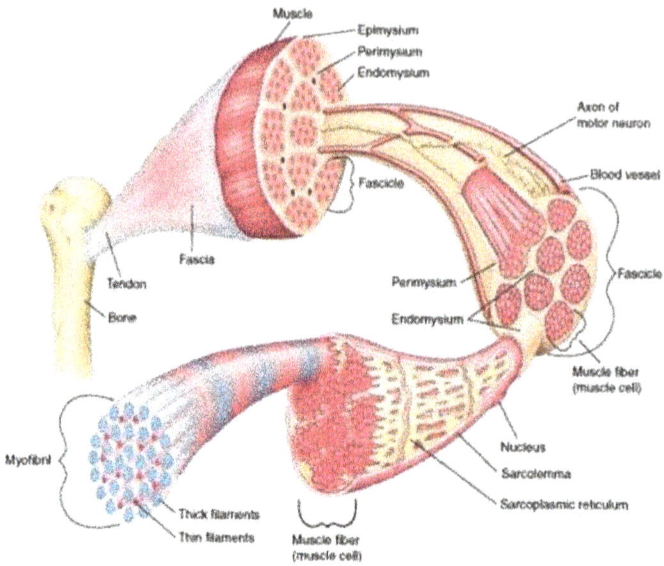

Figure 1 The ultrastructural arrangement of skeletal muscle. From: https://www.pinterest.com/explore/human-muscular-system/.

location and functional demand of the muscle. The size of the myofibres and (or) fascicles are visible when a muscle is cut on a transverse plane (Fig. 1), as when the butcher prepares steaks or chops. Meat texture then is what consumers would consider as the fineness of a cut surface. The texture of muscles can vary from fine (small myofibres/fascicle) to a coarse rough structure. The texture of the meat is influenced by the amount of connective tissue and marbling as well as by post-mortem breakdown of myofibrillar protein. Since myofibrillar protein degradation is related to meat tenderness, meat surface texture may serve as an indicator of tenderness.

In pork, colour and texture are the two significant factors influencing consumers' perception of pork quality (Qiao et al., 2007a). It is important to understand the many factors that can affect pork colour and texture, because these factors influence purchase decisions. Furthermore, those in charge of marketing pork must be informed regarding assessment methods available to monitor, evaluate and categorize pork colour and texture.

In this chapter, we will explore the biological and environmental factors that affect colour and texture in pig meat. Additionally, we will introduce and discuss existing and potential methods for assessment of the colour and texture attributes of pig meat.

2 Physicochemical factors effecting the conversion of muscle to meat

It is important to consider the chemical and physical (physicochemical) components that lead to differentiation of meat from living muscle in order to fully understand the nature of pig meat colour and texture.

2.1 High-energy phosphates

The two principal myofibrillar proteins responsible for muscle contraction are myosin (primary protein of the thick filament) and actin (primary protein component of the thin filament). Myosin possesses two binding sites on its large globular (S1) head: one site binds adenosine triphosphate (ATP) and has a myofibrillar actomyosin ATPase (*m*ATPase) that regulates the speed of contraction by the rate of ATP hydrolysis (degradation). The other site binds actin to allow for the subsequent contraction stroke. Muscle protein contraction is dependent on the synthesis, degradation and availability of high-energy phosphate compounds. ATP, adenosine diphosphate (ADP) and creatine phosphate (CP) are the ultimate energy currency for muscle contraction. Within the myofibre, creatine phosphate provides a ready reserve of high-energy phosphate that is used to rephosphorylate ADP to reform ATP during bouts of muscle contraction. Contraction begins when ATP binds within the globular head of myosin and is degraded by the *m*ATPase enzyme. The *m*ATPase is termed a *slow* enzyme because during the process of ATP hydrolysis ADP and its associated P*i* remain associated within the globular head. Once the P*i* is released, the binding cleft on myosin forms a weak bond with the globular actin protein of the thin filament. The subsequent release of ADP results in the power stroke (contraction) and a rigor (actomyosin; AM) bond is formed (Huxley and Simmons, 1971). It is important to note that ATP is responsible for both the formation of the AM rigor bond and its release. The AM rigor bond is released when ATP re-binds myosin which causes a shift in the arrangement of the protein structure of the globular head allowing myosin to release from actin and reactivate for another contraction. The reason why this is biologically important is because ATP must be present in sufficient quantity for contraction *and* relaxation of muscle to occur.

The pathways for glycolysis and the tricarboxylic acid (TCA) cycle produce the bulk of ATP within the myofibre. The source of substrate for engaging glycolysis and the TCA cycle is glucose, which is stored in muscle as the branched molecule glycogen. Liver and muscle are the two primary storage sites for glycogen. Because of its larger total mass, muscle contains three to four times more glycogen than the liver (Mayes and Bender, 2003). Glycogen from the liver functions to regulate and maintain blood sugar levels between meals while muscle glycogen is only utilized within the myofibre to produce ATP for myofibrillar contraction. Muscle glucose cannot exit the myofibre and enter blood circulation once it is stored as glycogen in the muscle cell.

Individual glucose units cleaved from muscle glycogen begin the conversion of muscle to meat. With the execution of exsanguination in the slaughter process, the cessation of blood circulation shifts muscle metabolism from aerobic to anaerobic. When muscle contracts in an anaerobic environment, glycogen disappears and lactic acid becomes the principal end product of glycolysis (Mayes and Bender, 2003). Under aerobic conditions lactic acid does not accumulate, because it is cleared from the muscle via the circulatory system and metabolized by the liver; or lactate may not accumulate in aerobic conditions at all because the final endpoint of aerobic glycolysis is pyruvate which is metabolized via the TCA cycle. During anaerobic post-mortem metabolism, one molecule of glucose will generate approximately 3 moles of ATP providing the high-energy phosphates necessary for post-mortem muscle contraction. Creatine phosphate is rapidly depleted under these conditions, yet depending on the conditions, post-mortem levels of ATP may be maintained as a result of anaerobic glycolysis. Accumulation of lactic acid in post-mortem muscle reduces the localized pH and muscle is converted to meat. Conversion of glycogen to lactic acid will continue to lower muscle pH until the glycogen (or ATP stores) are depleted or until the *m*ATPase ceases to function as a result of low intramuscular pH.

2.2 Calcium

Calcium is stored in the sarcoplasmic reticulum (Fig. 1) that surrounds the myofibrils of the muscle cell. Muscle contraction (myosin binding to actin) is regulated by Ca^{2+} levels inside the muscle cell sarcoplasm. Contraction is initiated from an efferent nervous signal which triggers the release of Ca^{2+} from the sarcoplasmic reticulum. Calcium ions bind to troponin C (a regulatory protein associated with the thin filament) causing a structural change in the actin globular protein that exposes the myosin-binding site. To cease contraction (relaxation), Ca^{2+} must be sequestered again within the sarcoplasmic reticulum.

High levels of sarcoplasmic Ca^{2+} in post-mortem muscle will also influence activity of Ca^{2+}-dependent enzymes such as phosphorylase kinase which is associated with the release of glucose from its storage molecule glycogen (Louis et al., 1993) and the calpain proteolytic system which plays a direct role in post-mortem protein denaturation that results in improved tenderness (Koohmaraie, 1992).

2.3 Muscle physiology

Everything that influences the growth and development of living muscle will play a role in the appearance and palatability of the final fresh meat product. The physiology of living muscle is influenced by many factors that begin as early as the sperm fertilizing the egg. Genetics, production environment, nutrition and health, as well as muscle function, anatomical location and myofibre metabolic activity impact muscle development and the conversion of muscle to meat.

The fibre-type classification of muscle fibres is indicative of the metabolic and anatomical responsibilities of any given muscle (Moss et al., 1995). The protein myosin is the largest member of the contractile apparatus in skeletal muscle fibres. The metabolic activity for differentiation between muscle fibre types is largely determined by the classification and concentration of myosin isoform that constitutes the muscle cell. The speed of myosin ATP hydrolysis (*m*ATPase activity) dictates the contractile speed of the entire myofibre, motor unit and muscle. The myosin isoform *m*ATPase is characterized by its ability to function at either acid or alkaline metabolic conditions and are associated with muscle fibres that possess the enzymes of oxidative or glycolytic metabolism.

For many years, meat scientists referred to muscle fibre types as *red* or *white* which correspond to the more current, broad classification of slow twitch/fatigue resistant (red fibres) and fast twitch/fast fatigue (white fibres). The nomenclature was derived from the darker (more red) visual appearance of muscles possessing a greater proportion of fatigue-resistant fibres versus the lighter pigmented meat of muscles containing a larger proportion of fast fatigue fibres. A real-world example of these phenomena is the serving of a whole roasted turkey for dinner. The host may ask the guest, '*Do you prefer dark or white meat?*' For the domestic turkey, the muscles of locomotion (legs and thighs) are considered *dark* meat while the breast muscle is considered *white* meat. The differences in colour are attributed to the differences in muscle fibre type as a result of the function of the muscles. Type I muscle fibres that have a slow *m*ATPase activity, are oxidative, and constitute slow, fatigue-resistant motor units. The make-up of myofibres present in muscles of locomotion constitute a higher percentage of Type I muscle fibres that have greater aerobic endurance and therefore possess a higher concentration of mitochondria and myoglobin (the red pigment protein). The higher myoglobin content is the reason the leg and thigh muscles are darker in colour (see Section 2.4). Type IIB myofibres, on the other hand, possess a fast *m*ATPase, a low aerobic (oxidative) capacity and a high glycolytic metabolism. The modern domestic turkey produced in the United States has been genetically selected for rapid muscle gain and larger pectoralis (breast) muscle. These specific genetic selection criteria have resulted in larger myofibre diameter associated with the type II muscle fibres. The breast muscle of the domestic turkey is metabolically suited for short bursts of power, it possesses a higher concentration of glycogen, fewer mitochondria and less myoglobin (which makes it subsequently lighter in colour).

The visual colour differences across different muscles are less dramatic in mammalian species; however, muscle differences associated with the relative proportion of fast or slow twitch muscle fibres profoundly influence the conversion of muscle to meat and the ultimate meat quality. Ryu and Kim (2005) and Kim et al. (2010) studied the relationship between muscle fibre characteristics (type, size and quantity), post-mortem metabolic rate and pig meat quality of the loin muscle. They found pork colour L* (lightness) value was correlated inversely with the number of type I and IIA myofibres. In other words, fewer type I and IIA fibres resulted in pork possessing L* readings closer to white (where L* of zero = black and L* of 100 = white). It could be said that the physiological increase in the percentage of type IIB fibre and a decrease in the percentages of type I and IIA fibres increased 'lightness' of the pork, which is negatively perceived at the marketplace.

Conversion of living muscle to meat depends on the physiology of individual muscle fibres responsible for driving the post-mortem biochemical activity and the reduction of intramuscular pH to a range from 5.3 to 6.7 (Pearson, 1987). The ultimate acidity of the meat will depend on the ability of the *m*ATPase to function at low pH conditions and the post-mortem glycogen content of the myofibre. Muscle fibre types IIB and IID have been reported to be more stable under acid conditions (Pette and Staron, 1990). Fast muscle types exhibit a glycolytic metabolism that is necessary for rapid and strong bursts of power. These fibre types will propel glycogenolysis to lactic acid formation longer under the anaerobic conditions of post-mortem muscle. In the living muscle, slow twitch type I and fast twitch type IIA muscle fibres are capable of sustained use and are found in high concentration in motor units capable of prolonged or sustained use. Myofibres classified as fatigue resistant are capable of maintaining a strong *m*ATPase activity at low pH levels; however, glycogen is rapidly depleted in these fibres and post-mortem activity cannot be maintained due to lack of glucose substrate that is necessary for post-mortem generation of ATP via anaerobic glycolysis. The post-mortem activity of these muscles will cease more quickly because without ATP, myosin cannot bind to actin to generate a power stroke, nor can it release from actin once bound. Post-mortem depletion of ATP stores within the muscle is the first step in the formation of the permanent actomyosin (rigor) bond. Essen-Gustavsson et al. (1992) reported that type I (slow, oxidative, fatigue-resistant) and type IIA (fast, higher oxidative capacity) muscle fibres from pig longissimus muscle (LM) were depleted of glycogen at slaughter, while type IIB (fast, lower oxidative capacity, higher glycolytic) fibres still retained glycogen. Muscles possessing a high concentration of fast IIB muscle fibres have the potential for a rapid reduction in post-mortem pH (given their high glycolytic potential). A rapid post-mortem pH decline while the carcass temperature is high can result in structural changes and changes in oxidative stability of the pigment protein myoglobin and also the proteins that comprise the myofilaments. These structural changes will influence the colour, texture and quality of the fresh meat product. For a complete scientific review of these changes, see Kim et al. (2014).

2.4 Myofilament structure in post-mortem muscle

Moss et al. (1995) provided an excellent review of the contractile properties of skeletal muscle. The muscle that humans consume as meat is considered *striated* because of the alternating dark and light bands that can be seen easily along the longitudinal cross section of a muscle fibre viewed under the polarized light by an electron microscope. These dark and light bands are a result of the overlapping thick and thin filaments (Fig. 1) and the contractile unit's boundary noted as the Z-line. The post-mortem condition of the thick and thin filament and extent of the myofilament overlap (distance between Z-lines) will influence the visual perception of colour and the physical expression of meat tenderness and juiciness.

Several post-mortem factors can contribute to increased light scattering and the presentation of pig meat that is perceived as pale by the human eye. Kim et al. (2014) provided a comprehensive review of research examining the physiological mechanism for increased light scattering and presentation of pale pig meat. A rapid post-mortem pH decline can result in early cessation of *m*ATPase activity resulting in reduced myofilament spacing. Also, the pH may decline to a level near the isoelectric point of actin and myosin. The amount of 'free water' maintained within meat depends on the amount of space between the myofilaments. Expulsion or retention of moisture is largely dependent on the amount of space between the thick and thin filaments and the hydrophilic attraction of water molecules to the charges on the myofilaments. Thus, as the available space between thick and thin filaments and (or) the isoelectric point is reduced, water is expelled (NPPC, 2000). Further, Offer (1991) reported that rapid decline of pH at a higher muscle temperature caused denaturation (shrinkage) of the myosin S1 heads and a compression of the myofibrillar lattice causing the expulsion of free water held between the myofibrils. The reduction in lateral spacing between the thick and thin filaments will influence the scattering and absorbance of light (appearing paler in colour) and increase the *softness* associated with PSE lean.

As described by Kim et al. (2014), muscle/meat colour depends upon the concentration and chemical state of myoglobin and the extent of light scattering. The chemical and structural state of the myoglobin globular protein can denature and unfold under post-mortem conditions of rapid pH decline associated with elevated carcass temperatures. This denaturation exposes the central haem iron to greater auto-oxidation and more rapid formation of an unacceptable colour during retail display. Also, denatured myoglobin and other sarcoplasmic proteins may settle onto the myofilament lattice and result in a greater scattering of incident light, which is visualized as paler coloured meat (Offer and Knight, 1988).

3 Ante-mortem factors affecting pork colour

3.1 Myoglobin

It is well documented that myoglobin is the principal water-soluble (sarcoplasmic) protein responsible for post-mortem meat colour changes. A thorough, comprehensive publication regarding the complete function of myoglobin in meat and meat processing

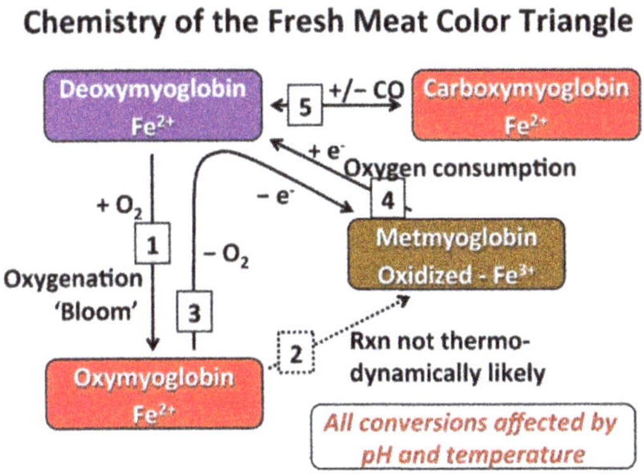

Figure 2 Fresh meat colour triangle displaying the pigment chemistry for the formation of deoxymyoglobin (DMb), carboxymyoglobin (COMb), oxymyoglobin (OMb) and metmyoglobin (MMb). Reproduced with permission from the American Meat Science Association (AMSA, 2012).

is available from the American Meat Science Association (AMSA, 2012). Myoglobin (Fig. 2) is chemically similar to the blood protein haemoglobin, in that both contain iron at the centre of a porphyrin ring structure and both are responsible for binding/carrying oxygen. As described above in the case of the American domestic turkey, if one muscle seems more red, it is most likely because it contains more myoglobin and therefore more iron. In living muscle, there are many factors which affect pork myoglobin concentration such as genetics, species, myofibre type and age of the animal (all described in greater detail below). Indeed, the state of post-mortem myoglobin is the most important factor in the expression of acceptable pork colour. The physiological function of myoglobin is to transport oxygen in muscle fibres. When chops or roasts are cut, myoglobin meets atmospheric oxygen and the colour will change from purple red to bright red (Hui et al., 2001). If the cut lean surface is deprived of oxygen exposure (as is the case with meat packaged in oxygen impermeable packaging), the colour of the meat will appear purple red. On the other hand, in the open-air markets, which are more common globally, the colour of the meat in the open-air markets appears bright red due to the freshness of processing and the abundant oxygen atmosphere. That said, if the cut lean surface is exposed to oxygen for a prolonged period, bright red oxymyoglobin will convert to tan or brown metmyoglobin.

The chemical state of myoglobin will ultimately determine if pig meat is acceptable or unacceptable to the consumer. Figure 2 provides a summary of the pigment chemistry as described in detail in AMSA (2012).

3.2 Genetics and species

Myoglobin concentration may differ between pig breeds or genetic line. Brewer et al. (2002) reported that genetic line plays a significant role in affecting the colour of pork loin chops. Certain genetic traits that influence muscle growth can have a negative impact on pig meat quality. For example, the colour of pork is dramatically affected by the presence of a halothane allele. Commonly referred to as porcine stress syndrome (PSS) in swine or malignant hyperthermia in humans, the condition is manifested as a mutation of the ryanodine receptor (myofibrillar calcium-release channel). A normal ryanodine receptor (genotype *NN*) allows calcium to be released from its myofibrillar storage organelle, the sarcoplasmic reticulum. The calcium provides substrate to initiate myosin binding actin (as described above) resulting in muscle contraction. Homozygotes of the halothane gene (genotype *nn*) possess a defected ryanodine receptor that floods the sarcoplasm with calcium resulting in rigor-like muscle contractions, increased body temperature, rapid respiration and heart rate, and systemic acidosis. Carriers of the halothane gene (genotype *Nn*) will manifest an intermediate reaction between *NN* and *nn* swine (Sellier et al., 1998). The halothane gene is a by-product of genetic selection for heavy muscle hogs possessing low body fat. The expression of PSS has long been linked with the occurrence of pale, soft and exudative (PSE) pig meat. Swine expressing PSS at slaughter begins the conversion of muscle to meat with an elevated muscle temperature and a very rapid post-mortem pH decline which ultimately influences fresh pork colour and texture as described in Section 2.4. Eggert et al. (2002) investigated pork quality attributes including colour from two distinct genetic Nn and NN pig populations. The results showed that meat from Nn pigs were paler than NN counterparts.

A second major gene associated with pork colour is the Rendement Napole (RN) gene. Pig meat from RN swine has been termed 'acid meat', because it possesses a very low ultimate (48 h) intramuscular pH (<5.5) which influences pig meat colour and texture. In 2002, Brewer et al. compared the lean colour quality of seven different genetic lines which included Duroc, Pietrain (Halothane negative, NN), Pietrain (Halothane positive, nn), Berkshire, Hampshire (ryanodine positive; rn+), Hampshire (RN−) and one synthetic line (PIC USA, Franklin, KY). The results showed chops from Pietrain-nn and Hampshire (RN− gene) carcasses had the highest a* (redness) values while chops from Duroc, Pietrain-NN and Hampshire-RN− carcasses were least red. Similar research articles have investigated halothane genetic effects on pig meat colour attributes (Moelich et al., 2003; Carr et al., 2009; Fisher et al., 2000; Velarde et al., 2001; Fàbrega et al., 2002; Ohene-Adjei et al., 2003; Bridi et al., 2006; Barbut et al., 2008).

3.3 Diet

Human athletes and their trainers have long known that the nutrients consumed by the athlete will influence the physicochemical factors associated with muscle performance, energy metabolism and tissue composition. Likewise, the quantity and (or) type of nutrients consumed or withheld from the diets of meat-producing animals can influence the physicochemical factors associated with the conversion of living muscle to meat (as described above). The relationship between swine diet and physicochemical properties of meat quality are described by Berg (2000). It has been stated above that post-mortem anaerobic glycolysis drives post-mortem conversion of muscle to meat and greatly influences pig meat colour and firmness. Therefore, many dietary strategies look to diminish the content of myofibrillar glycogen, so less glucose substrate is available to generate intramuscular lactic acid that drives pH decline or nutrients are fed that buffer lactate accumulation and pH decline. Other feeding strategies add novel antioxidants (i.e. oregano and peppermint) or increase the dietary level of antioxidants such as vitamin E to prolong pork colour stability in the retail case. It has also been theorized that feeding antioxidants will delay the oxidation of myoglobin. Pettigrew and Esnaola (2001) and Ellis and McKeith (1999) provided excellent reviews of the influence of swine diet on pig meat quality.

3.4 Pre-slaughter stress

Even though Pettigrew and Esnaola (2001) and Ellis and McKeith (1999) outlined many feeding strategies that show potential to improve the colour and quality of pork, much more emphasis has been placed on genetic selection and minimizing pre-slaughter stress. Stressors that trigger a physiological response can be in the form of short term (such as a loud noise, unfamiliar environment, fighting or goading with an electric prod) or long term (sickness, dehydration or malnourishment). The physiological response to stress is mediated by the two opposing factions of the autonomic nervous system: the sympathetic and parasympathetic nervous systems. Response and recovery from a stressor involves a balance between these two systems.

The sympathetic nervous system mediates vigilance, arousal, activation and mobilization, and is typically characterized by the physiological response to fight, fright and flight. The catecholamines epinephrine and norepinephrine are released from the adrenal medulla in response to activation of the sympathetic nervous system. The sympathetic nervous system stimulates a cascade of reactions to rapidly generate energy in response to the stressor. Epinephrine stimulates glycogenolysis and lipolysis, whereby glucose is mobilized from glycogen stores in the liver and muscle and fatty acids are released from fat cells. Epinephrine also stimulates an increase in heart rate and (or) blood pressure and invokes peripheral vasodilatation to increase blood flow to the extremities for delivery of metabolites (such as oxygen and glucose) to the muscle to facilitate 'escape' from the stressor. Likewise, the increased blood flow removes metabolic waste products such as lactic acid (which induce fatigue) and provides for heat dissipation produced as a result of enhanced muscle metabolism. Furthermore, the sympathetic release of norepinephrine stimulates peripheral vasoconstriction. Temperature is a positive modulator of *chemo-mechanical coupling* of muscle contractile proteins. Elevated muscle temperature increases the rate of shortening, increases contractile force, increases maximum tension and enhances Ca^{2+} sensitivity of the contractile proteins. Peripheral vasoconstriction inhibits external heat loss during the initial response to stress allowing for muscles to 'warm up', maximizing muscle efficiency. Therefore, stressful activation of the sympathetic nervous system in the short-term prior to stunning and exsanguination of market swine amplifies post-mortem (anaerobic) glycolysis, lactate accumulation and pH decline, while elevating muscle temperature. It can be recalled that rapid pH decline and high muscle temperature are a recipe for poor pork quality. This condition is even further exacerbated in heavy muscle hogs possessing a high concentration of fast twitch, glycolytic myofibres (Berg, 2001).

4 Post-mortem factors affecting pork colour

4.1 Ageing

Post-mortem ageing of meat is done to increase the flavour and improve tenderness (Maltin et al., 2003). In the modern meat industry, there are two ageing methods that are practised: dry ageing and wet ageing. Meat ageing is not as popular with pork as with beef because the cost of the process for commodity pork is not as valuable as it is for beef. That said, recent research from Korea has investigated the influence of dry ageing on pork loin quality (Lee et al., 2016). The results showed that dry-aged pork loin received higher, more favourable scores for sensory parameters associated with pork loin colour. Similar research from Australia compared two different durations for ageing (2 and 7 days) on pork meat. Channon et al. (2016) found that seven days ageing had a more positive impact on loin colour than two days ageing. However, research by Ngapo et al. (2012a) found no differences in colour, pH or drip loss for pork loins aged from 2 to 12 days. It is apparent that more research is necessary to determine if the cost and expense of ageing fresh pork will manifest greater economic return to the pork industry.

4.2 Meat packaging and storage time

The packaging of meat for retail display is a science of its own. The type of packaging selected must allow the prospective consumer the ability to visualize the meat product and it must accentuate the visual properties that are most important to the consumer. We have already established that fresh meat colour is the most important factor for consumers' purchase decision and that the biochemical state of the meat pigment myoglobin is the driving factor of acceptable or unacceptable meat (pork) colour. Cornforth and Hunt (2008) published an authoritative report (white paper) commissioned by the American Meat Science Association regarding the impact of meat packaging on meat quality, microbiology and safety. In this highly cited publication, the authors identified three main methods of packaging used in retail: 1) polyvinyl chloride (PVC) film overwrap, 2) high-oxygen (O_2) modified atmosphere packaging (MAP) and 3) low-O_2 MAP with carbon monoxide (CO). We will summarize these three common packaging methods as described by Cornforth and Hunt (2008).

The method of PVC overwrap packaging is a very common and traditional method of displaying meat in retail cases, whereby steaks or roasts are placed in a tray and covered with thin PVC film. It is popular at retail because the PVC film and packaging are inexpensive, easy to use and packaging integrity is easily attained by heat sealing. Traditionally, PVC has a high-O_2 permeability which is favourable for the formation of the attractive bright red (oxymyoglobin) surface colour of meat (Landrock and Wallace, 1955). Although widely used, PVC overwrap has several disadvantages, the most important being that PVC-wrapped meat possesses a relatively short shelf life at retail (5–7 days) because of rapid metmyoglobin formation. As described above (Fig. 2), the metmyoglobin (browning) rises from below the exterior surface of the meat that has been exposed to low levels of O_2, as is the case with PVC-wrapped products. The acceptable oxymyoglobin colour has a shallow penetration and retailers are forced to discount or discard packages possessing greater that 40% metmyoglobin surface exposure.

Prolonged exposure to low-O_2 concentrations results in brown discolouration of retail meat within 5–7 days of display. That said, steaks and roasts packaged in high atmospheric O_2 MAP can retain an acceptable retail colour for 10–14 days. The gas concentrations of high-O_2 MAP contain 70–80% oxygen and 20–30% CO_2 (Eilert, 2005). The high-O_2 atmosphere allows for greater penetration from the surface of the meat for a deeper oxymyoglobin layer that delays the surface exposure of metmyoglobin. The disadvantages of high-O_2 MAP are accelerated lipid oxidation and subsequent development of off-flavours and premature browning during cooking. Cornforth and Hunt (2008) state that premature browning of cooked meat could be a potential food safety hazard given that high-O_2 MAP ground beef can appear fully cooked at a temperature as low as 57°C. The recommended final cooking temperature for ground beef is 71°C to destroy pathogens such as *Escherichia coli* O157: H7.

Recently, the development of low-O_2 packaging has been implemented in order to address the negative quality issues associated with high-O_2 packaging. Atmospheric gas concentrations of these MAP are essentially anaerobic and contain low levels of CO (0.4%), 20–30% CO_2 and the balance made up by nitrogen. It can be recalled from Fig. 2 that DMb exposed to CO in the absence of oxygen will turn red and can be used as an effective means to stabilize meat colour. Cornforth and Hunt (2008) provided several reasons why MAP that contain CO are superior to PVC or high-O_2 MAP.

1 Acceptable retail colour accompanied by microbial plate counts below spoilage levels for 28 days (ground beef) and 35 days (steaks and roasts).
2 More acceptable flavour due to less oxidation.
3 No bone darkening in retail products.
4 Eliminates the potential of premature browning during cooking.
5 The presence of CO_2 (in both low- and high-O_2 MAP) decreases the growth of spoilage organisms and pathogenic bacteria is superior to PVC packaging.
6 Meat tenderness may improve due to reduced protein oxidation and the extension of the shelf life allows more time for ageing-related tenderization.

All of these advantages of low-O_2 MAP are overshadowed by consumers' negative impression of the use of carbon monoxide in food packaging and the misguided impression that meat products may be spoiled even though they appear fresh.

The MAP techniques described above are most effective for meats with a high myoglobin concentration such as beef or lamb. The brown discolouration associated with PVC-wrapped meats can be avoided in meats with low myoglobin concentrations (pork and poultry) by anaerobic vacuum packaging. Vacuum-packaged pork or poultry will not form dark purplish hue such as those exhibited by vacuum-packed beef, because of their low myoglobin content and reduced formation of deoxymyoglobin.

4.3 Display lighting

AMSA (2012) stated that the human eye does not 'see' colour, but rather captures the wavelengths of light that are reflected by an object and thus without light, there is no colour. Humans can only perceive light in the visible spectrum of colours. Colours (and corresponding wavelength) visualized and perceived by humans are violet (400–425 nm), indigo (425–455 nm), blue (455–490 nm), green (490–575 nm), yellow (575–580 nm), orange (580–650 nm) and red (650–700 nm). AMSA (2012) described colour perception as follows: when light illuminates the meat surface, it will either be absorbed, reflected or scattered. Those wavelengths absorbed by the meat are not detected by the eye because the meat tissues retain those wavelengths. The red wavelength is reflected, detected by the eye and perceived by the brain. A light source that emits a higher proportion of the red wavelength would therefore be more desirable for illuminating meat because the red wavelength has shallow absorption and greater reflection than the other colours of the visible spectrum (Raines et al., 2009). Conditions as described above for the case of PSE pork cause light to 'scatter' and the colour pattern captured by the eye, and interpreted by the brain, is perceived as a pale meat colour.

In order to optimize the most desirable pork colour presentation to consumers, AMSA (2012) identified six conditions that must be considered.

1 **Light source** (illuminant). Light sources such as sunlight, fluorescent, incandescent, halogen, high-intensity discharge (HID) and (or) light-emitting diode (LED) differ greatly with each possessing a unique spectral composition.
2 **Differences between observers**. Customers visiting the retail meat case will possess different colour sensitivities.
3 **The size of the cut of meat**. Colour is often perceived as brighter/more vivid from larger cuts because more light is reflected back to the eye.
4 **Smoothness/roughness of the meat surface**. A flat cut surface made with a sharp knife is more desirable.
5 **Differences in background colouration**. Cuts of meat placed in white or bright coloured trays are often perceived to have dull colour. This is why pork is often packaged on black trays because cuts on a darker background appear brighter.
6 **Direction of illumination**. An incorrect angle of light can result in gloss (shine) that hinders the interpretation of meat colour. The lighting in a meat display case should be overhead, backlighting should be avoided.

Raines et al. (2009) provided an excellent resource for meat retailers regarding proper illumination for meat retail display. This easy-to-understand, two-page guide can be obtained for free at http://animalscience.psu.edu/extension/meat/pdf/LightGuideSept09.pdf.

5 Factors affecting pork texture

The texture of pork has traditionally been evaluated by means of subjective assessment of firmness such as the *'soft'* aspect of PSE or the *'firm'* aspect of dark, firm and dry quality classification categories. The firmness or coarseness of the visible cut surface of a meat cut is impacted by anatomical location (muscles of locomotion/power vs muscle of posture), muscle fascicle size, the maturity of connective tissue, the age of the animal at slaughter, intramuscular fat content, status of rigor or relaxation, post-mortem chilling rate and post-mortem ageing. Physically, meat texture is visualized by the size of the muscle bundles (fasciculi) and the associated thickness of the surrounding collagen connective tissue membrane (perimysium). The strength of the collagen membrane can be attributed to stable covalent bonds formed between adjacent tropocollagen molecules (Swatland, 1994). As an animal ages, more covalent cross-linkages form as the collagen network becomes stronger. This increase in collagen cross-linkages leads to a subsequent increase in collagen stability over time. This is the reason that age of the animal at slaughter is the most important factor influencing texture and the sensory characteristics of cooked meat tenderness.

The state of rigor or the length of the sarcomere can also have a major influence on visualized texture and meat tenderness. A short distance between the Z-lines of striated muscle is indicative of a maximum overlap of the thick and thin filament. In other words, the myosin globular heads associated with the thick filament maximize all possible binding sites on the thin filament. As more actomyosin bonds are formed, the muscle shortens, bringing the Z-lines closer together resulting in a short sarcomere and maximum overlap of contractile proteins. As stated above, if this muscle shortening is triggered by activation of the sympathetic (fight or flight) nervous system, the short sarcomere will become permanent with the formation of rigor mortis. Combined with elevated early post-mortem muscle temperature and rapid pH decline in pork muscle, the physiological circumstances result in a pale colour and poor water-holding capacity.

If post-mortem muscle is chilled too quickly, over-contraction of the thick and thin filament (to the extent the I-band is not visible) can occur. This phenomenon, referred to as cold shortening by Locker and Hagyard (1963), produces very tough meat.

The temperature used to cook meat can improve or exacerbate the influence of connective tissue relative to tenderness. Lower cooking temperatures (71–82°C) at higher humidity for an extended period of time convert collagen to gelatin, enhancing tenderness. Boles et al. (1991) showed higher cooked temperatures were significant, and correlated with less tender and less juicy pork. Crawford et al. (2010) studied pork quality attributes under three different cooked temperatures (62°C, 71°C and 79°C) and the results showed as cooked temperature increased, the pork loin tenderness shear force value (toughness) also increased (62°C = 24.9N, 71°C = 27.8N and 79°C = 31.7N).

Post-mortem processing techniques such as dry or wet ageing of meat plays a significant role in meat tenderness. Research conducted by Lee et al. (2016) found pork sensory textural features such as hardness, cohesiveness, gumminess, chewiness and resilience were significantly decreased in dry-aged pork ($P < 0.05$). Dry-aged pork loin required less force to mechanically shear through 1-cm-diameter steak core samples compared to the non-aged control steaks ($P < 0.05$), meaning that dry-aged pork loin were more tender than controls. In contrast, the springiness of dry-aged pork was significantly increased ($P < 0.05$) (Lee et al., 2016). In another study, Ngapo et al. (2012b) investigated the influence of different ageing durations (2, 4, 6, 8, 10 or 12 days) on the quality properties of pork. Ageing the pork improved tenderness as shown by a reduction in mechanically measured force. Shear force values at 2, 6, 8 and 10 days were greater (tougher) than 4 and 12 days. Juarez et al. (2011) evaluated differences between dry and wet ageing of pork loins. They found that both ageing methods achieved better instrumental and sensory tenderness and juiciness values.

6 Measurement of meat colour and texture

6.1 Subjective measurements

Traditionally, the colour and texture of pork were evaluated subjectively by a trained quality grader through visual or sensory assessment (Larmond, 1976). The evaluator evaluates pork quality according to a reference standard developed and published by a pork industry entity such as the National Pork Board (Fig. 3) in the United States or Japanese Color Standards (Nakai et al., 1975). These methods rely heavily on human experience and human ability to remain objective during evaluation. Hence, subjective evaluation is subject to human error and therefore may not be the most applicable means to assess pork quality in an industrial setting due to variability between evaluators, differences in lighting, speed of operation and onset of human fatigue.

6.2 Objective measurements

Colorimeters/spectrophotometers

Colorimeters are instruments used across the pork supply chain to obtain an unbiased, objective, quantifiable assessment of the colour of the product. We have described above how light influences the perception of colour, so these instruments standardize the

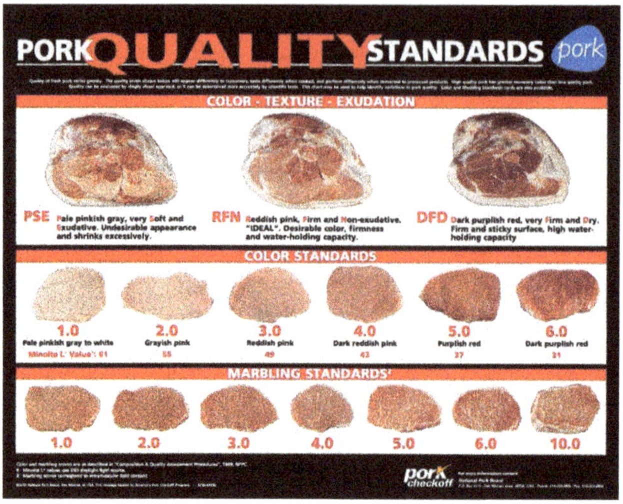

Figure 3 Pork quality reference standards for loin colour and marbling and ham firmness (NPB, 1999).

light source, the direction of the light source and measure the amount of reflected light wavelengths within the visual colour spectra. The instrument calculates and provides an output of colour values associated with three-dimensional colour space (Fig. 4). These values represent consumer perception of colour, which is based on a mixture of three attributes: 1) lightness, 2) hue (or hue angle) and 3) chroma (colour saturation, vividness) (Minolta, 1994). As illustrated in Fig. 4, hue angle refers to the 'true' nature of colour (Minolta, 1994), beginning at the positive a* axis and revolving 360° around the three-dimensional colour space, whereby 0° would be true red (positive a*), 90° would be true yellow (positive b*), 180° would be true green (negative a*) and 270° would be true blue (negative b*). With regard to pork colour, a smaller hue angle number would be closer to true red. A greater number, closer to true yellow, would be indicative of negative fat quality. In three-dimensional colour space, chroma becomes more 'vivid' as the colour point moves further from the three-dimensional centre outward. A colour point near the centre is perceived as a dull colour or grey (Minolta, 1994). For more information and colour pictures of three-dimensional colour space, see AMSA (2012).

Imaging technology - computer vision system

There has been an increased interest in research evaluating the use of imaging technology to objectively measure fresh pork colour and texture (Table 1) of which computer vision systems (CVS) are receiving the most attention. A typical CVS system consists of three parts: 1) an illumination device provides a constant, full-spectral source of lighting; 2) a charge coupled device (CCD) array camera which can acquire an image; and 3) a computer possessing image processing software to analyse the captured image (Fig. 5). Modern CVS have the potential to accurately assess pork quality at line speeds in an automated industrial setting. To make this means of assessment a reality in the pork processing plant, researchers strive to identify meaningful surface features from pork sample images. For example, Sun et al. (2016a) investigated RGB (red, green, blue), HSI (hue, saturation, intensity) and L*a*b* colour features obtained from images of fresh pork

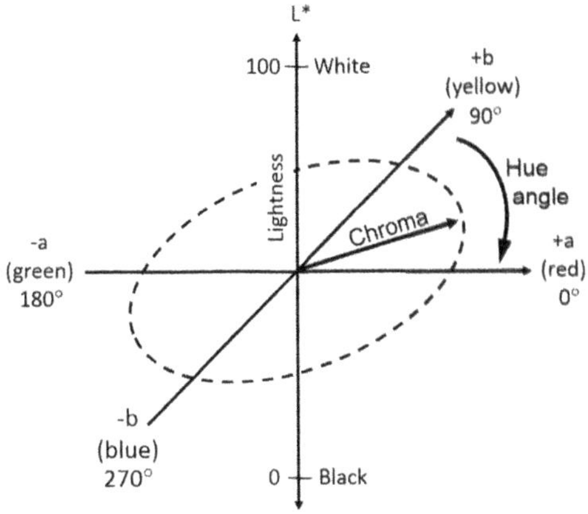

Figure 4 Illustration of three-dimensional colour space for CIE L* (lightness), a* (red/green), b* (yellow/blue), and chroma (colour saturation, vividness) and hue (true red).

chops. Prediction equations were developed using Minolta colorimeter (Chroma Meter CR 2000, Minolta Camera, Osaka, Japan) values as the dependent value. Eighteen image colour features (means and standard deviation of R, G, B, H, S, I, L*, a*, b* values) were used as independent values extracted from the various combinations of the three colour spaces. The vision system accounted for 83% of the variability in fresh pork chop quality as established by industry standards for pork quality. Similarly, Girolami et al. (2013) reported that CVS illumination was more comparable to the actual colour of meat than colours generated from colorimeter for L*, a* and b* values. Girolami et al. (2013) concluded that the translucency of the meat influenced the L*, a* and b* values reported from the colorimeter device and resulted in colour values that were not truly representative of the meat colour. Liao et al. (2010) used CVS to predict pork chop colour and intramuscular fat (IMF) on a conveyor belt at online speed (0.25 m/s). They concluded that the CVS had full potential to evaluate pork chop colour and quality at line speed. Overall, CVS is a promising method for predicting pork quality attributes, especially colour. That said, prediction accuracy must be improved and proper prediction models must be verified before these systems can be implemented at industry line speeds.

Near-infrared spectroscopy

Another popular objective method to predict pork quality is near-infrared spectroscopy (NIRS) technology. The principle of NIRS is to detect colour targets in the near-infrared region of the electromagnetic spectrum (from about 700 nm to 2500 nm), which is beyond the upper limit of the red wavelength in the visible colour spectra. Because of the extreme range of NIRS, researchers will often combine NIRS analysis with readings in the visible spectra (VIS; 400–700 nm). These data features are then utilized for generating prediction equations relative to pork quality attributes such as pH, colour, cooking loss, intramuscular fat (IMF) and mechanical toughness (Warner-Bratzler Shear Force; WBSF).

Although NIRS have the potential to predict pork quality attributes (Table 1), the technique possesses some limitations for implementation in a modern pork processing environment. First, NIRS is unable to visually detect the entire pork loin surface due to the

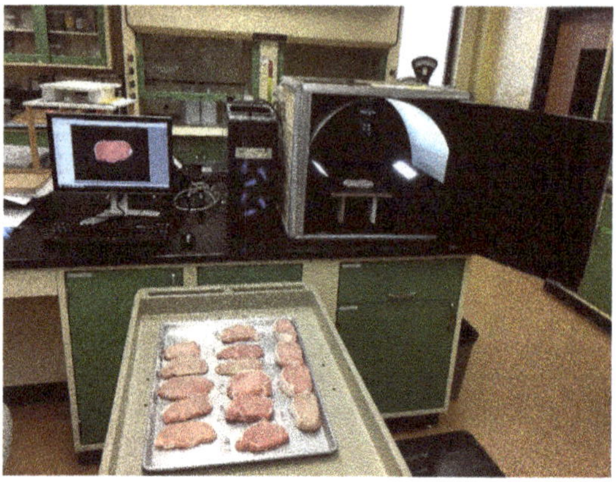

Figure 5 Computer vision system for assessment of pork colour quality. Picture courtesy of X. Sun, North Dakota State University.

limited size of the spatial sensor aperture. The NIRS sensor can only collect the spectral information that is directly under the sensor. Second, prediction equation accuracy and repeatability require further testing and validation to satisfy industry expectations. Future investigations using NIRS as a means of assessing fresh pork colour and textural quality should focus on increasing the field of view relative to the sensor detection area function, improving prediction model accuracy and automating the technology to meet the speed requirement for use in an online industrial application (Balage et al., 2015).

Hyperspectral imaging system

Research of hyperspectral imaging techniques is gaining attention for its technical ability for programmable automated target detection and the ability to recognize and analyse the composition of the target of interest (Table 1). Hyperspectral imaging systems (HIS) integrate spectroscopic and imaging techniques as a means to identify different component features and then distinguish the spatial distribution within the tested sample. HIS produces a two-dimensional spatial array of vectors which represents the spectrum at each pixel location. From these data, the HIS generates a three-dimensional data set that consists of two spatial dimensions and one spectral dimension, which is known as the data cube or hypercube. Compared with the traditional image analysis methods, the advantages of HIS include minimal sample preparation, non-destructive feature detection and visualizing spatial distribution of numerous chemical compositions simultaneously.

Hyperspectral imaging has some disadvantages which would directly influence its use for industrial pork quality detection. Use of the hyperspectral imaging technique requires time for image acquisition and analysis, therefore, HIS would have limited application in an online system. Also, hyperspectral images contain a substantial amount of data, including vast accumulation of redundant information, and pose considerable computational and data storage challenges. That said, if data features could be identified that are common across the board for either high-quality or low-quality pork, the system could be programmed to target those specific features and reduce the amount of redundancy in the data collected.

7 Summary of recent research: assessment of pork colour and texture using imaging technology

Table 1 Summary of research investigating assessment of pork colour and texture using imaging technology in the past ten years

Name	Technology[1]	Results and conclusion
Moelich et al. (2003), Sun et al. (2016b)	CVS	Computer vision method through the development of support vector machine (SVM) reported strong potential as a means to predict pork colour scores.
Chmiel et al. (2011, 2016)	CVS	Shows potential to classify the pig meat obtained from *semimembranosus* muscle as RFN (normal) and PSE (exudative) meat.
Girolami et al. (2013)	CVS	CVS method validated measurements that reproduced a colour assessment similar to the true colour of the meat obtained from colorimeter.
Sun and Peng (2016)	NIRS	NIRS technique within 400–1000 nm wavelength range demonstrated reasonable accuracy for non-destructive assessment of pork quality attributes from a portable system that operated in real-time.
Balage et al. (2015)	NIRS	The partial least squares regression (PLSR) and VIS/NIRS-based models for IMF and WBSF of pork quality were inaccurate; supplementary research is required to identify better prediction equations for these meat features.
Wang et al. (2015)	NIRS	Reflectance spectra of 60 chilled pork samples were collected from a dual-band VIS/NIRS system which covered 350–1100 nm and 1000–2600 nm wavelength band. The results showed that the PLSR model based on spectral information was more accurate when using dual-band spectra rather than single-band spectra. The correlation coefficient of validation (Rv) were 0.9469, 0.9495, 0.9180, 0.9054 and 0.8789 for L*, a*, b*, pH value and cooking loss, respectively.
Zhang et al. (2013)	NIRS	A VIS/NIRS in the range of 350–1100 nm was used to collect the spectral data for fresh pork samples. The correlation coefficient (root mean square error) of the validation set for water content and pH were 0.810 (0.653) and 0.803 (0.098), respectively. The research shows that the real-time non-destructive detection system based on VIS/NIRS can be efficient to predict the quality of fresh meat.
Xie et al. (2015)	NIRS/HIS	A PLSR model was evaluated using HIS technology for predicting pork chop meat quality. The coefficients of determination for prediction (Rp^2) for L*, cooking loss, b*, drip loss and a* were 0.907, 0.845, 0.814, 0.762 and 0.716, respectively. The study showed HIS had significant potential to predict quality of pork for the industry.
Kapper et al. (2012a,b)	NIRS	NIRS prediction equations were developed to predict drip loss% and L* of pork longissimus dorsi muscle samples. The NIRS equations for colour a*, b* and pH were not applicable for the prediction of pork quality for commercially slaughtered pigs.

(Continued)

Table 1 (*Continued*)

Name	Technology[1]	Results and conclusion
Pu et al. (2015)	VIS/NIRS/ HIS	VIS, NIRS and HIS were used to classify fresh and frozen-thawed pork. Features were extracted from hyperspectral images in the region of 400–1000 nm. Six feature wavelengths (400, 446, 477, 516, 592 and 686 nm) were identified. Image textural features of the principal component images from hyperspectral images were obtained and the results indicated that the classification accuracy could be improved by combining spectral features with textural features. Also, the fusion of critical spectral and textural features had better potential than single spectral extraction in classifying fresh and frozen-thawed pig meat quality.
Ma et al. (2015)	HIS	Fresh and frozen-thawed pork longissimus dorsi meats were classified by VIS-NIR hyperspectral imaging. Eight optimal wavelengths (624, 673, 460, 588, 583, 448, 552 and 609 nm) were selected by successive projection algorithm (SPA). The results showed that the system could identify if pork loin had been previously frozen with 97.73% accuracy.
Tao and Peng (2014)	HIS	A HIS was investigated for evaluating pork meat tenderness and *Escherichia coli* contamination. Hyperspectral imaging in the spectral range of 400–1100 nm were collected on 31 pork longissimus dorsi muscle samples. The researchers concluded that HIS has potential for rapid determination of pork meat tenderness and *E. coli* contamination and that their technique could monitor meat quality and safety.
Pu et al. (2014)	VIS/NIR/HIS	In this study, wavelet textural analysis was applied to HIS in the visible and near-infrared (VIS/NIR) region (400–1000 nm) for differentiation between fresh and frozen-thawed pork. The spectral data of acquired hyperspectral images were analysed using partial least squares (PLS) regression. Five wavelengths (462, 488, 611, 629 and 678 nm) were selected as the feature wavelengths by the PLS model. The results showed the possibility of developing a fast and reliable HIS for discrimination between fresh and frozen-thawed pork samples based on wavelet texture in the VIS/NIR wavelength range.
Barbin et al. (2012a,b)	HIS	In this study, a HIS technique was investigated for objective determination of pork quality attributes. Hyperspectral images in the near-infrared region (900–1700 nm) were acquired from the pork longissimus dorsi muscle. Colour reflectance (L), pH and drip loss of pork meat could be predicted with determination coefficients (R-CV(2)) of 0.93, 0.87 and 0.83, respectively. Results indicate that this technique is a possible tool for rapid assessment of pork quality. In the second study, wavelengths at 960, 1074, 1124, 1147, 1207 and 1341 nm were extracted for use as predictors of hyperspectral-based classification of three pork quality grades (pale/pinkish grey, soft and exudative (PSE), reddish pink, firm and non-exudative (RFN), dark purplish red, firm and dry (DFD)) by showing the longissimus dorsi samples.

(*Continued*)

Name	Technology[1]	Results and conclusion
Li et al. (2011)	HIS	HIS was used to predict pork freshness based on quality attributes such as total volatile basic nitrogen (TVB-N), pH value, L*, a* and b* colour parameters. Hyperspectral scattering images were collected from the pork surface at the range of 400–1100 nm. It was concluded that hyperspectral scattering technique is a valid tool for real-time, non-destructive detection of pork freshness using a combination of TVB-N, pH and colour parameters.
Liu et al. (2010)	HIS	In this study, a Gabor filter-based HIS technique was presented to develop an accurate system for pork quality level classification. The principal component analysis (PCA) was used to compress spectral features over the entire visible wavelength (400–1000 nm) into principal components (PCs). It was concluded that HIS with image texture features reached a statistically significant improvement for pork quality prediction.
Qiao et al. (2007a,b)	HIS	HIS was investigated for its potential for objective determination of pork quality attributes. The system extracted spectral and spatial characteristics simultaneously to determinate the quality attributes, drip loss, pH and colour of pork loin chops. Six feature band images were selected for predicting the drip loss (459, 618, 655, 685, 755 and 953 nm), pH (494, 571,637, 669, 703 and 978 nm) and colour (434, 494, 561, 637, 669 and 703 nm), respectively. The authors found that with the HIS, the drip loss, pH and colour of pork meat could be predicted with correlation coefficients of 0.77, 0.55 and 0.86, respectively. The second study found HIS to have potential to classify the reddish, firm and non-exudative (RFN) and reddish, soft and exudative (RSE) pork at 75–80% accuracy.

[1] CVS (computer vision system); NIRS (near-infrared spectroscopy); HIS (hyperspectral image system).

8 Conclusion

For pig meat, the most important aspect is its quality – which is affected by combined key factors like colour and texture. In this chapter, we investigated and discussed both ante-mortem and post-mortem factors that can affect pig meat colour and texture attributes, which have significant impacts to consumer's purchase decision. In addition, subjective and objective measurement methods for pig colour and texture attributes were also described.

9 Where to look for further information

Meat quality is important due to the high consumer demand for higher quality food products from the global market. Production of healthy, high quality meat products is one of the top issues to the modern meat industry. For pig meat, colour and texture are the two most important factors that can affect pork quality. Knowing how to use the advanced technologies as research tools to improve, evaluate and produce the higher quality of pig meat is necessary for pig meat industry in the future.

10 References

AMSA. (2012) AMSA Meat Color Measurement Guidelines. American Meat Science Association: Champaign, Illinois, USA 61820, pp. 1–135.

Balage, J. M., da Luz, E. S. S., Gomide, C. A. and et al. (2015) Predicting pork quality using Vis/NIR spectroscopy. *Meat Sci.* 108: 37–43.

Barbin, D., Elmasry, G., Sun, D. W. and et al. (2012a) Near-infrared hyperspectral imaging for grading and classification of pork. *Meat Sci.* 90: 259–68.

Barbin, D. F., ElMasry, G., Sun, D. W. and et al. (2012b) Predicting quality and sensory attributes of pork using near-infrared hyperspectral imaging. *Anal. Chim. Acta* 719: 30–42.

Barbut, S., Sosnicki, A. A., Lonergan, S. M. and et al. (2008) Progress in reducing the pale, soft and exudative (PSE) problem in pork and poultry meat. *Meat Sci.* 79: 46–63.

Berg, E. P. (2000) Swine nutrition, the conversion of muscle to meat, and pork quality. *In* Austin J. Lewis, L. Lee Southern (Eds), *Swine Nutrition*, 2nd edn. Boca Raton, CRC Press. 631–658.

Berg, E. P. (2001) Influence of stress on composition and quality of meat, poultry and meat products. *J. Anim. Sci.* 79: 204.

Boles, J., Parrish, F., Skaggs, C. and et al. (1991) Effect of porcine somatotropin, stress susceptibility, and final end point of cooking on the sensory, physical, and chemical properties of pork loin chops. *J. Anim. Sci.* 69: 2865–70.

Brewer, M. S., Jensen, J., Sosnicki, A. A. and et al. (2002) The effect of pig genetics on palatability, color and physical characteristics of fresh pork loin chops. *Meat Sci.* 61: 249–56.

Bridi, A. M., de Oliveira, A. R., Fonseca, N. A. N. and et al. (2006). Effect of halothane genotype, ractopamine and sex on pork meat quality. *Rev. Bras. Zootecn.* 35: 2027–33.

Carr, S. N., Hamilton, D. N., Miller, K. D. and et al. (2009) The effect of ractopamine hydrochloride (Paylean®) on lean carcass yields and pork quality characteristics of heavy pigs fed normal and amino acid fortified diets. *Meat Sci.* 81: 533–9.

Channon, H. A., D'Souza, D. N. and Dunshea, F. R. (2016) Developing a cuts-based system to improve consumer acceptability of pork: Impact of gender, ageing period, endpoint temperature and cooking method. *Meat Sci.* 121: 216–27.

Chmiel, M., Slowinski, M. and Dasiewicz, K. (2011) Lightness of the color measured by computer image analysis as a factor for assessing the quality of pork meat. *Meat Sci.* 88: 566–70.

Chmiel, M., Słowiński, M., Dasiewicz, K. and et al. (2016) Use of computer vision system (CVS) for detection of PSE pork meat obtained from m. semimembranosus. *LWT Food Sci. Technol.* 65: 532–6.

Cornforth, D. and Hunt, M. (2008) Low-oxygen packaging of fresh meat with carbon monoxide. *AMSA white paper series* 2(10), pp. 1–12.

Crawford, S. M., Moeller, S. J., Zerby, H. N. and et al. (2010) Effects of cooked temperature on pork tenderness and relationships among muscle physiology and pork quality traits in loins from Landrace and Berkshire swine. *Meat Sci.* 84: 607–12.

Eggert, J., Depreux, F., Schinckel, A. and et al. (2002) Myosin heavy chain isoforms account for variation in pork quality. *Meat Sci.* 61: 117–26.

Eilert, S. (2005) New packaging technologies for the 21st century. *Meat Sci.* 71: 122–7.

Ellis, M. and McKeith, F. (1999) Nutritional influences on pork quality. *Pork Fact Sheets*, pp. 1–8.

Essen-Gustavsson, B., Karlström, K. and Lundström, K. (1992) Muscle fibre characteristics and metabolic response at slaughter in pigs of different halothane genotypes and their relation to meat quality. *Meat Sci.* 31: 1–11.

Fàbrega, E., Manteca, X., Font, J. and et al. (2002) Effects of halothane gene and pre-slaughter treatment on meat quality and welfare from two pig crosses. *Meat Sci.* 62: 463–72. https://doi.org/10.1016/s0309-1740(02)00040-2

Fisher, P., Mellett, F. and Hoffman, L. (2000) Halothane genotype and pork quality. 1. Carcass and meat quality characteristics of three halothane genotypes. *Meat Sci.* 54: 97–105.
Girolami, A., Napolitano, F., Faraone, D. and et al. (2013) Measurement of meat color using a computer vision system. *Meat Sci.* 93: 111–18.
Hammond, J. (1932) *Growth and the Development of Mutton Qualities in the Sheep. A Survey of the Problems Involved in Meat Production.* Oliver and Boyd. Edinburgh, Lothian, 597.
Hui, Y. H., Nip, W.-K. and Rogers, R. (2001) *Meat Science and Applications*: Boca Raton, CRC Press.
Huxley, A. F. and Simmons, R. M. (1971) Proposed mechanism of force generation in striated muscle. *Nature* 233: 533–8.
Juarez, M., Caine, W. R., Dugan, M. E. R. and et al. (2011) Effects of dry-ageing on pork quality characteristics in different genotypes. *Meat Sci.* 88: 117–21.
Kapper, C., Klont, R. E., Verdonk, J. M. A. J. and et al. (2012a) Prediction of pork quality with near infrared spectroscopy (NIRS) 1. Feasibility and robustness of NIRS measurements at laboratory scale. *Meat Sci.* 91: 294–9.
Kapper, C., Klont, R. E., Verdonk, J. M. A. J. and et al. (2012b) Prediction of pork quality with near infrared spectroscopy (NIRS) 2. Feasibility and robustness of NIRS measurements under production plant conditions. *Meat Sci.* 91: 300–5.
Kim, G. D., Jeong, J. Y., Hur, S. J. and et al. (2010) The relationship between meat color (CIE L* and a*), myoglobin content, and their influence on muscle fibre characteristics and pork quality. *Korean J. Food Sci. Anim. Resour.* 30: 626–33.
Kim, Y. H. B., Warner, R. D. and Rosenvold, K. (2014) Influence of high pre-rigor temperature and fast pH fall on muscle proteins and meat quality: a review. *Anim. Prod. Sci.* 54: 375–95.
Koohmaraie, M. (1992) The role of Ca 2+-dependent proteases (calpains) in post mortem proteolysis and meat tenderness. *Biochimie* 74: 239–45.
Landrock, A. H. and Wallace, G. A. (1955) Discoloration of fresh red meat and its relationship to film oxygen permeability. *Food Technol.* 9: 194–6.
Larmond, E. (1976) Texture measurement in meat by sensory evaluation. *J. Texture Stud.* 7: 87–93.
Lee, C. W., Lee, J. R., Kim, M. K. and et al. (2016) Quality improvement of pork loin by dry aging. *Korean J. Food Sci. Anim. Resour.* 36: 369.
Li, Y. Y., Zhang, L. L., Peng, Y. K. and et al. (2011) Hyperspectral imaging technique for determination of pork freshness attributes. *Sensing for Agriculture and Food Quality and Safety Iii* 8027.
Liao, Y. T., Fan, Y. X., Wu, X. Q. and et al. (2010) On-line determination of pork color and intramuscular fat by computer vision. *Sensing for Agriculture and Food Quality and Safety Ii* 7676.
Liu, L., Ngadi, M. O., Prasher, S. O. and et al. (2010) Categorization of pork quality using Gabor filter-based hyperspectral imaging technology. *J. Food Eng.* 99: 284–93.
Locker, R. and Hagyard, C. (1963) A cold shortening effect in beef muscles. *J. Sci.Food and Agr.* 14: 787–93.
Louis, C., Rempel, W. and Mickelson, J. (1993) Porcine stress syndrome: Biochemical and genetic basis of this inherited syndrome of skeletal muscle. *Proceedings of 46th Annual Reciprocal Meat Conference, Lincoln, NE.* 89–96.
Ma, J., Pu, H. B., Sun, D. W. and et al. (2015) Application of Vis-NIR hyperspectral imaging in classification between fresh and frozen-thawed pork Longissimus Dorsi muscles. *Int. J. Refrig.* 50: 10–18.
Maltin, C., Balcerzak, D., Tilley, R. and et al. (2003) Determinants of meat quality: tenderness. *Proc. Nutr. Soc.* 62: 337–47.
Mancini, R. A. and Hunt, M. C. (2005) Current research in meat color. *Meat Sci.* 71: 100–21.
Mayes, P. A. and Bender, D. A. (2003) Metabolism of glycogen. *Harper's Illustrated Biochemistry*, p. 145.
Minolta, C. (1994) *Precise Color Communication: Color Control from Feeling to Instrumentation*: Minolta.
Moelich, E., Hoffman, L. and Conradie, P. (2003) Sensory and functional meat quality characteristics of pork derived from three halothane genotypes. *Meat Sci.* 63: 333–8.
Moss, R., Diffee, G. and Greaser, M. (1995) Contractile properties of skeletal muscle fibres in relation to myofibrillar protein isoforms. *Reviews of Physiology, Biochemistry and Pharmacology*, Volume 126. Springer, Berlin and Heidelberg, pp. 1–63.

Nakai, H., Saito, F., Ikeda, T. and et al. (1975) Standard models of pork-colour. *Bull. Natl. Inst. Anim. Ind.* 29: 69–74.

Ngapo, T. M., Riendeau, L., Laberge, C. and et al. (2012a) Marbling and ageing - Part 1. Sensory quality of pork. *Food Res. Int.* 49: 396–405.

Ngapo, T. M., Riendeau, L., Laberge, C. and et al. (2012b) Marbling and ageing - Part 1. Sensory quality of pork. *Food Res. Int.* 49: 396–405.

NPB. (1999) *Pork Quality Standards*. National Pork Board: Des Moines, IA, USA.

NPPC. (2000) Pork fat quality. In Berg, E. P. (Ed.), *Pork Carcass Composition and Quality Assessment Procedures*. National Pork Producers Council: Des Moines, IA.

Offer, G. (1991) Modeling of the formation of pale, soft and exudative meat - effects of chilling regime and rate and extent of glycolysis. *Meat Sci.* 30: 157–84.

Offer, G. and Knight, P. (1988) The structural basis of water-holding in meat. Part 1: General principles and water uptake in meat processing. In Lawrie, R. (Ed.), *Developments in Meat Science*, Vol 4. Elsevier Applied Science: London, pp. 63–171.

Ohene-Adjei, S., Ellis, M., McKeith, F. K. and et al. (2003) Halothane genotype, chilling and location within muscle: Effects on pork quality. *J. Muscle Foods* 14: 107–18.

Pearson, A. (1987) Muscle function and postmortem changes. In J. F. Price, B. S. Schweigert (eds.), *The Science of Meat and Meat Products*. Academic Press, Cambridge, USA, pp. 155–92.

Pette, D. and Staron, R. S. (1990) Cellular and molecular diversities of mammalian skeletal muscle fibres. *Reviews of Physiology, Biochemistry and Pharmacology*, Volume 116. Springer, Berlin and Heidelberg, pp. 1–76.

Pettigrew, J. and Esnaola, M. (2001) Swine nutrition and pork quality: A review. *J. Anim. Sci.* 79: E316–42.

Pu, H., Sun, D. W., Ma, J. and et al. (2015) Classification of fresh and frozen-thawed pork muscles using visible and near infrared hyperspectral imaging and textural analysis. *Meat Sci.* 99: 81–8.

Pu, H. B., Sun, D. W., Ma, J. and et al. (2014) Using wavelet textural features of visible and near infrared hyperspectral image to differentiate between fresh and frozen-thawed pork. *Food and Bioprocess Tech.* 7: 3088–99.

Qiao, J., Ngadi, M. O., Wang, N. and et al. (2007a) Pork quality and marbling level assessment using a hyperspectral imaging system. *J. Food Eng.* 83: 10–16.

Qiao, J., Wang, N., Ngadi, M. O. and et al. (2007b) Prediction of drip-loss, pH, and color for pork using a hyperspectral imaging technique. *Meat Sci.* 76: 1–8.

Raines, C. R., Hunt, M. C., Seyfert, M. and et al. (2009) *Meat Lighting Facts; A guide to selecting the best light for your meat product display*. Penn State University.

Ryu, Y. C. and Kim, B. C. (2005) The relationship between muscle fibre characteristics, postmortem metabolic rate, and meat quality of pig longissimus dorsi muscle. *Meat Sci.* 71: 351–7.

Sellier, P. (1998) Genetics of meat and carcass traits. In M. F. Rothschild and A. Ruvinsug (eds) *The genetics of the pig*. CAB International Wallingford, Oxon, UK, pp. 463–510.

Sun, H. W. and Peng, Y. K. (2016) A portable nondestructive real-time detection system for inspection of pork quality attributes using Vis/NIR spectral technique. *Sensing for Agriculture and Food Quality and Safety Viii* 9864.

Sun, X., Young, J., Liu, J. H. and et al. (2016a) Prediction of pork color attributes using computer vision system. *Meat Sci.* 113: 62–4.

Sun, X., Young, J. M., Liu, J. H. and et al. (2016b) Predicting pork color scores using machine vision and support vector machine technologies. *J. Anim. Sci.* 94: 67. https://doi.org/10.2527/msasas2016-143

Swatland, H. J. (1994) *Structure and Development of Meat Animals and Poultry*: CRC Press.

Tao, F. F. and Peng, Y. K. (2014) A method for nondestructive prediction of pork meat quality and safety attributes by hyperspectral imaging technique. *J. Food Eng.* 126: 98–106.

Velarde, A., Gispert, M., Faucitano, L. and et al. (2001) Effects of the stunning procedure and the halothane genotype on meat quality and incidence of haemorrhages in pigs. *Meat Sci.* 58: 313–19.

Wang, W. X., Peng, Y. K., Li, Y. Y. and et al. (2015) Nondestructive detection of pork quality based on dual-band VIS/NIR spectroscopy. *Sensing for Agriculture and Food Quality and Safety Vii* 9488.
Xie, A. G., Sun, D. W., Xu, Z. Y. and et al. (2015) Rapid detection of frozen pork quality without thawing by Vis-NIR hyperspectral imaging technique. *Talanta* 139: 208–15.
Zhang, H. Y., Peng, Y. K., Zhao, S. W. and et al. (2013) Development of VIS/NIR spectroscopic system for real-time prediction of fresh pork quality. *Sensing for Agriculture and Food Quality and Safety V* 8721.

Chapter 8

Nutritional composition and the value of pig meat

Lauren E. O'Connor and Wayne W. Campbell, Purdue University, USA

1 Introduction

Pork, meat that comes from hogs or domesticated swine, and pork products are commonly consumed worldwide. Pork is sometimes referred to as 'the other white meat', a slogan once used by the National Pork Board, but is a red meat because of its myoglobin content (United States Department of Agriculture (USDA), 2013). The most popular cuts of pork available for retail purchase are from the shoulder, loin and side of the animal (National Pork Board, 2013). The National Pork Board reports the consumption of bacon, ham, sausage, pork chops, roasts (loin and tenderloin) and ribs most frequently based on an internal segmentation study conducted in March 2015.

For this chapter, the nutritional profiles of the most frequently consumed pork and pork products as well as the most recent consumption and recommended intakes of total pork are presented. A brief background is provided about dietary guidance in the United States regarding pork and total red meat consumption and how the effects of pork consumption are investigated within the human diet. Summaries of the scientific literature relevant to the effects of pork consumption on various aspects of human health (weight management and body composition, cardiometabolic diseases, cancer and sleep health) are provided by describing research studies in which humans consumed varying amounts and types of pork and pork products.

http://dx.doi.org/10.19103/AS.2017.0030.08

2 Nutritional content of pork

'Meats' are the second most commonly consumed protein source in the United States, following poultry (Phillips et al., 2015). Pork protein, like all animal-based proteins, is a high-quality complete protein (meaning that it is an adequate source of all nine essential amino acids) which does not contribute a disproportionate amount of energy to the diet of people living in the United States (Phillips et al., 2015, Murphy et al., 2011, O'Neil et al., 2012). O'Neil et al. (2012) reported that pork, ham and bacon contribute ~6% and ~2% to total protein and energy intake, respectively, while frankfurters, sausages and lunch meats contribute ~4% and ~3% of total protein and energy intake, respectively. Murphy et al. (2011) reported that fresh pork contributes 27–31% to total protein intake, while contributing <10% to total energy intake. These percentages are calculated from the National Health and Nutrition Examination Survey (NHANES) data, which is a compilation of research studies combining survey data (i.e. 24-hr dietary recalls) and physical examinations designed to assess the health and nutritional status of the population of the United States. The USDA Food Composition Database (https://ndb.nal.usda.gov) contains individual nutrition information about commercially available pork meats and pork products.

Commercially available cuts of pork have varying macronutrient profiles depending on the percentage of lean muscle tissue, intramuscular fat and external trimmable fat (Williams, 2007). The protein and fat contents of the most frequently consumed pork and pork products can vary from ~12–77% to 18–86% of total energy, respectively. Pork does not naturally contain carbohydrates. As the amount of protein increases, the amount of fat and energy decreases (see Fig. 1; data are adapted from the USDA Food Composition Database). The proportion of saturated fat varies from 22% to 36% (or ~1–13 g of saturated

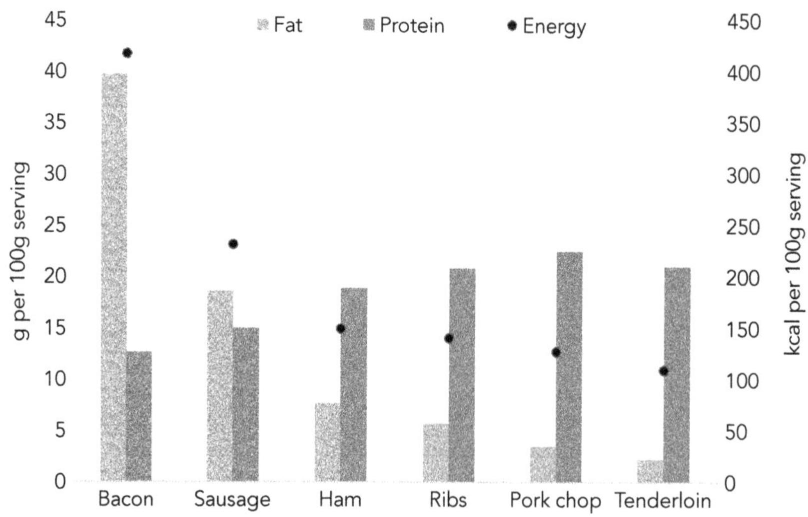

Figure 1 Fat, protein and energy contents of the most frequently consumed pork meats and pork products. Nutrition composition data are adapted from the USDA Food Composition Database (https://ndb.nal.usda.gov); the most frequently consumed pork meats were identified by the National Pork Board.

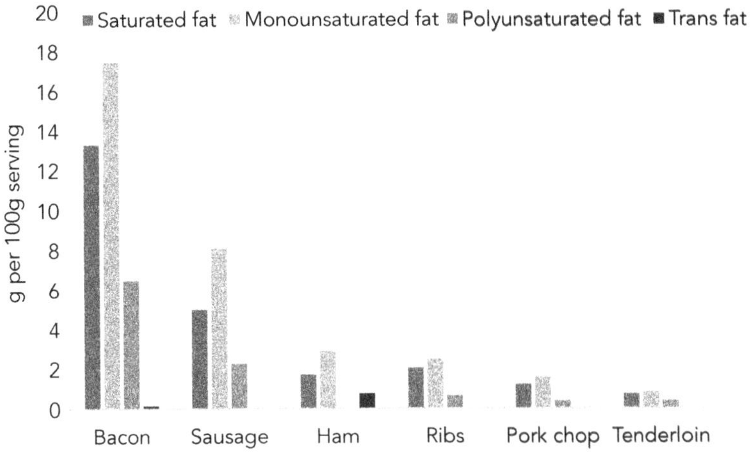

Figure 2 Long-chain fatty acid contents of the most frequently consumed pork meats and pork products. Nutrition composition data are adapted from the USDA Food Composition Database (https://ndb.nal.usda.gov); the most frequently consumed pork meats were identified by the National Pork Board.

fat per 100 g serving; see Fig. 2; data are adapted from the USDA Food Composition Database) of the total fat content. Stearic acid is the main saturated fatty acid found in pork, comprising ~30% of saturated fat and ~12% of total fat, similar to other animal-based protein sources such as beef, lamb, milk and cheese (Beef Checkoff, 2007). About 35–50% and ~10% of pork's total fatty acid content is comprised of monounsaturated fatty acids and polyunsaturated fatty acids, respectively, with trace amounts of *trans* fats (Araujo de Vizcarrondo et al., 1998).

Over the past 20 years, commercially available cuts of pork have become leaner containing 16% and 27% less total and saturated fat, respectively (Pork Checkoff, 2011, Williams J et al., 2011). This fat reduction was due to advancements in animal breeding and feeding as well as an increase in fat trimming by retail establishments. A serving of pork tenderloin is now as lean as a skinless chicken breast and there are seven cuts of pork that meet the USDA guidelines of a lean protein source (Pork Checkoff, 2012). Lean meats are defined as containing <10 g of total fat, <4.5 g of saturated fat and <95 mg of cholesterol (USDA, 2015).

Although fresh pork is naturally low in sodium, sodium is a common preservative added to pork to prolong shelf life and enhance flavour (<50 mg vs >650 mg in a 100 g serving of fresh pork chop vs cured bacon; see Fig. 3; data are adapted from the USDA Food Composition Database). Pork and pork products can be purchased as fresh unprocessed cuts (no preservation techniques except refrigeration or freezing) (American Institute for Cancer Research, 2007) or processed meats (preserved by smoking, curing, salting and/or the addition of chemical preservatives) (USDA, 2015). Frankfurters, sausages and luncheon meats are the fourth highest source of sodium found in the diet of those living in the United States, according to 2003–2006 NHANES data, behind table salt, bread products and cheese (O'Neil et al., 2012). As more sodium is added to pork products, potassium concentrations decrease. Potassium concentrations are highest in fresh pork (~400 mg vs ~200 mg per 100 g serving of pork tenderloin vs bacon, respectively; data are adapted

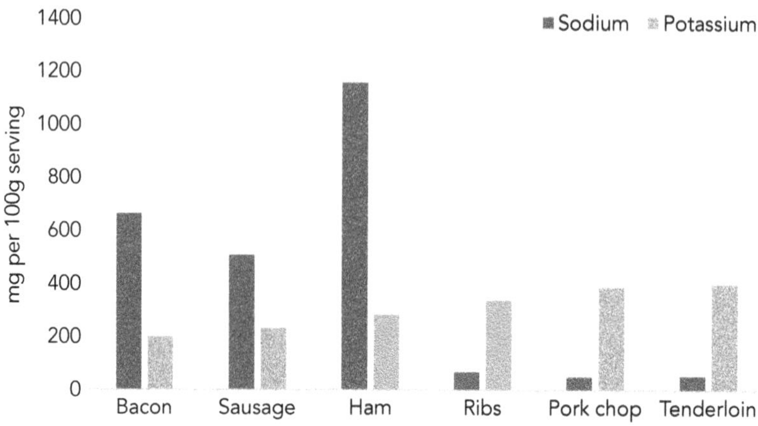

Figure 3 Sodium and potassium contents of the most frequently consumed pork meats and pork products. Nutrition composition data are adapted from the USDA Food Composition Database (https://ndb.nal.usda.gov); the most frequently consumed pork meats were identified by the National Pork Board.

from the USDA Food Composition Database). Pork and pork products contribute various other micronutrients to the diet of those living in the United States including vitamin B_{12}, niacin, vitamin B_6, iron, zinc, phosphorous, riboflavin and selenium (Williams, 2007, Murphy et al., 2011).

Non-nutritive preservatives are also added to process or preserve pork meat. Nitrites are commonly added to inhibit microbial growth of *Clostridium botulinum*, add colour and preserve shelf life by delaying lipid peroxidation, or rancidity (Heinz et al., 2010). Bacon and ham have higher concentrations of nitrites (ppm) compared to other processed meats (Nunez De Gonzalez et al., 2012). Some processed pork products use meat extenders or meat fillers to lower the cost of production. Meat extenders and meat fillers are non-meat protein and carbohydrate supplements, respectively, added to meats to provide more bulk at a lower cost. Examples of these additives include flours, starches, soy concentrates, water and mechanically deboned meats or non-muscle parts of the animal carcass (Heinz et al., 2010). These additives are included in the ingredients list on the nutrition facts label but are not yet quantified or available in the USDA database.

Other non-nutritive compounds, not included in the nutrition facts label or ingredients list, can be released during cooking. High-heat cooking, such as grilling, produces advanced glycation end products which are highly oxidant glycotoxins. There is a higher concentration of advanced glycation end products in animal-based foods than carbohydrate-based foods such as fruits, vegetables, whole grains and dairy products before and after cooking (Uribarri et al., 2010).

There are gradients of fat, protein, sodium and potassium levels throughout the most commonly consumed pork meat and pork products but most other micronutrients remain relatively constant. Although the amount of fat can vary, the fatty acid content of pork and pork products remains relatively constant as well. This variety allows consumers to choose quality protein sources that are most appropriate for their diet and/or palate. The incorporation of pork as a whole food, and its nutritive and non-nutritive components, into the human diet and the effects on human health will be discussed in later sections.

3 Pork consumption and dietary guidance in the US

3.1 Pork consumption in the US

Pork is the most commonly consumed meat in the world, per capita. China and the European Union are the largest total producers and consumers of pork, 5x and 2x that of the United States, respectively (USDA, 2016). In 2015, annual consumption of pork per United States capita was 50 pounds compared to 54 and 106 pounds of beef and poultry, respectively (National Chicken Council, 2016). The amount of pork consumed per capita has remained relatively stable since 1960; beef and total red meat have steadily declined while poultry consumption increased (see Fig. 4) (Daniel et al., 2011, USDA, 2016). Every two weeks 1.5–3.2 servings, on average, of fresh pork are consumed per household in the United States, surpassed by beef (mostly ground beef) and chicken consumptions, according to the National Pork Board's internal tracking study (November 2015, unpublished).

Pork, and other food and nutrient, consumption in the United States is tracked mainly by consumer reporting in NHANES. The average intakes of unprocessed and processed total red meat, beef and pork by adults living in the United States, according to 2009–2010 NHANES data, are shown in Fig. 5. The average total pork intake in the United States per capita is estimated to be 29 g per day, according to 2003–2006 NHANES data provided by the National Pork Board, or 24 g per day based on total pork disappearance data and USDA consumption patterns (USDA, 2015). The average intake of fresh, processed and lean total pork per day is estimated to be 8 g, 22 g and 11 g per day, respectively, according to 2003–2006 NHANES data provided by the National Pork Board.

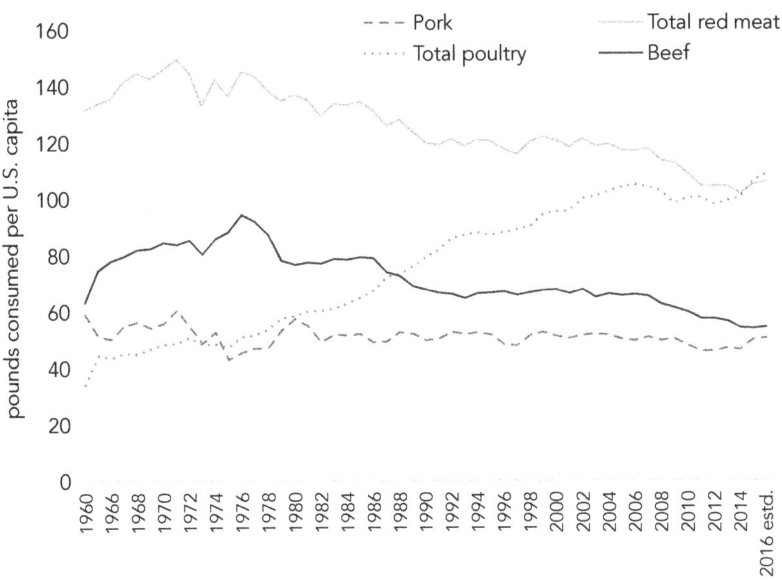

Figure 4 Red meat and poultry consumption trends over time in the United States. Data are adapted from National Chicken Council: http://www.nationalchickencouncil.org/about-the-industry/statistics/per-capita-consumption-of-poultry-and-livestock-1965-to-estimated-2012-in-pounds/.

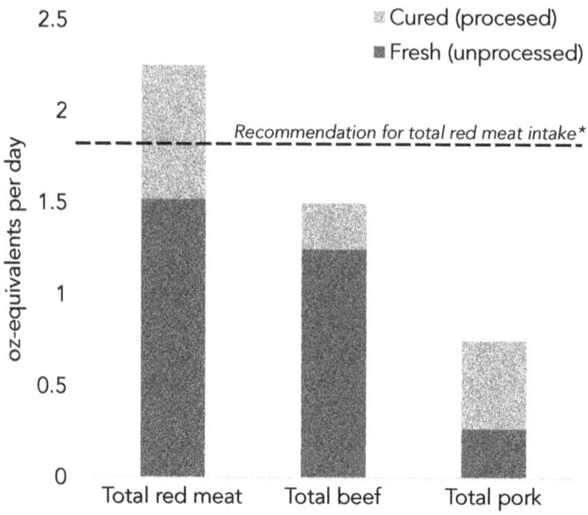

Figure 5 Daily total red meat and pork consumption compared to recommended intake in the United States. Intake data are adapted from What We Eat in America, NHANES 2009–2010 and Food Patterns Equivalents Database 2009–2010; *Total red meat recommendation is from the Healthy U.S.-Style Pattern reported in the Scientific Report of the 2015 Diétary Guidelines Advisory Committee and the 2015–2020 Dietary Guidelines for Americans.

While beef and total red meat consumption have decreased over time in the United States, pork consumption has remained steady but lower compared to other meats. As will be discussed in the next section, it is recommended for humans to limit total red meat consumption due to its relatively high concentrations of saturated fat and sodium. The National Pork Board developed the slogan 'the other white meat' in 1987 around the time in which red meat became a public health concern. This successful marketing strategy may have contributed to the steady pork consumption patterns compared to the decrease in beef and overall red meat and the increase in poultry consumption.

3.2 Dietary guidance in the US

Inclusion of pork in the human diet allows consumers to vary their protein choices to obtain the benefits of a higher protein diet and pork offers a variety of nutrients when consumed within energy needs (Freeland-Graves et al., 2013). The acceptable macronutrient distribution range for protein is 10–35% of total energy intake. Specific protein recommendations are also commonly made based on an individual's body weight; the recommended dietary allowance for protein is 0.8 g of protein per kg of body mass per day for men and women 19+ years of age (Institute of Medicine, 2005).

The Dietary Guidelines for Americans (DGA) is a publically available government document (https://health.gov/dietaryguidelines) updated every five years since 1980 by the United States Department of Health and Human Services and the Department of Agriculture. The DGAs provide 'science-based advice on promoting health and reducing the risk of major chronic diseases through a healthy diet and regular physical activity'

(USDA, 2015). The DGAs target generally healthy people aged 2+ years for health promotion and disease prevention via dietary recommendations. A Dietary Guidelines Advisory Committee of nationally recognized experts is appointed every five years to inform the federal government of the available scientific evidence pertaining to nutrition and human health. These documents are also publically accessible and are one of many resources used to shape the DGAs.

The 2015–2020 DGA present three healthy eating patterns, or a combination of foods recommended for consumption to promote health. The key recommendations of these eating patterns emphasize consumption of fruits, vegetables, whole grains, non/low-fat dairy products, oils and a variety of lean protein sources while limiting saturated fats, *trans* fats, added sugars and sodium. The DGA emphasize that protein sources, particularly meats, should be consumed in lean forms. Processed meats and poultry can be consumed as long as sodium, saturated fats, added sugars and total energy intake are within the limits of the desired eating pattern (USDA, 2015). Details about the Healthy U.S.-Style Eating Pattern, the Healthy Mediterranean-Style Eating Patterns and the Healthy Vegetarian Eating Pattern can be found in the 2015–2020 DGA.

Over the years, the DGAs suggest limiting red meat due to the relatively high concentrations of total and saturated fat as well as sodium and cholesterol. The 2015–2020 DGA state that the evidence regarding the effects of dietary cholesterol on human health is weak so there is no longer a recommended limit for dietary cholesterol intake as there was in the past (<300 mg cholesterol per day in the 2010–2015 DGA). As well, the DGAs shifted concerns about total fat to focus more on the types of fat consumed, specifically to limit saturated fat below 10% of total energy intake. Currently in the United States, saturated fat consumption is ~1–3% over the recommended amount. It is also recommended to keep sodium intake below 2300 mg per day but currently in the United States this recommendation is exceeded by 1000+mg per day (USDA, 2015).

A specific recommended intake level for red meat subtypes, such as pork and pork products, has never been established. The 2015–2020 DGA do not specify what a serving size of red meat is but according to the American Heart Association, a cooked serving of red meat is 2–3 oz. The 2010–2015 DGA suggest 1.4–1.8 oz. of red meat per day in a 2000-calorie omnivorous diet; this is equivalent to ~0.6–0.7 servings of red meat per day or ~4–5 servings per week. However, in the 2015–2020 DGA red meat is grouped into an overarching category of 'meat, eggs and poultry' (recommended amount is 26 oz. per week) resulting in an ambiguous recommendation pertaining to red meat, inclusive of pork. See Table 1 for a summary of dietary guidance pertaining to total red meat intake.

By limiting red meat consumption, the speculation is that it will help limit saturated fat and sodium consumption in the diet of people living in the United States. Protein foods, inclusive of pork and pork products, contribute about 15% and 14% of total saturated fat and sodium intake, respectively, in the diet of people living in the United States, while mixed dishes containing pork products (such as pizza, sandwiches and soups) contribute about half of these nutrients. As shown in Fig. 5, the United States' population consumes on average 1 oz. of red meat in excess of what is recommended (see Table 1 for various recommendations). As well, red meat consumption overall has noticeably declined since the 1960s but obesity rates and the associated comorbidities are still rising in the United States (USDA, 2015, Lim et al., 2012).

Table 1 Guidance ranges for total red meat consumption in the United States

Source	ounces/day	calculated servings/day	calculated servings/week
2010–2015 Dietary Guidelines for Americans DASH diet	1.4	0.6	4.2
2010–2015 Dietary Guidelines for Americans USDA Food Pattern	1.8	0.7	4.9
Scientific Report of the 2015 Dietary Guidelines Advisory Committee	1.7	0.7	4.9
2015 Dietary Guidelines for Americans, all patterns[¥]	n/a	n/a	n/a
World Cancer Research Fund	<1.6	0.6	4.2

Servings are calculated based on 2.5 oz. from the American Heart Association's recommendation of 2–3 oz. per serving of red meat; the most recent DGAs do not specify a clear red meat serving size.

4 Studying the influence of nutrition on human health

Throughout the next sections of this chapter we will discuss the scientific literature available regarding the effects of consuming pork and pork products in the human diet on the risk of obesity and related comorbidities, that is, cardiometabolic diseases, cancer, and sleep health. The types of studies being presented will consist of observational and experimental human data. Observational study designs follow a group of people over time tracking their behaviours and associating those behaviours to end outcomes like disease onset or death. Randomized controlled trials (RCTs) are an experimental approach to isolate the effects of consuming a food or nutrient on health outcomes.

Both observational and experimental study designs offer valuable evidence about dietary choices and human health but the conclusions do not always agree. The correlations from observational studies are insightful but can be confounded by uncontrollable factors and are unable to evaluate cause and effect. Conducting RCTs allow scientists to assess causality by random comparison to a non-experimental control group. These design features reduce bias and the effect of confounding factors to ensure that the variable of interest is isolated. However, RCTs rarely assess endpoints (i.e. disease onset or death) as do observational studies; therefore, RCTs assess intermediate disease risk factors (i.e. blood lipids or lipoproteins or blood pressures). See references Booth and Tannock (2014), Barton (2000) and Hebert et al. (2016) for more detailed comparisons of these study designs.

Regardless of study design, nutrition scientists can assess the effect of individual nutrients, individual foods or overall eating patterns on human health. Humans do not consume individual nutrients, rather we consume nutrient-containing foods within some sort of an eating pattern that meets personal needs. This hierarchy of consumption (individual nutrients, whole foods and eating patterns) will be used to represent how pork consumption influences human health, specifically 1) weight control and body composition, 2) cardiometabolic health, 3) cancer and 4) sleep health. In places where limited evidence is available specifically for pork, we will supplement with data pertaining to total red meat intake and higher protein diets.

5 Effects of pork consumption on weight control and body composition

Over two-thirds of the adult population in the United States is overweight or obese in addition to 17% of the population's children (Ogden et al., 2014). Obesity leads to the development of preventable comorbidities such as cardiovascular disease, type 2 diabetes, and some cancers. Along with the increased risk for disease is the large financial burden of obesity; the medical cost of obesity is the United States was $147 billion in 2008 (Centers for Disease Control and Prevention, 2015). Weight control via diet and physical activity can reduce or prevent obesity and its related comorbidities. Weight control is achieved via energy balance which is the difference of calories consumed (energy intake) and calories expended (energy expenditure). Higher protein diets, inclusive of pork, may aid in weight control and result in an improved body composition by reducing fat mass.

A higher protein diet, which can be achieved by consuming pork, can control appetitive sensations and lessen the energy intake side of the energy balance equation. Protein is a satiating macronutrient, causing humans to feel fuller longer, resulting in a prolonged period between meals. This is partially explained by protein having a slower transit time through the digestive tract than carbohydrates or fats and decreased appetite sensations in the hypothalamus of the central nervous system. Protein ingestion causes the peripheral nervous system to reduce ghrelin secretion (an appetite-stimulating gut hormone) and increase cholecystokinin, glucagon-like peptide 1 and peptide YY secretion (appetite-suppressing gut hormones) (Fromentin et al., 2012, Halton and Hu, 2004). These objective measurements of appetite are used in human research and are often complimented by subjective appetite questionnaires completed using a 100 mm visual analogue scale. See Carreiro et al. (2016) for more information about the effects of protein consumption on appetite.

To measure appetite and satiety in humans, it is common to use acute feeding study designs. Acute feeding studies allow researchers to take subjective and/or objective measurements of appetite following ingestion of a test meal throughout a predetermined time frame. In an acute feeding trial of 13 overweight/obese men, Leidy et al. (2010) measured subjective appetite ratings as well as objective concentrations of gut hormones. The study consisted of four randomized testing days with test meals containing either 14% or 25% of total energy from protein, including pork, with either 3 or 6 eating occasions. The 25% protein diet led to a 6% increase in fullness, as reported by the subjects, and a 20% increase in peptide YY concentrations, regardless of eating frequency. These results suggest that the higher protein meals containing pork increase short-term satiety (or fullness) in overweight/obese men regardless of how often meals are consumed.

In order to aid in weight control, feelings of satiety need to be prolonged past an acute one-day feeding trial. In a longer 12-week RCT, middle-aged men chronically consumed 14% or 25% of total energy from protein, including pork, with a 750 kcal energy deficit. Appetitive sensations were measured over 15 hours at different time points throughout the study. Men consuming the 25% versus 14% protein diet reported greater fullness throughout the day (511 ± 56 vs. 243 ± 54 mm per 15 hours area under the curve from subject responses to a 100 mm visual analogue scale), and fewer preoccupations or thoughts about food (17 ± 5 vs. 30 ± 5mm). Although this demonstrates a statistically significant increase in satiety, the authors do not mention whether the increase in satiety was great enough to result in more successful weight loss or dietary compliance (Leidy et al., 2011).

In a similar study design, 46 overweight/obese women consumed a diet containing either 30% or 18% of total energy from protein, inclusive of pork, with a 750 kcal energy deficit for 12 weeks. The 30% protein diet resulted in a 6% increase in perceived pleasure, regardless of the energy deficit, compared to a 6% decrease in pleasure in the 18% protein diet group. As well, a 10% versus 27% reduction in fullness occurred in the 30% versus 18% protein diet group, respectively, with the energy restriction. There was no difference in the amount of weight loss between groups since the energy deficit was the same (Leidy et al., 2007). Both of these studies not only support the use of pork as a satiating source of protein, as seen in the Leidy et al. (2010) acute feeding study, but also show that a higher protein diet containing pork is effective in prolonging satiety during energy restriction in men and women. Similar results are seen with consumption of a higher protein energy-restricted diet inclusive of other red meat types (Layman et al., 2003, Johnston et al., 2004).

The previously mentioned studies were tightly controlled feeding studies in which the subjects followed a prescribed diet. However, these results cannot be directly translated into how pork influences energy intake when subjects are asked to plan their own diets. When conducting dietary interventions, researchers can assess 1) efficacy with a tightly controlled feeding study or 2) effectiveness and feasibility in a more free-living condition; however, it is difficult to achieve both in one study. In a controlled feeding study, subjects are asked not to consume other foods or beverages outside of what is prescribed/provided; therefore, the weight loss is the same in the control and experimental group, regardless of their satiety levels. When subjects are consuming pork and eating *ad libitum*, or as much and how often they desire, it provides external validity for real-life applications.

The above studies do not provide information regarding how the differential appetitive responses translate into changes in *ad libitum* energy intake. Murphy et al. (2012) conducted a study to assess how incorporating pork into an otherwise habitual *ad libitum* diet would affect weight control. They provided the intervention group 1 kg of pork per week to incorporate into their habitual diet while the control group continued eating their habitual diet with no increase in pork consumption. Even though there was no instructed dietary energy restriction, the pork consuming group lost 0.8 ± 0.3 kg compared to the control group that gained 0.4 ± 0.4 kg at the end of six months. The pork consuming group reported consuming less energy throughout six months while energy expenditure via daily activity was the same between both groups. We can speculate that the decreased energy intake is due to greater satiety, but no appetitive responses were measured to confirm this. More research is needed about *ad libitum* energy intake with increased pork consumption.

These studies suggest that consuming pork decreases appetitive sensations in overweight/obese men and women. This in turn, assumingly, decreases energy intake and positively influences weight control. However, during diet-induced weight loss ~25% of total body mass lost is lean mass (total body mass minus total body fat mass, accounting for muscle, bones and organs) (Weinheimer et al., 2010). Retention of lean body mass is especially important in ageing populations because they are at risk for sarcopenia, or the age-related loss of skeletal muscle, with or without obesity. Sarcopenia is predictive of a higher mortality rate in individuals 73+ years old (Bunout et al., 2011).

A higher protein diet (1.0 g of protein per kg of body mass per day; 0.2 g more than the normal recommendation) is recommended during diet-induced weight loss to help retain lean body mass (Villareal et al., 2005). In the previously mentioned study by Leidy et al. (2007), consumption of the 30% protein diet resulted in losing 1.3 kg less lean mass than when consuming the 18% protein diet. Men in a similar study lost 1.1 kg less lean body mass with 25% versus 15% consumption of total energy from protein during a 750 kcal energy deficit (Tang et al., 2013a). Both of these studies used pork to achieve the protein levels of the diet. A secondary analysis of these two RCTs showed that consumption of the higher protein diets resulted in less lean body mass loss in the trunk and legs but showed no difference in regional changes in fat mass (Tang et al., 2013b). This suggests that consuming a higher protein diet obtained by the consumption of pork aids in retention of lean mass, mostly by conserving the weight-bearing muscles of the lower body, regardless of sex.

Consuming a higher protein diet during energy restriction-induced weight loss to retain lean mass is a well-studied area of nutrition research. A recent meta-analysis assessed the effects of consuming ≥ versus <1.0 g of protein per kg of body mass per day or ≥ versus <25% of total energy from protein on diet-induced changes in body composition in adults 50+ years old. Eight and three out of the twenty RCTs used red meat and specifically pork, respectively, as a means to achieve the desired protein levels in the diet. Although total body mass lost did not differ between groups, Kim et al. showed that subjects consuming a higher protein diet retained more lean body mass (0.45–0.83 kg) and lost more fat mass (0.53–0.57 kg) (Kim et al., 2016).

The results from Kim et al. (2016) are consistent with a previous meta-analysis by Wycherley et al. (2012) that assessed body composition changes while consuming either a high-protein/low-fat diet or a normal-protein/low-fat diet in adults aged ≥18 years old. Consumption of a higher protein diet resulted in more total body and fat mass loss (-0.79 and −0.87 kg, respectively) and greater retention of lean mass (+0.43 kg) than the normal-protein diet. Seven and four out of the sixteen RCTs included in this meta-analysis indicated the consumption of red meat and specifically pork, respectively, to increase the protein content of the diet. Both of these meta-analyses support that a higher protein diet, which can be achieved by the consumption of pork and other red meats, helps adults of all ages obtain a favourable body composition profile during energy restriction-induced weight loss.

Higher protein diets are successful in longer-term weight loss programmes as well. Over a 6-month dietary intervention period, total body mass loss and total fat mass loss almost doubled with consumption of a 25% versus 12% protein diet (Due et al., 2004, Skov et al., 1999). Continuation of a higher protein intake (18% of total energy) after initial body mass loss helps avoid body mass regain (Lejeune et al., 2005). This successful long-term energy restriction in combination with a higher protein intake may be explained by the practicality and easy adoption of high-protein diets (Paddon-Jones et al., 2008) and the increase in perceived pleasure and satiety as mentioned earlier. These researchers did not specify the main protein sources used in their study diets. There is a need for long-term weight loss and weight maintenance studies specific to pork consumption.

During weight loss, the body reduces resting energy expenditure and becomes more efficient at using and storing energy. Lean body mass is more metabolically active than fat mass and accounts for ~70% of total resting energy expenditure throughout the day (Brozek and Grande, 1955). Therefore a retention of lean mass, as shown with the consumption of higher protein diets containing pork, can influence the other side of the energy balance equation by increasing energy expenditure. A subgroup of the Wycherley et al. (2012) meta-analysis showed a smaller reduction in resting energy expenditure when consuming a higher protein diet during diet-induced weight loss. One of these four studies specified the consumption of pork to obtain a higher protein diet. Higher protein diets also help retain lean body mass by increasing fat break down (lipolysis) and increasing muscle protein synthesis (Petzke et al., 2014). There is limited research about how pork specifically affects lipolysis and/or muscle protein synthesis, but other red meat (such as 113 g beef) can stimulate muscle protein synthesis in adults (Symons et al., 2007).

There are other ways in which higher protein diets increase energy expenditure. The thermic effect of food, or diet-induced thermogenesis, is the amount of energy expended to digest, store and dispose of nutrients after they are consumed. Ingestion of protein-rich foods such as pork and pork products increases diet-induced thermogenesis by ~15–20% more than carbohydrate or fat ingestion due to more complex pathways of protein metabolism and less storage (Westerterp, 2004). The extra energy expended during this process contributes to daily total energy expenditure but is unlikely to be substantial enough to influence changes in weight (Wolfe, 2006).

The evidence regarding the effects of pork consumption as part of a higher protein diet on energy balance is limited. As shown by acute and longer-term feeding studies, higher protein diets achieved by consuming pork increase satiety. More research is needed to determine whether this increase in satiety is great enough to decrease *ad libitum* energy intake without a prescribed energy restriction. As well, the increase in energy expenditure with a higher protein diet, in the absence of an energy restriction, is not enough to contribute to weight loss. However, a higher protein diet achieved by the consumption of pork helps attenuate the decrease in resting energy expenditure that accompanies weight loss. This is accomplished due to greater retention of lean body mass during diet-induced weight loss while consuming a higher protein diet.

6 Effects of pork consumption on cardiometabolic health

Cardiometabolic health is a measure of the risk for cardiovascular disease and diabetes which is determined by measurable health parameters and biomarkers such as blood lipid and lipoprotein concentrations (total cholesterol, LDL-cholesterol, HDL-cholesterol and triglycerides), blood glucose and insulin concentrations, and blood pressures. Cardiovascular disease is attributable to 610 000 deaths each year in the United States (Centers for Disease Control and Prevention, 2015) and 1.4 million diabetes diagnoses occur each year in the United States (American Diabetes Association, 2016). Cardiovascular disease-related deaths are 1.7 times more likely in adults with diabetes than those without diabetes (American Diabetes Association, 2016). A small proportion of the population of the United States follows a lifestyle that promotes cardiometabolic health. Greater than 50% of the adult population has dyslipidaemia and/or hypertension, which are major modifiable risk factors for cardiovascular disease (Mozaffarian et al., 2015), and ~27% of the population is pre-diabetic (American Diabetes Association, 2016). This section will describe how consuming pork, total red meat and the nutritional components in these meats can affect cardiometabolic health parameters.

Saturated fat is a nutritional component of pork that is inconsistently associated with cardiometabolic diseases in observational studies (Siri-Tarino et al., 2010, de Souza et al., 2015, Feskens et al., 1995). These inconsistencies may be explained by variations in study designs or analyses as well as the dependent variables of interest. Experimentally, consuming saturated fat raises blood LDL-cholesterol concentrations (LDL: the 'bad' or pro-atherosclerotic blood lipid transporter) but some saturated fats stabilize high-density lipoprotein cholesterol concentrations (HDL: the 'good' or anti-atherosclerotic blood lipid transporter) (Eckel et al., 2014).

Stearic acid, a long-chain saturated fatty acid, is the main saturated fatty acid found in pork accounting for ~30% of total saturated fat (Beef Checkoff, 2007). High stearic acid diets (8–17% of total energy) compared to diets high in other saturated fatty acids lower or stabilize blood lipoproteins, and improve total cholesterol:HDL ratio. Therefore, replacement of other saturated fatty acids with stearic acid in the human diet does not negatively influence cardiovascular disease risk factors. Similar effects are seen when replacing monounsaturated fats but not polyunsaturated fats with stearic acid (Hunter et al., 2010). These different characteristics of saturated fatty acids may contribute to the inconsistent correlations between saturated fat consumption and human cardiometabolic health. However, the health effects of the relatively low concentrations of stearic acid in pork (~12% of total fat) are unknown.

Newer evidence is suggesting that moderate saturated fat consumption does not influence diabetes risk (Morio et al., 2016). Higher consumption of saturated fat versus poly- and monounsaturated fats lessens the clearance of glucose from the bloodstream, causing higher blood glucose levels which over time increases insulin resistance (Kim et al., 2015). An increase in insulin resistance causes the beta-cells of the pancreas to produce more insulin to combat the high concentrations of blood glucose, eventually causing beta-cell failure at later stages of the disease (National Institute of Diabetes and Digestive and Kidney Disease, 2009). Full-fat dairy consumption, an animal-based protein relatively high in saturated fat and stearic acid, is associated with a decrease in diabetes risk (Huth and Park, 2012, Rice, 2014) but little is known specific to pork fat consumption.

From a whole food perspective, red meat (inclusive of pork) has historically been associated with an increased risk for cardiovascular disease (Kaluza et al., 2012, Chen et al., 2013), adult-onset type 2 diabetes (Fung et al., 2004, van Dam et al., 2002) and related mortality (Abete et al., 2014, Wang et al., 2016) throughout the observational literature. This notion is recently being challenged due to inherent limitations of observational research (Klurfeld, 2015). Commonly, observational studies group red meat with processed meat during the collection of dietary intake data. Grouping these two variables is not an accurate representation of red meat since the amount of sodium, fat and preservatives in these foods varies, ultimately influencing cardiometabolic disease risk differently (see Fig. 1–2). Micha et al. were the first to meta-analyse observational cohort data with fresh unprocessed red meat and processed meat as separate variables. The results showed that consuming 50 g of processed meat per day increased the risk for developing cardiovascular disease and diabetes by 42% and 19%, respectively, while consuming 100 g of unprocessed red meat per day showed no increased risk, despite the doubling in serving size (Micha et al., 2010). A continuation of this research supports a stronger association between processed meat consumption and a weak or no association with unprocessed red meat consumption and cardiometabolic diseases (Larsson and Orsini, 2014, Micha et al., 2012).

No experimental RCTs compare the cardiometabolic health risks of unprocessed versus processed meat consumption but many assess the cardiometabolic health risks of total red meat consumption. O'Connor et al. (2017) recently meta-analysed RCTs which compared ≥ versus <0.5 (a common recommended serving size amount) and ≥ versus <1 serving (about the average consumption per United States capita) of total red meat per day. The consumption of ≥0.5 servings of red meat per day did not influence improvements in blood lipids or lipoprotein concentrations (total cholesterol, LDL, HDL, total cholesterol:HDL and triglycerides) or blood pressures, therefore not negatively influencing cardiovascular disease risk factors. Out of the 24 articles assessed in this meta-analysis, eight included fresh pork as a source of protein and three used fresh pork as the dominant or only animal flesh-based protein source in the diet. One study diet included processed meat, 15 study diets used unprocessed meat exclusively and information was unavailable for the remaining eight study diets. This meta-analysis did not include fasting glucose or insulin concentrations, creating a need for future research about the causality between red meat, specifically pork, consumption and diabetes risk.

Two studies included by O'Connor et al. (2017) incorporated pork as the only animal flesh-based protein source consumed by subjects. The research subjects consumed 140 g and 200 g of pork for 12 and 4 weeks, respectively, within the context of their habitual diet compared to the control group who consumed their habitual diets without any pork. There were no changes in blood lipids or lipoproteins in either study, regardless of the predominant protein source; concentrations of blood glucose and insulin were not measured (Flynn et al., 1982, Horrocks and Yeo, 1999). Murphy et al. (2012), mentioned previously, incorporated 135 and 97 g/day of fresh lean pork for men and women, respectively, in an otherwise unchanged habitual diet. This intervention also showed no changes in blood pressures, blood lipids or lipoproteins, or blood glucose and insulin concentrations over time. Therefore, substituting fresh pork for other animal flesh-based protein sources in the context of subjects' habitual diets does not adversely affect cardiometabolic disease risk factors.

The Dietary Approaches to Stop Hypertension (DASH) diet is designed to be low in meats containing relatively high concentrations of saturated fat and cholesterol (Sacks, 1995). Specifically, the DASH dietary pattern in the 2010–2015 and 2015–2020 DGA suggests <0.5 servings of red meat per day. Sayer et al. (2015) hypothesized that the DASH diet would retain effectiveness whether lean poultry/fish or lean pork was the main protein source. In a 5-week RCT, 19 men and women with elevated blood pressures consumed a DASH diet with either chicken/lean fish or lean pork as 55% of their total protein intake. Improvements in clinical blood pressures and blood lipid–lipoprotein concentrations were the same with consumption of both diets, regardless of whether chicken/fish or pork was the predominant protein source consumed. Blood glucose and insulin concentrations were unchanged with both diets (Sayer et al., 2015). Similarly, Leidy et al. (2007) and Tang et al. (2013a), mentioned previously, reported no differences in glucose improvements in the weight loss diets regardless of whether higher amounts of pork were consumed. Several other studies support these results which indicate that pork and other types of red meat including beef, lamb or kangaroo can be incorporated into healthy diets and show similar improvements in blood lipids or lipoprotein concentrations (Aadland et al., 2015, Davidson et al., 1999, Gascon et al., 1996, Wolmarans et al., 1999, Beauchesne-Rondeau et al., 2003, Sinclair et al., 1987) and blood pressures (Nowson et al., 2009, Wolmarans et al., 1999) compared to various other protein sources.

The inconsistent and frequently changing evidence on saturated fat consumption mentioned previously may inadvertently push consumers to purchase processed meats rather than fresh unprocessed meats because there is a variety of lean processed meats available. Some processed pork products may be lower in total and saturated fat than fresh unprocessed pork but they have higher amounts of added preservatives like sodium and nitrites. The 400% sodium content increase of processed meat can account for more than two-third of the observed associations between processed meat consumption and cardiovascular disease due to increased blood pressure (Micha et al., 2010, Micha et al., 2012). Nitrites, another common meat preservative, cause endothelial cell dysfunction (the inner lining of blood vessels) which promotes coagulation, inflammatory cytokine release and reactive oxygen species production, all of which advance atherosclerotic lesion development and maturation (Forstermann, 2008). Nitrites can also increase insulin resistance (Kim et al., 2015).

Cooking mechanisms raise concerns for cardiometabolic diseases as well. High-heat cooking techniques produce advanced glycation end products which are associated with an increased insulin resistance when high blood concentrations are detected (Kim et al., 2015). Advanced glycation end products also reduce vascular endothelium response via decreased vasodilation and increase oxidative stress and inflammation, therefore increasing the risk for cardiovascular complications (White and Collinson, 2013, Ulrich and Cerami, 2001). Very little human experimental data are available about the cardiometabolic effects of advanced glycation end products, especially specific to pork and pork products.

Results from observational studies and RCTs pertaining to total red meat consumption and cardiometabolic disease are not consistent. Historically, red meat consumption was associated with increased incidents of cardiometabolic diseases but recent evidence suggests that the level of meat processing confounds these results. There are no RCTs to further investigate this comparison between processed and unprocessed red meats but RCTs of total red meat, including pork, support that higher red meat consumption results in no negative effects on cardiovascular disease risk factors; however, data are limited pertaining to diabetes risk. The cardiometabolic effects of consuming preservatives (types and amounts) and cooking mechanisms, especially in pork, is not clear and more research is needed in this area.

7 Effects of pork consumption on other indices of human health

7.1 Effects of pork consumption on cancer

Colon cancer is the second leading non-sex-specific cancer killer in the United States, according to the Centers for Disease Control and Prevention (2016). In December 2015, the International Agency for Research on Cancer classified processed meats as carcinogenic to humans for colorectal cancer as a Group 1 carcinogen, a category inclusive of cigarette smoking and asbestos. Fresh red meat was classified as 'probably carcinogenic to humans' for colorectal cancer as a Group 2 carcinogen, a category inclusive of DDT and lead. Other positive associations for stomach and pancreatic and prostate cancer were found as well for processed meats and red meat, respectively (Bouvard et al., 2015).

Most of the evidence supporting the report from the International Agency for Cancer Research is from observational cohort data. For example, Chen et al. showed a 17% and 18% increased risk for colorectal cancer for every 100 g of unprocessed red meat and 50 g of processed meat consumed per day from ten observational cohort studies (Chan et al., 2011). When statistical models are adjusted for confounding factors such as smoking, medical history and body weight compared to a crude statistical model, the associations seen between red meat consumption and cancer risk are weakened (Alexander et al., 2015, Ollberding et al., 2012). Carr et al. (2015) concluded from 19 observational cohort studies that beef and lamb increased colorectal cancer relative risk by 11% and 24%, respectively, but there was no association seen for pork consumption. However, this meta-analysis was published after the release of the International Agency for Cancer Research report in which subtypes of red meat are not addressed.

There are various possible mechanisms in which red meat consumption increases colorectal cancer but not all are demonstrated in humans yet (Boada et al., 2016). The bioavailable haem-iron in fresh red meat can initiate cancer development via an increase in lipid peroxidation which can produce reactive oxygen species, damage cell membranes and cause DNA mutations (Marnett, 1999, Bastide et al., 2015). This is a greater concern for beef/lamb consumption rather than pork since pork contains about half the amount of iron as beef/lamb. This may explain the stronger associations between beef/lamb and colorectal cancer compared to pork consumption shown by Carr et al (2015).

As mentioned in the previous section regarding cardiometabolic disease, preservatives and high-heat cooking methods of meat cause concern for cancer development as well. Nitrite additives form carcinogenic nitrosamines in the stomach. Smoking and curing as preservation techniques creates *N*-nitroso compounds and polyacyclic aromatic amines and increases *N*-nitroso compounds and lipid peroxidation concentrations in human stools (Pierre et al., 2013). High-heat cooking creates mutagenic compounds such as heterocyclic amines and polycyclic aromatic hydrocarbons which can cause DNA damage (Sinha et al., 1994).

A desired next step in biomedical research would be to conduct an experimental human RCT to assess the effects of consuming different red meat subtypes on cancer risk and/or development, but this is a difficult task. Studying cancer prevention in humans is not deemed ethical because you cannot feed humans a food that is hypothesized to cause cancer. It is important to be wary of the limitations of biomedical science and heed the best evidence available when establishing policies pertaining to dietary guidance. However, more details are needed from observational and animal studies to further assess red meat subtypes and consumption amounts in order to provide appropriate and applicable dietary guidance to the United States' population.

7.2 Effects of pork consumption on sleep health

'Sleep health' is a new term regarding how the nation sleeps and what factors influence it. According to the National Sleep Foundation (https://sleep.org/), about two-thirds of adults report that their sleep health is poor and that they experience an issue with sleep almost every night (i.e. waking too early, feeling lethargic the next morning or snoring). Poor sleep quality and short sleep duration is associated with obesity (Wu et al., 2014) and cardiometabolic disease (Cappuccio et al., 2010, 2011).

In the past, the focus of nutrition and sleep-related research has primarily been on how poor sleep quality and short sleep duration effects dietary choices, energy intake and body weight. However, the relationship between diet and sleep could be bidirectional. It is unknown how dietary choices or macronutrient consumption influence indexes of sleep. The Scientific Report of the 2015 Dietary Guidelines Advisory Committee recognizes this lack of evidence as a need for future research to determine how dietary choices can influence sleep health.

Two 12-week randomized controlled feeding trials used the Pittsburgh Sleep Quality Index to assess changes in sleep parameters during weight loss with varying protein intakes. The Pittsburgh Sleep Quality Index is a subjective questionnaire that calculates a global sleep score about the previous month (Buysse, 1989). Lower global sleep score values indicate better sleep health. The first study compared 10%, 20% or 30% of total energy from protein achieved by pork and beef versus soy and legumes during a 750 kcal energy deficit. Higher protein intake lowered (improved) global sleep scores in subjects, regardless of the source of protein, with post-intervention global sleep scores of 10% (6.0 ± 0.4 arbitrary unites (au)) vs 20% (5.0 ± 0.4 au) versus 30% protein (5.4 ± 0.6 au).

In the second study, 44 overweight men and women consumed 0.8 versus 1.5 g of protein per kg body weight per day during weight loss for 16 weeks; the diet contained pork but the increase in protein was achieved mainly by a milk protein isolate. The higher protein group improved their global sleep score but the normal-protein group did not (post-intervention global sleep scores for the higher- vs. normal-protein groups were 4.0 ± 0.6 and 5.9 ± 0.5 au, respectively) (Zhou et al., 2016). Both of these studies used only a subjective measure of sleep quality. The scale of the global sleep quality questionnaire ranges from 0 to 21 au, therefore a 1–2 au difference between groups may not be clinically meaningful. However, the post-intervention global sleep scores of the higher protein group in the second study fell below 5.0, which is the threshold for 'good' sleep health.

Although the improvements in global sleep score were small in these RCTs, both show the same directional changes regardless of protein source. More research is needed with both subjective and objective measurements of sleep (such as actigraphy, melatonin production and brain activity) to determine if consumption of higher protein diets achieved by the inclusion of pork can improve sleep health, and therefore overall health, in adults.

8 Summary and conclusions

Total red meat and beef intakes have decreased over time, but pork consumption has remained relatively stable in the United States. The nutritional composition of pork and pork products varies depending on the leanness and processing level of the meat. Pork is a high-quality protein source rich in an array of nutrients but also has nutritive and non-nutritive components such as saturated fat, sodium and nitrites that should be consumed with caution. The nutritive and non-nutritive components of pork can contribute to human health in positive and negative ways.

Consumption of processed meats, inclusive of pork, is an emerging topic in nutrition research. Observational data suggest that processed meats are more detrimental to human health than unprocessed red meats but evidence is limited. Little is known about how specific additives and preservatives, other than sodium, in processed meat can affect human health. There is a need for experimental human RCTs to assess the effects of processed meat on cardiometabolic risk outcomes, specifically for pork because it contains higher concentrations of nitrites compared to other processed meats.

Pork consumption as part of a higher protein diet increases satiety levels and has favourable influences on body composition changes during diet-induced weight loss. The effects of pork consumption on cardiometabolic outcomes are inconsistent among the observational and experimental evidence, with observational data suggesting a positive association of higher red meat consumption, inclusive of pork, on incidences of cardiometabolic morbidities and mortalities. However, this association is weakened when processing level is considered. Experimental evidence supports no negative effect of consuming red meat, inclusive of pork, on cardiovascular disease risk factors. More experimental research is needed to assess the causality between total red meat and red meat subtype consumption on indexes of diabetes risk.

Limited evidence is available regarding the consumption of pork as part of a higher protein diet or total red meat intake on sleep health and cancer. The effect of dietary protein on sleep health is an emerging topic and more research is needed to support a possible positive effect of higher protein diets, inclusive of pork, on health. Additionally, red and processed meats were recently established, based on observational data, as carcinogenic to humans. However, when red meat subtypes are assessed, pork consumption is not associated with increased cancer risk. More evidence is needed to provide guidance on this issue but it is difficult to experimentally study cancer prevention and development in humans.

9 Where to look for further information

For more information regarding:

Dietary Guidelines for Americans: https://health.gov/dietaryguidelines/
American Heart Association: http://www.heart.org/HEARTORG/
American Diabetes Association: http://www.diabetes.org/?referrer=https://www.google.com/
Centers for Disease Control and Prevention: https://www.cdc.gov/
National Cancer Institute: https://www.cancer.gov/
United States Department of Agriculture: https://www.usda.gov/
National Pork Board: Pork Checkoff: http://www.pork.org/
National Cattlemen's Beef Association: http://www.beefusa.org/

10 References

Aadland, E. K., Lavigne, C., Graff, I. E., Eng, O., Paquette, M., Holthe, A., Mellgren, G., Jacques, H. and Liaset, B. 2015. Lean-seafood intake reduces cardiovascular lipid risk factors in healthy subjects: Results from a randomized controlled trial with a crossover design. *Am. J. Clin. Nutr.*, 102, 582–92.

Abete, I., Romaguera, D., Vieira, A. R., Lopez de Munain, A. and Norat, T. 2014. Association between total, processed, red and white meat consumption and all-cause, CVD and IHD mortality: A meta-analysis of cohort studies. *Br. J. Nutr.*, 112, 762–75.

Alexander, D. D., Weed, D. L., Miller, P. E. and Mohamed, M. A. 2015. Red meat and colorectal cancer: A quantitative update on the state of the epidemiologic science. *J. Am. Coll. Nutr.*, 34, 521–43.

American Diabetes Association. 2016. Statistics about Diabetes. Available: http://www.diabetes.org/diabetes-basics/statistics/ [Accessed 2016].

American Institute for Cancer Research. 2007. *Food, Nutrition, Physical Activity, and the Prevention of Cancer: A Global Perspective*. Washington, DC: AICR.

Araujo de Vizcarrondo, C., Carrillo de Padilla, F. and Martin, E. 1998. Fatty acid composition of beef, pork, and poultry fresh cuts, and some of their processed products. *Arch. Latinoam. Nutr.*, 48, 354–8.

Barton, S. 2000. Which clinical studies provide the best evidence? The best RCT still trumps the best observational study. *BMJ*, 321, 255–6.

Bastide, N. M., Chenni, F., Audebert, M., Santarelli, R. L., Tache, S., Naud, N., Baradat, M., Jouanin, I., Surya, R., Hobbs, D. A., Kuhnle, G. G., Raymond-Letron, I., Gueraud, F., Corpet, D. E. and Pierre, F. H. 2015. A central role for heme iron in colon carcinogenesis associated with red meat intake. *Cancer Res.*, 75, 870–9.

Beauchesne-Rondeau, E., Gascon, A., Bergeron, J. and Jacques, H. 2003. Plasma lipids and lipoproteins in hypercholesterolemic men fed a lipid-lowering diet containing lean beef, lean fish, or poultry. *Am. J. Clin. Nutr.*, 77, 587–93.

Beef Checkoff. 2007. Stearic acid- a unique saturated fat. Available: http://www.beefnutrition.org/CMDocs/BeefNutrition/StearicAcid.pdf [Accessed 2016].

Boada, L. D., Henriquez-Hernandez, L. A. and Luzardo, O. P. 2016. The impact of red and processed meat consumption on cancer and other health outcomes: Epidemiological evidences. *Food Chem. Toxicol.*, 92, 236–44.

Booth, C. M. and Tannock, I. F. 2014. Randomised controlled trials and population-based observational research: partners in the evolution of medical evidence. *Br. J. Cancer*, 110, 551–5.

Bouvard, V., Loomis, D., Guyton, K. Z., Grosse, Y., Ghissassi, F. E., Benbrahim-Tallaa, L., Guha, N., Mattock, H., Straif, K. and International Agency for Research on Cancer Monograph Working, G. 2015. Carcinogenicity of consumption of red and processed meat. *Lancet Oncol.*, 16, 1599–600.

Brozek, J. and Grande, F. 1955. Body composition and basal metabolism in man: Correlation analysis versus physiological approach. *Hum. Biol.*, 27, 22–31.

Bunout, D., de la Maza, M. P., Barrera, G., Leiva, L. and Hirsch, S. 2011. Association between sarcopenia and mortality in healthy older people. *Australas. J. Ageing*, 30, 89–92.

Buysse, D. J., Reynolds, III, C. F., Monk, T. H., Berman, S. R. and Kupfer, D. J. 1989. The Pittsburgh Sleep Quality Index: A new instrument for psychiatric practice and research. *Psychiatry Res.*, 28, 193–213.

Cappuccio, F. P., D"Elia, L., Strazzullo, P. and Miller, M. A. 2010. Quantity and quality of sleep and incidence of type 2 diabetes: A systematic review and meta-analysis. *Diabetes Care*, 33, 414–20.

Cappuccio, F. P., Cooper, D., D'elia, L., Strazzullo, P. and Miller, M. A. 2011. Sleep duration predicts cardiovascular outcomes: A systematic review and meta-analysis of prospective studies. *Eur. Heart J.*, 32, 1484–92.

Carr P. R., Walter, V., Brenner, H., Hoffmeister, M. 2015. Meat subtypes and their associations with colorectal cancer: Systematic review and meta-analysis. *Int. J. Cancer*, 138, 293–302.

Carreiro, A. L., Dhillon, J., Gordon, S., Higgins, K. A., Jacobs, A. G., Mcarthur, B. M., Redan, B. W., Rivera, R. L., Schmidt, L. R. and Mattes, R. D. 2016. The Macronutrients, Appetite, and Energy Intake. *Annu. Rev. Nutr.*, 36, 73–103.

Centers for Disease Control and Prevention. 2015. Adult Obesity Facts. Available: https://www.cdc.gov/obesity/data/adult.html [Accessed 2016].

Centers for Disease Control and Prevention. 2015. *Heart Disease Facts*. Available: http://www.cdc.gov/heartdisease/facts.htm [Accessed 2016].

Centers for Disease Control and Prevention. 2016. Colorectal (Colon) Cancer. Available: http://www.cdc.gov/cancer/colorectal/index.htm [Accessed 2016].

Chan, D. S., Lau, R., Aune, D., Vieira, R., Greenwood, D. C., Kampman, E. and Norat, T. 2011. Red and processed meat and colorectal cancer incidence: Meta-analysis of prospective studies. *PLoS One*, 6, e20456.

Chen, G. C., Lv, D. B., Pang, Z. and Liu, Q. F. 2013. Red and processed meat consumption and risk of stroke: A meta-analysis of prospective cohort studies. *Eur. J. Clin. Nutr.*, 67, 91–5.

Daniel, C. R., Cross, A. J., Koebnick, C. and Sinha, R. 2011. Trends in meat consumption in the USA. *Public Health Nutr.*, 14, 575–83.

Davidson, M. H., Hunninghake, D., Maki, K. C., Kwiterovich Jr., P. O. and Kafonek, S. 1999. Comparison of the effects of lean red meat vs lean white meat on serum lipid levels among free-living persons with hypercholesterolemia: A long-term, randomized clinical trial. *Arch. Intern. Med.*, 159, 1331–8.

De Souza, R. J., Mente, A., Maroleanu, A., Cozma, A. I., Ha, V., Kishibe, T., Uleryk, E., Budylowski, P., Schunemann, H., Beyene, J. and Anand, S. S. 2015. Intake of saturated and trans unsaturated fatty acids and risk of all cause mortality, cardiovascular disease, and type 2 diabetes: Systematic review and meta-analysis of observational studies. *BMJ*, 351, h3978.

Due, A., Toubro, S., Skov, A. R. and Astrup, A. 2004. Effect of normal-fat diets, either medium or high in protein, on body weight in overweight subjects: A randomised 1-year trial. *Int. J. Obes. Relat. Metab. Disord.*, 28, 1283–90.

Eckel, R. H., Jakicic, J. M., Ard, J. D., De Jesus, J. M., Miller, N. H., Hubbard, V. S., Lee, I. M., Lichtenstein, A. H., Loria, C. M., Millen, B. E., Nonas, C. A., Sacks, F. M., Smith, S. C., Svetkey, L. P., Wadden, T. A. and Yanovski, S. Z. 2014. 2013 AHA/ACC Guideline on Lifestyle Management to Reduce Cardiovascular Risk A Report of the American College of Cardiology/American Heart Association Task Force on Practice Guidelines. *J. Am. Coll. Cardiol.*, 63, 2960–84.

Feskens, E. J., Virtanen, S. M., Rasanen, L., Tuomilehto, J., Stengard, J., Pekkanen, J., Nissinen, A. and Kromhout, D. 1995. Dietary factors determining diabetes and impaired glucose tolerance. A 20-year follow-up of the Finnish and Dutch cohorts of the Seven Countries Study. *Diabetes Care*, 18, 1104–12.

Flynn, M. A., Naumann, H. D., Nolph, G. B., Krause, G. and Ellersieck, M. 1982. Dietary 'meats' and serum lipids. *Am. J. Clin. Nutr.*, 35, 935–42.

Forstermann, U. 2008. Oxidative stress in vascular disease: Causes, defense mechanisms and potential therapies. *Nat. Clin. Pract. Cardiovasc. Med.*, 5, 338–49.

Freeland-Graves, J. H., Nitzke, S., Academy of, Nutrition & Dietetics. 2013. Position of the academy of nutrition and dietetics: Total diet approach to healthy eating. *J. Acad. Nutr. Diet.*, 113, 307–17.

Fromentin, G., Darcel, N., Chaumontet, C., Marsset-Baglieri, A., Nadkarni, N. and Tome, D. 2012. Peripheral and central mechanisms involved in the control of food intake by dietary amino acids and proteins. *Nutr. Res. Rev.*, 25, 29–39.

Fung, T. T., Schulze, M., Manson, J. E., Willett, W. C. and Hu, F. B. 2004. Dietary patterns, meat intake, and the risk of type 2 diabetes in women. *Arch. Intern. Med.*, 164, 2235–40.

Gascon, A., Jacques, H., Moorjani, S., Deshaies, Y., Brun, L. D. and Julien, P. 1996. Plasma lipoprotein profile and lipolytic activities in response to the substitution of lean white fish for other animal protein sources in premenopausal women. *Am. J. Clin. Nutr.*, 63, 315–21.

Halton, T. L. and Hu, F. B. 2004. The effects of high protein diets on thermogenesis, satiety and weight loss: A critical review. *J. Am. Coll. Nutr.*, 23, 373–85.

Hebert, J. R., Frongillo, E. A., Adams, S. A., Turner-Mcgrievy, G. M., Hurley, T. G., Miller, D. R. and Ockene, I. S. 2016. Perspective: Randomized controlled trials are not a panacea for diet-related research. *Adv. Nutr.*, 7, 423–32.

Heinz, G., P., Hautzinger, P. 2010. Meat Processing Technology for Small- to Medium-scale Producers. Available: http://www.fao.org/docrep/010/ai407e/ai407e00.htm [Accessed 2016].

Horrocks, L. A. and Yeo, Y. K. 1999. Docosahexaenoic acid-enriched foods: Production and effects on blood lipids. *Lipids*, 34 Suppl, S313.

Hunter, J. E., Zhang, J. and Kris-Etherton, P. M. 2010. Cardiovascular disease risk of dietary stearic acid compared with trans, other saturated, and unsaturated fatty acids: A systematic review. *Am. J. Clin. Nutr.*, 91, 46–63.

Huth, P. J. and Park, K. M. 2012. Influence of dairy product and milk fat consumption on cardiovascular disease risk: A review of the evidence. *Adv. Nutr.*, 3, 266–85.

Institute of Medicine. 2005. *Dietary Reference Intakes for Energy, Carbohydrate, Fiber, Fat, Fatty Acids, Cholesterol, Protein, and Amino Acids*. Washington, DC: National Academies Press.

Johnston, C. S., Tjonn, S. L. and Swan, P. D. 2004. High-protein, low-fat diets are effective for weight loss and favorably alter biomarkers in healthy adults. *J. Nutr.*, 134, 586–91.

Kaluza, J., Wolk, A. and Larsson, S. C. 2012. Red meat consumption and risk of stroke: A meta-analysis of prospective studies. *Stroke*, 43, 2556–60.

Kim, J. E., O'Connor, L. E., Sands, L. P., Slebodnik, M. B. and Campbell, W. W. 2016. Effects of dietary protein intake on body composition changes after weight loss in older adults: A systematic review and meta-analysis. *Nutr. Rev.*, 74, 210–24.

Kim, Y., Keogh, J. and Clifton, P. 2015. A review of potential metabolic etiologies of the observed association between red meat consumption and development of type 2 diabetes mellitus. *Metabolism*, 64, 768–79.

Klurfeld, D. M. 2015. Research gaps in evaluating the relationship of meat and health. *Meat Sci.*, 109, 86–95.
Larsson, S. C. and Orsini, N. 2014. Red meat and processed meat consumption and all-cause mortality: A meta-analysis. *Am. J. Epidemiol.*, 179, 282–9.
Layman, D. K., Boileau, R. A., Erickson, D. J., Painter, J. E., Shiue, H., Sather, C. and Christou, D. D. 2003. A reduced ratio of dietary carbohydrate to protein improves body composition and blood lipid profiles during weight loss in adult women. *J. Nutr.*, 133, 411–7.
Leidy, H. J., Carnell, N. S., Mattes, R. D. and Campbell, W. W. 2007. Higher protein intake preserves lean mass and satiety with weight loss in pre-obese and obese women. *Obesity (Silver Spring)*, 15, 421–9.
Leidy, H. J., Armstrong, C. L., Tang, M., Mattes, R. D. and Campbell, W. W. 2010. The influence of higher protein intake and greater eating frequency on appetite control in overweight and obese men. *Obesity (Silver Spring)*, 18, 1725–32.
Leidy, H. J., Tang, M., Armstrong, C. L., Martin, C. B. and Campbell, W. W. 2011. The effects of consuming frequent, higher protein meals on appetite and satiety during weight loss in overweight/obese men. *Obesity (Silver Spring)*, 19, 818–24.
Lejeune, M. P., Kovacs, E. M. and Westerterp-Plantenga, M. S. 2005. Additional protein intake limits weight regain after weight loss in humans. *Br. J. Nutr.*, 93, 281–9.
Lim, S. S., Vos, T., Flaxman, A. D., Danaei, G., Shibuya, K., Adair-Rohani, H. et al. 2012. A comparative risk assessment of burden of disease and injury attributable to 67 risk factors and risk factor clusters in 21 regions, 1990–2010: A systematic analysis for the Global Burden of Disease Study 2010. *Lancet*, 380, 2224–60.
Marnett, L. J. 1999. Lipid peroxidation-DNA damage by malondialdehyde. *Mutat. Res.*, 424, 83–95.
Micha, R., Wallace, S. K. and Mozaffarian, D. 2010. Red and processed meat consumption and risk of incident coronary heart disease, stroke, and diabetes mellitus: A systematic review and meta-analysis. *Circulation*, 121, 2271–83.
Micha, R., Michas, G. and Mozaffarian, D. 2012. Unprocessed red and processed meats and risk of coronary artery disease and type 2 diabetes–an updated review of the evidence. *Curr. Atheroscler. Rep.*, 14, 515–24.
Morio, B., Fardet, A., Legrand, P. and Lecerf, J. M. 2016. Involvement of dietary saturated fats, from all sources or of dairy origin only, in insulin resistance and type 2 diabetes. *Nutr. Rev.*, 74, 33–47.
Mozaffarian, D., Benjamin, E. J., Go, A. S., Arnett, D. K., Blaha, M. J., Cushman, M., De Ferranti, S., Despres, J. P., Fullerton, H. J., Howard, V. J., Huffman, M. D., Judd, S. E., Kissela, B. M., Lackland, D. T., Lichtman, J. H., Lisabeth, L. D., Liu, S., Mackey, R. H., Matchar, D. B., Mcguire, D. K., Mohler, E. R., III, Moy, C. S., Muntner, P., Mussolino, M. E., Nasir, K., Neumar, R. W., Nichol, G., Palaniappan, L., Pandey, D. K., Reeves, M. J., Rodriguez, C. J., Sorlie, P. D., Stein, J., Towfighi, A., Turan, T. N., Virani, S. S., Willey, J. Z., Woo, D., Yeh, R. W., Turner, M. B., American Heart Association Statistics, C. & Stroke Statistics, S. 2015. Heart disease and stroke statistics–2015 update: A report from the American Heart Association. *Circulation*, 131, e29–322.
Murphy, K. J., Thomson, R. L., Coates, A. M., Buckley, J. D. and Howe, P. R. 2012. Effects of eating fresh lean pork on cardiometabolic health parameters. *Nutrients*, 4, 711–23.
Murphy, M. M., Spungen, J. H., Bi, X. and Barraj, L. M. 2011. Fresh and fresh lean pork are substantial sources of key nutrients when these products are consumed by adults in the United States. *Nutr. Res.*, 31, 776–83.
National Chicken Council. 2016. *Per capita consumption of poultry and livestock, 1965 to estimated 2016, in pounds.* Available: http://www.nationalchickencouncil.org/about-the-industry/statistics/per-capita-consumption-of-poultry-and-livestock-1965-to-estimated-2012-in-pounds/ [Accessed 2016].
National Institute of Diabetes and Digestive Kidney Disease. 2009. Prediabetes and insulin resistance. Available: https://www.niddk.nih.gov/health-information/diabetes/types/prediabetes-insulin-resistance [Accessed 2016].
National Pork Board. 2013. Pork's most popular cuts. Available: https://www.porkretail.org/filelibrary/Retail/NPB%20Nomenclature%20cut%20chart%20041913%20HR.pdf [Accessed 2016].

National Pork Board. 2015. Internal Tracking Study, unpublished.

Nowson, C. A., Wattanapenpaiboon, N. and Pachett, A. 2009. Low-sodium dietary approaches to stop hypertension-type diet including lean red meat lowers blood pressure in postmenopausal women. *Nutr. Res.*, 29, 8–18.

Nunez De Gonzalez, M. T., Osburn, W. N., Hardin, M. D., Longnecker, M., Garg, H. K., Bryan, N. S. and Keeton, J. T. 2012. Survey of residual nitrite and nitrate in conventional and organic/natural/uncured/indirectly cured meats available at retail in the United States. *J. Agric. Food Chem.*, 60, 3981–90.

O'Connor, L. E., Kim, J. E., Campbell, W. W. 2017. Total red meat intake of >/=0.5 servings/d does not negatively influence cardiovascular disease risk factors: a systemically searched meta-analysis of randomized controlled trials. *Am. J. Clin. Nutr.*, 105(1), 57–69. Jan.

Ogden, C. L., Carroll, M. D., Kit, B. K. and Flegal, K. M. 2014. Prevalence of childhood and adult obesity in the United States, 2011–2012. *JAMA*, 311, 806–14.

Ollberding, N. J., Wilkens, L. R., Henderson, B. E., Kolonel, L. N. and Le Marchand, L. 2012. Meat consumption, heterocyclic amines and colorectal cancer risk: The Multiethnic Cohort Study. *Int. J. Cancer*, 131, E1125–33.

O'Neil, C. E., Keast, D. R., Fulgoni, V. L. and Nicklas, T. A. 2012. Food sources of energy and nutrients among adults in the US: NHANES 2003–2006. *Nutrients*, 4, 2097–120.

Paddon-Jones, D., Westman, E., Mattes, R. D., Wolfe, R. R., Astrup, A. and Westerterp-Plantenga, M. 2008. Protein, weight management, and satiety. *Am. J. Clin. Nutr.*, 87, 1558S–1561S.

Petzke, K. J., Freudenberg, A. and Klaus, S. 2014. Beyond the role of dietary protein and amino acids in the prevention of diet-induced obesity. *Int. J. Mol. Sci.*, 15, 1374–91.

Phillips, S. M., Fulgoni, 3rd, V. L., Heaney, R. P., Nicklas, T. A., Slavin, J. L. and Weaver, C. M. 2015. Commonly consumed protein foods contribute to nutrient intake, diet quality, and nutrient adequacy. *Am. J. Clin. Nutr.* 101, 1346S-1352S.

Pierre, F. H., Martin, O. C., Santarelli, R. L., Tache, S., Naud, N., Gueraud, F., Audebert, M., Dupuy, J., Meunier, N., Attaix, D., Vendeuvre, J. L., Mirvish, S. S., Kuhnle, G. C., Cano, N. and Corpet, D. E. 2013. Calcium and alpha-tocopherol suppress cured-meat promotion of chemically induced colon carcinogenesis in rats and reduce associated biomarkers in human volunteers. *Am. J. Clin. Nutr.*, 98, 1255–62.

Pork Checkoff. 2011. Balancing Animal Proteins on the New American Plate. Available: http://www.meat-ims.org/wp-content/uploads/2013/07/ADA_RD_fact_sheet.pdf [Accessed 2016].

Pork Checkoff. 2012. Pork Inspirations Issue 3. Available: https://www.porkretail.org/filelibrary/Retail/Pork%20Inspirations%20%233%20080912.pdf [Accessed 2016].

Rice, B. H. 2014. Dairy and cardiovascular disease: A review of recent observational research. *Curr. Nutr. Rep.*, 3, 130–8.

Sacks, F. M., Obarzanek, E., Windhauser, M. M., Svetkey, L. P., Vollmer, W. M., Mccullough, M., Karanja, N., Lin, P. H., Steele, P., Proschan, M. A., Evans, M. A., Appel, L. J., Bray, G. A., Vogt, T. M., Moore, T. J. and Investigators, D. 1995. Rationale and design of the Dietary Approaches to Stop Hypertension trial (DASH). A multicenter controlled-feeding study of dietary patterns to lower blood pressure. *Ann. Epidemiol.*, 5, 108–18.

Sayer, R. D., Wright, A. J., Chen, N. and Campbell, W. W. 2015. Dietary Approaches to Stop Hypertension diet retains effectiveness to reduce blood pressure when lean pork is substituted for chicken and fish as the predominant source of protein. *Am. J. Clin. Nutr.*, 102, 302–8.

Sinclair, A. J., O'Dea, K., Dunstan, G., Ireland, P. D. and Niall, M. 1987. Effects on plasma lipids and fatty acid composition of very low fat diets enriched with fish or kangaroo meat. *Lipids*, 22, 523–9.

Sinha, R., Rothman, N., Brown, E. D., Mark, S. D., Hoover, R. N., Caporaso, N. E., Levander, O. A., Knize, M. G., Lang, N. P. and Kadlubar, F. F. 1994. Pan-fried meat containing high levels of heterocyclic aromatic amines but low levels of polycyclic aromatic hydrocarbons induces cytochrome P4501A2 activity in humans. *Cancer Res.*, 54, 6154–9.

Siri-Tarino, P. W., Sun, Q., Hu, F. B. and Krauss, R. M. 2010. Meta-analysis of prospective cohort studies evaluating the association of saturated fat with cardiovascular disease. *Am. J. Clin. Nutr.*, 91, 535–46.

Skov, A. R., Toubro, S., Ronn, B., Holm, L. and Astrup, A. 1999. Randomized trial on protein vs carbohydrate in ad libitum fat reduced diet for the treatment of obesity. *Int. J. Obes. Relat. Metab. Disord.*, 23, 528–36.

Symons, T. B., Schutzler, S. E., Cocke, T. L., Chinkes, D. L., Wolfe, R. R. and Paddon-Jones, D. 2007. Aging does not impair the anabolic response to a protein-rich meal. *Am. J. Clin. Nutr.*, 86, 451–6.

Tang, M., Armstrong, C. L., Leidy, H. J. and Campbell, W. W. 2013a. Normal vs. high-protein weight loss diets in men: Effects on body composition and indices of metabolic syndrome. *Obesity (Silver Spring)*, 21, E204–10.

Tang, M., Leidy, H. J. and Campbell, W. W. 2013b. Regional, but not total, body composition changes in overweight and obese adults consuming a higher protein, energy-restricted diet are sex specific. *Nutr. Res.*, 33, 629–35.

Ulrich, P. and Cerami, A. 2001. Protein glycation, diabetes, and aging. *Recent. Prog. Horm. Res.*, 56, 1–21.

United States Department of Agriculture. 2010. *Report of the Dietary Guidelines Advisory Committee on the Dietary Guidelines for Americans, 2010.* Available: http://www.cnpp.usda.gov/DGAs2010-DGACReport.htm [Accessed 2016].

United States Department of Agriculture. 2011. Livestock and Poultry: World Markets and Trade, Annual Report. Available: http://usda.mannlib.cornell.edu/usda/fas/livestock-poultry-ma//2010s/2011/livestock-poultry-ma-10–14-2011.pdf [Accessed 2016].

United States Department of Agriculture; Food and Safety Inspection Service. 2013. *Fresh Pork from Farm to Table.* Available: http://www.fsis.usda.gov/wps/portal/fsis/topics/food-safety-education/get-answers/food-safety-fact-sheets/meat-preparation/fresh-pork-from-farm-to-table/CT_Index [Accessed 2016].

United States Department of Agriculture. 2015. *2015–2020 Dietary Guidelines for Americans.* Washington, DC: US Government Printing Office. Available: https://health.gov/dietaryguidelines/2015/guidelines/[Accessed 2016].

United States Department of Agriculture. 2015. Factors affecting U.S. pork consumption. Available: http://www.ers.usda.gov/media/326138/ldpm13001_1_.pdf [Accessed 2016].

United States Department of Agriculture. 2015. Scientific Report of the 2015 Dietary Guidelines Advisory Committee (Advisory Report). Available: https://health.gov/dietaryguidelines/2015-scientific-report/PDFs/Scientific-Report-of-the-2015-Dietary-Guidelines-Advisory-Committee.pdf [Accessed 2016].

United States Department of Agriculture. 2016. Livestock and poultry: world markets and trade. Available: http://apps.fas.usda.gov/psdonline/circulars/livestock_poultry.pdf [Accessed 2016].

Uribarri, J., Woodruff, S., Goodman, S., Cai, W., Chen, X., Pyzik, R., Yong, A., Striker, G. E. and Vlassara, H. 2010. Advanced glycation end products in foods and a practical guide to their reduction in the diet. *J. Am. Diet. Assoc.*, 110, 911–16 e12.

Van Dam, R. M., Willett, W. C., Rimm, E. B., Stampfer, M. J. and Hu, F. B. 2002. Dietary fat and meat intake in relation to risk of type 2 diabetes in men. *Diabetes Care*, 25, 417–24.

Villareal, D. T., Apovian, C. M., Kushner, R. F., Klein, S., American Society For, N. and Naaso, T. O. S. 2005. Obesity in older adults: Technical review and position statement of the American Society for Nutrition and NAASO, The Obesity Society. *Obes. Res.*, 13, 1849–63.

Wang, X., Lin, X., Ouyang, Y. Y., Liu, J., Zhao, G., Pan, A. and Hu, F. B. 2016. Red and processed meat consumption and mortality: Dose-response meta-analysis of prospective cohort studies. *Public Health Nutr*, 19, 893–905.

Weinheimer, E. M., Sands, L. P. and Campbell, W. W. 2010. A systematic review of the separate and combined effects of energy restriction and exercise on fat-free mass in middle-aged and older adults: Implications for sarcopenic obesity. *Nutr. Rev.*, 68, 375–88.

Westerterp, K. R. 2004. Diet induced thermogenesis. *Nutr. Metab. (Lond)*, 1, 5.

White, D. L. and Collinson, A. 2013. Red meat, dietary heme iron, and risk of type 2 diabetes: The involvement of advanced lipoxidation endproducts. *Adv. Nutr.*, 4, 403–11.

Williams, J. R., Howe, J., Trainer, D., Snyder, C., Boillot, K., Lofgren, P., et al. 2011. Nutritional changes in fresh pork cuts between 1991–2005. Available: https://www.ars.usda.gov/ARSUserFiles/80400525/Articles/IFT2006_PorkComp.pdf [Accessed 2016].

Williams, P. 2007. Nutritional composition of red meat. *Nutr. Diet.*, 64, S113–S119.

Wolfe, R. R. 2006. The underappreciated role of muscle in health and disease. *Am. J. Clin. Nutr.*, 84, 475–82.

Wolmarans, P., Laubscher, J. A., Van Der Merwe, S., Kriek, J. A., Lombard, C. J., Marais, M., Vorster, H. H., Tichelaar, H. Y., Dhansay, M. A. and Benade, A. J. 1999. Effects of a prudent diet containing either lean beef and mutton or fish and skinless chicken on the plasma lipoproteins and fatty acid composition of triacylglycerol and cholesteryl ester of hypercholesterolemic subjects. *J. Nutr. Biochem.*, 10, 598–608.

Wu, Y., Zhai, L. and Zhang, D. 2014. Sleep duration and obesity among adults: A meta-analysis of prospective studies. *Sleep Med.*, 15, 1456–62.

Wycherley, T. P., Moran, L. J., Clifton, P. M., Noakes, M. and Brinkworth, G. D. 2012. Effects of energy-restricted high-protein, low-fat compared with standard-protein, low-fat diets: A meta-analysis of randomized controlled trials. *Am. J. Clin. Nutr.*, 96, 1281–98.

Zhou, J., Kim, J. E., Armstrong, C. L., Chen, N. and Campbell, W. W. 2016. Higher-protein diets improve indexes of sleep in energy-restricted overweight and obese adults: Results from 2 randomized controlled trials. *Am. J. Clin. Nutr.*, 103, 766–74.

Part 3

Sustainability

Chapter 9

Assessing the environmental impact of swine production

G. J. Thoma, University of Arkansas, USA

1 Introduction

Mitigation of, and adaptation to, climate change is necessary for the continued success of modern agriculture. Agriculture is one of the most pervasive human activities on the planet, and thus is a significant source of greenhouse gas (GHG) emissions. Significant efforts are being undertaken globally to assess and understand the impact that managed agro-ecosystems have on the environment, including global climate change (Foley et al., 2011; Tilman et al., 2011; Gerber et al., 2013; Macleod et al., 2013). System-scale models provide a mechanism for integrating scientific knowledge, management practices, and environmental and economic impacts of agro-ecosystems. While agriculture, and the swine production industry in particular, has reduced environmental impacts during the past 50 years due to productivity gains (Xin et al., 2010; Capper, 2011; Boyd et al., 2012), there are increasing demands on our resource base, and we need robust tools to continue to support the best decisions in the face of these challenges. The tools described in this chapter describe approaches to system-scale productivity and environmental management for the swine sector.

The meat production sector of the agricultural market has been under increasing scrutiny from a segment of the public due to perceived impacts of production practices

http://dx.doi.org/10.19103/AS.2017.0030.09

on our natural resources and environment quality. The livestock sector competes for scarce resources, such as land, water and energy, and has an impact on air, water and oil quality because of its emissions (de Vries and de Boer, 2010). On a global scale, pork has accounted for an approximately constant 36.5% of the world's meat production between 2005 and 2014. China is the world's largest producer (48% of total pig meat in 2014), followed by the United States and European Union which produce 35% of pig meat (FAO, 2016). Between 2005 and 2014 total pig meat production, on a carcass weight equivalent basis, increased from 94.3 to 115 million metric tons, while during the same period total meat production increased from 255 to almost 318 million metric tons. The United States, Canada and European Union together account for over 80% of pig meat exports (USDA FAS, 2016). Until 2014 the United States was the largest exporter of pork, contributing approximately 34–27% as Canadian and European production increased more rapidly, with European exports overtaking US exports in 2015.

Pigs are raised in a wide variety of systems, ranging from smallholder backyard production to large-scale intensive systems (Macleod et al., 2013; McGlone, 2013). Each of these systems will have different environmental impacts, depending upon scale and location. These impacts can affect both water and air quality, and some impacts will be local in nature (eutrophication from nutrient loss) while others will be global in nature (GHG emissions). Life cycle assessment (LCA) is a methodological framework for assessing environmental impacts of goods and services taking a life cycle perspective. That is, providing a full account of emissions and impacts beginning from the original extraction of resources from nature, to processing (e.g. fertilizers and fuel), and agricultural production, followed by food processing, distribution, retail, consumption and finally end-of-life disposal (International Orgainzation for Standardization, 2006). LCA enables the identification of hot spots within the supply chain, those activities with significant contributions to impacts, which represent targets for improvements. It also provides a transparent assessment of trade-offs among the impacts, both between stages in the supply chain and between different impact categories. LCA can also be a benchmarking activity, to document current environmental performance and provide a defensible and documentable metric for demonstrating continual improvement.

As the demand for high-quality protein increases globally due to the confluence of an increasing population and growing middle class (Godfray et al., 2010; Melorose, Perroy and Careas, 2015), critical assessment of the sustainability of our food production system is increasingly important. Typically, LCA of food products is based on one of the following functional units: live weight at the farm gate, carcass weight at the processor gate or the weight consumed, often accounting for food loss in the supply chain. However, there is an increasing awareness that this approach has some shortcomings, and that nutritional characteristics should be included for a full assessment of the environmental consequences of food production and consumption choices (Fulgoni, Keast and Drewnowski, 2009; Heller, Keoleian and Willett, 2013; van Dooren et al., 2014; Van Kernebeek et al., 2014; Stylianou et al., 2015; Gustafson et al., 2016).

This chapter presents a review of environmental sustainability impacts of swine production, focused at the farm level because the majority of environmental impacts occur by this stage of the supply chain. Two case studies comparing the European and US swine production and the adoption of gestation pens to replace gestation stalls are presented to demonstrate the application of LCA.

2 Environmental emissions and impacts at farm level: GHG emissions

The livestock production supply chain, including swine production, contributes to environmental impacts (Steinfeld et al., 2006; Macleod et al., 2013). From a full supply chain perspective, impacts up to the farm gate generally contribute over 70% of the total supply chain impact. Feed production and manure management are the two largest contributors for non-ruminant livestock. As the industry consolidates and confined animal operations become more common, there are concerns that the environmental consequences of production may increase (Burkholder et al., 2007; Hribar, 2010). This section considers farm-level emissions of GHGs. The next two sections will then turn to water and air emissions, the spread of pathogenic microbes and antibiotic resistance, and pharmaceutical emissions.

The most significant GHG releases associated with agricultural production are carbon dioxide (CO_2), methane (CH_4) and nitrous oxide (N_2O). CO_2 is emitted from combustion of fossil fuels, both on-farm and from production of electricity. CH_4 emissions from swine production arise primarily from the anaerobic microbial processes occurring during manure handling (Petersen et al., 2013). N_2O emissions primarily occur during the nitrification-denitrification cycle in soils. There are two important N_2O sources: direct emissions from transformation of nitrogen fertilizers applied to cropland (both inorganic and manure based), and indirect emissions associated with ammonia volatilization and subsequent deposition onto the soil where it enters the nitrification-denitrification cycle and can be converted to N_2O (Dong et al., 2006).

The management of the pork production system influences GHG emissions. Basset-Mens and van der Werf (2005) compared three production systems in France. Their LCA was a cradle-to-farm gate study with a functional unit of live weight at slaughter. The first of the three systems evaluated was a 'good agricultural practice' (GAP) scenario, which involves intensive production where pigs are raised in confinement on slatted floors in buildings. The second system was an organic (OA) scenario, where pigs are born and reared outdoors until weaning (42 days of age), then moved to open-front straw-litter buildings at low animal density. The final production scenario, red label (RL), corresponds to a specific quality label in France. This scenario (RL) is similar to OA; however, piglets are weaned at 28 days of age and reared under specified standards aimed at providing a premium product. The authors reported the LCA data using two functional units: 1 kg of pig produced and one ha of land surface used. Interestingly, the three production scenarios rank differently in their global warming potentials (GWPs) depending on choice of the functional unit. When the functional unit is 1 kg of pig produced (live weight at slaughter), GAP, RL and OA values were 2.3, 3.46 and 3.97 kg CO_2e per ha, respectively. However, when the functional unit was 1 ha of land surface used, the ranking changed order: OA, GAP and RL values were 4022, 4236 and 5510 kg CO_2e ha^{-1}, respectively. These results clearly demonstrate that production system and functional unit can influence the results of LCA and the interpretation of those results. In another study, also based on these production scenarios, the same authors investigated the effect of uncertainty in the data on the impact assessment results and subsequent decisions recommended for changes to the system. They report that uncertainties in field emissions associated with crop production are the most significant in consideration of GWP. However, for other environmental impact metrics (acidification and eutrophication) they report that emissions from the barn dominate the uncertainty in the results. Not surprisingly, they report that for eutrophication there is a strong dependence on the farm's specific watershed characteristics (Basset-mens and van der Werf, 2005; Basset-mens et al., 2006, 2007).

LCA consistently indicates that feed production and manure management are the dominant contributors to GHG emissions in pork production (Basset-mens and van der Werf, 2005; Dalgaard and Halberg, 2005; Basset-mens et al., 2007; Dalgaard, Halberg and Hermansen, 2007; SNC-LAVALIN Agro and Ciraig, 2009; Lammers et al., 2010; Jones and Cherruault, 2011; Reckmann, Traulsen and Krieter, 2013). Unfortunately, direct comparisons among the different studies are problematic. Not only are the functional units different, the data analysis models used by the authors are not always clearly defined. For example, Dalgaard et al. (2007) reported that the GHG equivalent of producing grain for feed was approximately 2 kg CO_2e/kg live weight at the farm gate while Eriksson et al. (2005) reported values from 0.55 to 0.74 kg CO_2e/kg pig growth. Reckmann et al. (2013) report 3.22 kg CO_2e/kg live weight for typical production in Germany, while Noya et al. (2017) report 3.42kg CO_2e/kg live weight for pigs produced in Spain. Besides the difference in functional unit, the Eriksson study is based on different system boundaries that exclude sows and farrowing (Eriksson et al., 2005). Clearly, attempts to convert these values to the same units of measure will increase uncertainty, and may result in invalid conclusions. The wide range of management systems used in the swine industry adds to the complexity of performing comparisons among studies.

In a Swedish study, Eriksson et al. (2005) investigated the impact of feed choice on the environmental impact of growing-finishing pig production using LCA methodology. In this study, the functional unit was 1 kg of pig growth, as an average of the growth from 29 to 115 kg. The rations evaluated were the use of imported soybean meal, locally grown peas and rapeseed cake, or peas and rapeseed cake supplemented with synthetic amino acids. The overall system included crop production and the processes involved in production of growing pigs including manure handling. Similar to the findings of Dalgaard et al. (2007), Eriksson et al. (2005) reported GHG emissions from feed production was greater than that of the pig housing system, including the manure emissions. This equated to 60–66% of the GWP for feed production and 34–40% for pig housing sub-system. Manure storage, specifically methane release under anaerobic conditions, was the source of greatest importance in the pig housing sub-system. The result of feeding pigs with soybean meal was 1.5 kg CO_2e/kg pig growth (with a significant contribution from transportation of the imported feed), followed by 1.4 kg CO_2e/kg pig growth for pigs that were fed peas and rapeseed cake supplemented with synthetic amino acids. Pigs fed with locally grown peas and rapeseed cake had the lowest impact with 1.3 kg CO_2e/kg pig growth. Pigs fed on diets supplemented with synthetic amino acids had an overall positive effect on the impact categories of eutrophication, and lower emission of ammonia may also contribute to lowering GHG emissions, as ammonia can give rise to indirect nitrous oxide formation (Eriksson et al., 2005; Apple et al., 2013; Ogino et al., 2013; Garcia-Launay et al., 2014; Maxwell et al., 2015).

Vergé et al. (2009) using the IPCC tier 2 protocol for GHG calculations reported that the Canadian swine industry increased its total GHG emissions by 54% from 1981 to 2001, attributable to the dramatic increase in swine production. At the same time, because of increased efficiency in swine production, the quantity of GHG emissions on a live weight basis as reduced from approximately 3 kg CO_2e kg^{-1} live market weight to 2.3 kg CO_2e kg^{-1}. Hog production in Alberta was evaluated by LCA and the GHG emissions were estimated to be 4 kg CO_2e kg^{-1} of meat (ready for shipping from the processing facility). The main N_2O emissions in the pork production life cycle arise from applied fertilizer N and crop residue in some cases (SNC-LAVALIN Agro and Ciraig, 2009; Vergé et al., 2009).

A US study (Thoma et al., 2011; Castellini et al., 2012) reported that the contribution of emission burdens for each stage was approximately: a) 9.8%: sow barn (including feed and manure handling); b) 53.3%: nursery to finish (including feed and manure handling); c) 7.0%: processing (5.65%) and packaging (1.34%); d) 7.3%: retail (electricity and refrigerants); e) 22.7%: the consumer (refrigeration, cooking and methane from food waste in landfill). A more detailed breakdown of the supply chain emissions is shown in Fig. 1 and 2.

Groen et al. (2016) demonstrate that feed conversion efficiency, methane emissions from manure management and crop yields are the most sensitive parameters affecting

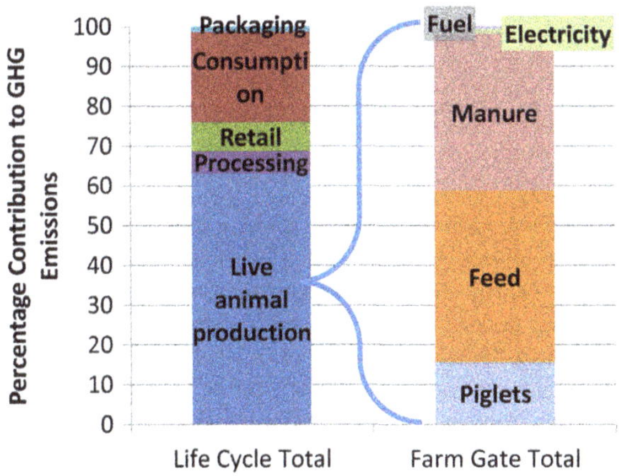

Figure 1 Life cycle contribution analysis for GHG emissions from US swine production and consumption. Food waste/loss included. Reproduced from Thoma et al. (2011).

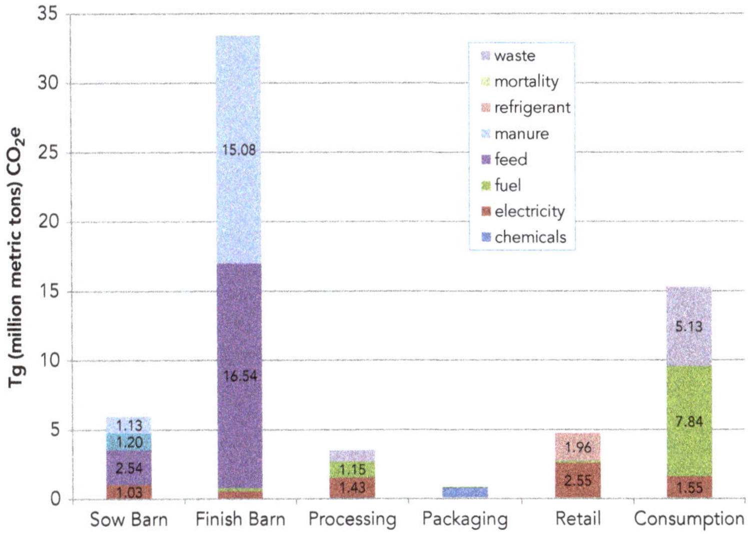

Figure 2 US domestic swine sector cumulative contribution to GHG emissions in the United States. Adapted from Castellini et al. (2012).

GHG emissions from swine production systems in the Netherlands. These results are likely applicable in most large-scale modern swine production systems. The implications of this work are that these factors represent the most promising points for intervention in the system with regard to opportunities for mitigation of GHG emissions. Numerous other studies support the importance of feed production and manure emissions as key drivers of emission in swine production (Perez, 2009; Reckmann, Traulsen and Krieter, 2012; Macleod et al., 2013; Reckmann and Krieter, 2015; Noya et al., 2016).

3 Environmental emissions and impacts at farm level: emissions to water and air

3.1 Emissions to water

Emissions

Environmental concerns with regard to swine production and water quality arise from two principal sources: manure nutrient run-off from crop production and manure management. Broadly speaking nutrient loss contributes to both terrestrial and aquatic eutrophication and acidification. In majority of continental freshwater systems phosphorus is the limiting nutrient, and therefore emissions of phosphorus to aquatic ecosystems contribute mainly to freshwater eutrophication, while nitrogen is generally regarded as the limiting nutrient in marine systems (Howarth and Marino, 2006; Conley et al., 2009). There are, of course, locations in which freshwater systems are nitrogen limited. In addition, numerous geospatially specific characteristics collectively determine the fate of applied nutrients. These factors include the soil type, slope and aspect as well as soil management including the type and timing of nutrient application. It is for these reasons that watershed-specific knowledge is necessary to fully evaluate the effects of nutrient emissions on these ecosystems. Tomer et al. (2016) documented differences in the edge-of-field phosphorous losses for two fields receiving inorganic fertilizer or swine manure, but otherwise managed identically, over a 11-year period. The fields were under a 3-year corn-corn-soy rotation and one field received swine manure for each corn planting. The majority of phosphorous loss was associated with moderate rainfall events (30–60 mm rainfall) with resulting average annual run-off rates 1.8 and 1.05 kg ha^{-1} yr^{-1} for manure and inorganic treatments, respectively.

While the main emissions to surface waters arise from run-off after field application of manure as fertilizer, there may also be emissions to groundwater associated with on-farm storage (DeSutter, Pierzynski and Ham, 2005). Elevated levels of emissions were reported up to 1 m beneath a 25-year-old decommissioned swine lagoon in Kansas ranging from 2000 to 400 mg kg^{-1} from between 0.05 and 1 m beneath the lagoon bottom (DeSutter, Pierzynski and Ham, 2005). Burkholder et al. (2007) summarize several studies in which nitrate concentrations in surface and subsurface drainage up to 30 mg/L, even when swine effluent is applied at recommended rates. These levels can exceed those known to cause algal blooms.

Mitigation options

Management practices are known to be effective at controlling nutrient losses from fields and thus decreasing eutrophication potential in surface and marine waters (Sharpley and

Wang, 2014). Specifically, because the primary pathway that swine production leads to eutrophication is through nutrient run-off following field application, the 4Rs are an excellent framework for understanding mitigation potential. These are right source, right time, right place and right amount. The last three are relevant for manure, as the source is specified. The timing of application may have two dimensions: short-term and mid-term. Short-term timing is relevant with regard to rainfall, as it is well documented that the longer the period between manure application and rain, the lower the run-off of nutrients (Sharpley and Wang, 2014; Tomer et al., 2016). Mid-term timing refers to application of nutrients at the time that will lead to optimal plant uptake while reducing the chances for losses. Interestingly, a recent meta-analysis did not find a correlation between timing (fall vs. spring) of nutrient application and also demonstrated increasing loss with precipitation up to a certain application rate, beyond which the fraction of applied N lost plateaued (Christianson and Harmel, 2015). The authors attributed this plateauing to a dominating effect of over application. However, a study by Adeli et al. (2003) showed increased Bermuda grass productivity in a comparison of swine manure application in September compared to October suggesting that late season application was poorly timed to crop needs.

Manure management options have been compared using LCA (Lopez-Ridaura et al., 2009; Prapaspongsa et al., 2010; Petersen et al., 2013; García-González et al., 2016). Prapaspongsa et al. (2010) compared several combinations of treatment, storage and land application and demonstrated that the primary impact categories of concern were global warming, aquatic eutrophication, respiratory inorganics and terrestrial eutrophication based on relative contribution to single score impacts from multiple impact assessment methodologies. They report that technology combinations that maximize GHG reduction while conserving nutrient content are the most effective at mitigating impacts, for example combining anaerobic digestion with energy recovery and deep injection for field application. García-González et al. (2016) showed that numerous, both on-farm and centralized off-farm, treatment options, including mechanical separation, digestion, composting all reduced GHG emissions between 53 and 90% compared to the baseline scenario.

In addition to manure management, ration manipulation has also been suggested as a means to reduce nutrient excretion, and thus to ultimately reduce emissions to the environment. Several authors have evaluated the effect of reduced crude protein in the ration, and shown that significant reductions in excretion of nitrogen can be achieved, which subsequently lower the loss of nitrogen to the environment (Mosnier et al., 2011; Garcia-Launay et al., 2014; Webb et al., 2014; Burek et al., 2015). Nonetheless, it remains important to maintain the feed quality under these management scenarios. In a study evaluating effects of lowering crude protein and substituting crystalline amino acids, Apple et al. (2013) observed improvement in growth performance at lower levels of substitution, but decreased growth performance and increased back fat at the highest levels of amino acid substitution. Results of this study were also consistent with other findings that reducing the crude protein content of grower and finisher diets elevates intramuscular fat content from 14 to 65%, whereas the reductions in intramuscular fat content or marbling anecdotally associated with feeding ractopamine are more than likely a response to dietary lysine levels in excess of 0.8%. In all these studies, diets were formulated on an ME basis and as soybean meal is reduced in diets, the Lys/NE is actually decreased which may explain some of the increase in fat deposition in pigs fed with ME-based diets formulated by decreasing soybean meal and including high levels of synthetic amino acids.

3.2 Emissions to air

Emissions

The major emissions to air from swine production at farm level are GHGs, ammonia and odorous volatile compounds, hydrogen sulphide in particular (Lim et al., 2003; Ni et al., 2007; Schauberger et al., 2013; Philippe and Nicks, 2015; Phillips et al., 2016). Ammonia emissions directly contribute to both eutrophication and acidification, and in addition play a role in secondary particulate formation, with potential impacts to human health. Deposition of these ammonia emissions to the soil enter the nitrification-denitrification cycle and may be converted to nitrous oxide, resulting in an indirect contribution to GHG emissions (Arogo, Westerman and Heber, 2004; Dong et al., 2006). In general, ammonia emissions are affected by several factors including the number of animals and their weight; housing type coupled with manure storage and management; crude protein content and nitrogen excretion rates; and environmental conditions (Arogo, Westerman and Heber, 2004).

Phillips et al. (2016) reported ammonia emissions, per animal unit (AU = 500 kg live weight) of 39.1 to 52.2 g AU^{-1} day^{-1} in winter and summer, respectively, from deep-litter sheds with or without external stockpiling, representing more than a 60% decrease in emissions compared to conventional housing with anaerobic lagoons. Philippe and Nicks (2013) reported increased ammonia emissions with increasing temperature and ventilation – increasing from 12.8 to 14.6 g NH_3 $pig^{-1}day^{-1}$, as temperature increased from 17 to 28°C, which is similar to the findings of Phillips et al. (2016).

Philippe et al. (2007) reported on ammonia and gas emissions from different fattening pig housing systems and showed significant increases of emissions from animals housed on deep straw beds compared to slatted floors. They reported approximately double the emissions per pig per day for ammonia and nitrous oxide and roughly similar emissions for methane and carbon dioxide. Accounting for the average weight and animal population in each system over a four month grow out period, the slatted floor system resulted in approximately 46 g NH_3 AU^{-1} day^{-1}, while the litter system resulted in approximately 97 g NH_3 AU^{-1} day^{-1}. These results are similar to those of a study reported by Nicks et al. (2004), for fattening pigs raised on deep litter with straw compared to sawdust; they reported 12–13 g NH_3 pig^{-1} day^{-1}.

Several studies have also evaluated the emission of odorous compounds associated with swine production facilities (Pelletier et al., 2004; Schauberger et al., 2012, 2013; Kafle and Chen, 2014; Webb et al., 2014). Not surprisingly, both the number of pigs and ventilation rate have a significant effect on the concentration of particles, and compounds released to the air. While LCA does not generally consider odour in its impact assessment (Peters et al., 2014), the release of odorous compounds can have an effect on animal well-being, and therefore on production efficiency, which is an important factor in LCA. Furthermore, facility siting and community acceptance are often strongly related to these emissions (Hribar, 2010; Schauberger et al., 2012), thus a full sustainability assessment should at least qualitatively consider odour emissions.

Mitigation options

A number of technical solutions exist for reduction of air emissions associated with swine production (Ndegwa et al., 2008; Webb et al., 2014; Van der Heyden, Demeyer and Volcke, 2015). Swine manure applied to land by injection techniques significantly reduces ammonia emissions: 65–80% reduction for trailing shoe and open slot injection machines, respectively, and up to 90% reduction by immediate incorporation by plough; however, even a few days delay in ploughing significantly reduces the effective emission abatement (Webb et al., 2010). Ammonia emissions are also reduced from housing systems with animals fed on lower crude protein in the ration (Hayes et al., 2004; Le, Aarnink and Jongbloed, 2009; Webb et al., 2014; Hou, Velthof and Oenema, 2015). Several studies have assessed the use of biofilters, scrubbers and other technologies for mitigation of air emissions (Ndegwa et al., 2008; Van der Heyden, Demeyer and Volcke, 2015; Liu et al., 2016).

4 Environmental emissions and impacts at farm level: pathogenic microbes, antibiotic resistance and pharmaceuticals

4.1 Emissions

In some regions there are additional concerns regarding the impact associated with emissions of pathogens, including antibiotic-resistant microbes, and pharmaceuticals (Hribar, 2010; Landers et al., 2012; EPA-OW and U.S. Environmental Protection Agency, 2013; Laxminarayan et al., 2013). There has been growing concern for number of years that the use of antibiotics on a routine basis in livestock production is increasing pressures towards larger numbers of antibiotic-resistant human pathogens (Koike et al., 2007; Thorne, 2007; Hribar, 2010; Landers et al., 2012; Center for Disease Dynamics Economics & Policy, 2015; Van Boeckel et al., 2015). It is anticipated that global antimicrobial use in food production will increase by 67% between 2010 and 2030, with significant increases in China (Van Boeckel et al., 2015). Thus, despite the increase in oversight, such as the Veterinary Feed Directive (U.S. Food and Drug Administration, 2013), effective 1 January 2017 in the United States, and regulations that ban the use of growth-promoting antimicrobial use in the European Union (*Regulation (EC) No 1831/2003*, 2003), it is anticipated that continued use of antibiotics for therapeutic treatment of unhealthy animals will continue to place pressure on livestock producers to carefully manage antibiotic usage (Van Boeckel et al., 2015). It should also be noted that this use of antibiotics for treatment of disease is not only important from environmental (by imparting resource use efficiency by returning animals' healthy status) and economic (mortality and morbidity reduce efficiency and therefore profit margins; Stone et al., 2011) perspectives, it is also an animal welfare issue (Struve, 2011).

Tong et al. (2009), in a study targeting 13 common antimicrobial compounds used in China, reported ng L^{-1} levels in ground and surface water and ng to mg L^{-1} levels in swine wastewater, while He and others reported the presence of antibiotic-resistant genes as well as antibiotics and soils receiving treated swine manure as fertilizer (He et al., 2016). Mackie et al. (2006) and Koike et al. (2007) reported on the presence of tetracycline and tetracycline-resistant genes in groundwater adjacent to swine lagoons in the United States. Awad et al. (2014) reported tetracycline and sulfonamide concentrations in water, sediment and soil near a swine manure composting facility, ranging from 2 to 200 ng L^{-1} for tetracyclines, but generally below 10 ng/L for sulfonamides. The release of veterinary antibiotics to the environment arises from the fact that they are not completely metabolized by the animal, and thus between 30 and 90% of the administered antibiotics are excreted in the faeces (Awad et al., 2014).

The USDA reported in 2006 and 2012 that over 90% of *Escherichia coli* isolates from swine manure were resistant to at least one antimicrobial agent (tetracycline), and approximately one-third of isolates were resistant to Sulfisoxazole (Services and USDA Veterinary Services, 2009; USDA, 2016). Sayah et al. (2005) reported *E. coli* isolates from swine faecal samples from had higher levels of multidrug resistance than isolates from human or wildlife sources in a Michigan watershed. Haley and others reported on the prevalence and antimicrobial susceptibility of *Salmonella* in swine faecal samples (Haley et al., 2012). The overall level of antimicrobial resistance increased between 2000 and 2006, with 20.4% and 38.1% of isolates being susceptible to all antimicrobials in 2006 and 2000, respectively. However, they reported that the proportion of *Salmonella* isolates resistant to at least three antimicrobials was similar in both years, just greater than 50%. Similar results have been observed in China where 41% of *Salmonella* isolates were susceptible to all antimicrobials tested, 48% were resistant to at least one antimicrobial and over 34% were resistant to three or more antimicrobials (Kuang et al., 2015).

Luby and others studied the transport of antibiotic resistance genes, using quantitative polymerase chain reaction techniques, in drained fields either receiving or not receiving swine manure from animals fed with tylosin (Luby, Moorman and Soupir, 2016). They reported significant differences in resistant gene concentrations in drainage water from fields receiving swine manure – interestingly reporting no difference between tillage practices on the transport of genes. While the concentrations of the resistance genes dropped to background levels by the following season, the authors suggested that the increased release of the genes from the field application sites poses increased risk off-site (Luby, Moorman and Soupir, 2016).

4.2 Mitigation options

Vonderohe et al. (2015) reported on a comparative study of animals grown with and without antibiotics. Seven hundred and twenty-four mixed sex pigs were placed in 11 rooms. Control animals received diets containing antibiotics and were treated with injectable antibiotics when deemed necessary. Antibiotic-free animals did not receive injectable antibiotics and were fed on diets with antibiotic alternatives (acidification plus Zn; direct-fed microbials; oregano). Sick animals that were unresponsive to antibiotic alternatives were removed from the experiment. Weight gain was measured for each dietary phase, and daily records of mortality and morbidity were taken. Nursery pigs from the control group consumed more feed and gained weight more rapidly than antibiotic-free animals, and thus had a similar feed conversion ratio (FCR). Both groups had similar performance in the grower phase such that the control pigs were slightly heavier at the beginning of the finisher phase. During the finisher phase the antibiotic-free pigs grew 3% faster ($P < 0.01$) and had a 6% better FCR ($P < 0.001$). The net result was no overall effect on average daily gain between treatments, but there was a trend for increased feed consumption and slightly lower FCR for control animals compared to antibiotic-free animals were removed from the study ($n = 30$, 8.3%) compared to control animals ($n = 11$, 3.0%). Thus, the economics should be carefully weighed, as fewer animals may be marketed from antibiotic-free systems. Gong et al. (2014) reviewed a number of studies in which Chinese herbal formulations are substituted for antibiotics in swine rations and reported numerous instances of beneficial effects. Further work in this area appears to be justified given the potential for acceptable performance of animals fed antibiotic alternatives.

5 Environmental emissions throughout the life cycle

Figure 3 presents a schematic of system boundaries commonly used in LCA. Generally, more complete, and fully ISO-compliant studies encompass the full system – from field to fork. However, many LCAs focus on either cradle-to-farm gate or processor gate. Wealth management is included as a potential secondary function of the production system because it is common in many smallholder situations in developing countries (Herrero et al., 2009; Raney et al., 2009; Weiler et al., 2014). A recent review by McAuliffe et al. (2016) provided an excellent summary of LCA applied to pig production systems. The review covers studies across numerous geographies, some focusing on a cradle-to-farm gate analysis and others focusing on specific technologies, for example manure management or precision feeding. Table 1 presents a summary of primarily cradle-to-farm gate studies. A significant challenge in agricultural LCA is that different practitioners can choose different functional units, system boundaries, allocation procedures as well as other methodological choices and may have access to different data sets. This situation makes comparisons among published LCA challenging. The categorization provided by McAuliffe et al. (2016) grouped the studies under review, broadly, by feed production, animal husbandry and manure management, moves a step in the direction of enabling comparisons among studies.

There are relatively few studies which encompass the full cradle-to-grave life cycle of pork production and consumption (Blonk et al., 2008; Kingston, Fry and Aumonier, 2009; Castellini et al., 2012). For carbon footprint, Castellini and Kingston reported between 68 and 72% of the full life cycle impacts occur by the farm gate, while Blonk et al. reported approximately 88% of impacts occur by the farm gate, as shown in Fig. 4.

Similarly, Kingston et al. (2009) reported that the acidification and eutrophication contribution for the post-farm gate supply chain accounts for only 4% of each impact category, with the primary contribution arising from feed production and manure management.

6 Case studies

6.1 Gestation management

Gestation stalls and pens: Use of gestation stalls for housing pregnant sows has been criticized by animal welfare groups because the stalls restrict the animals' movement (Lammers et al., 2007; McGlone, 2013). Although Europe, Australia and New Zealand as well as several states in the United States have banned the use of sow gestation stalls, their use remains widespread throughout much of the rest of the world. There remain arguments in the United States regarding the shift in production away from gestation stalls, and it is important to understand that even small production changes might have large environmental impacts. If changes to management practices are made based on a single criterion, it could result in undesirable outcomes for other metrics. Therefore, assessment of impact trade-offs associated with change to production practices is necessary to ensure

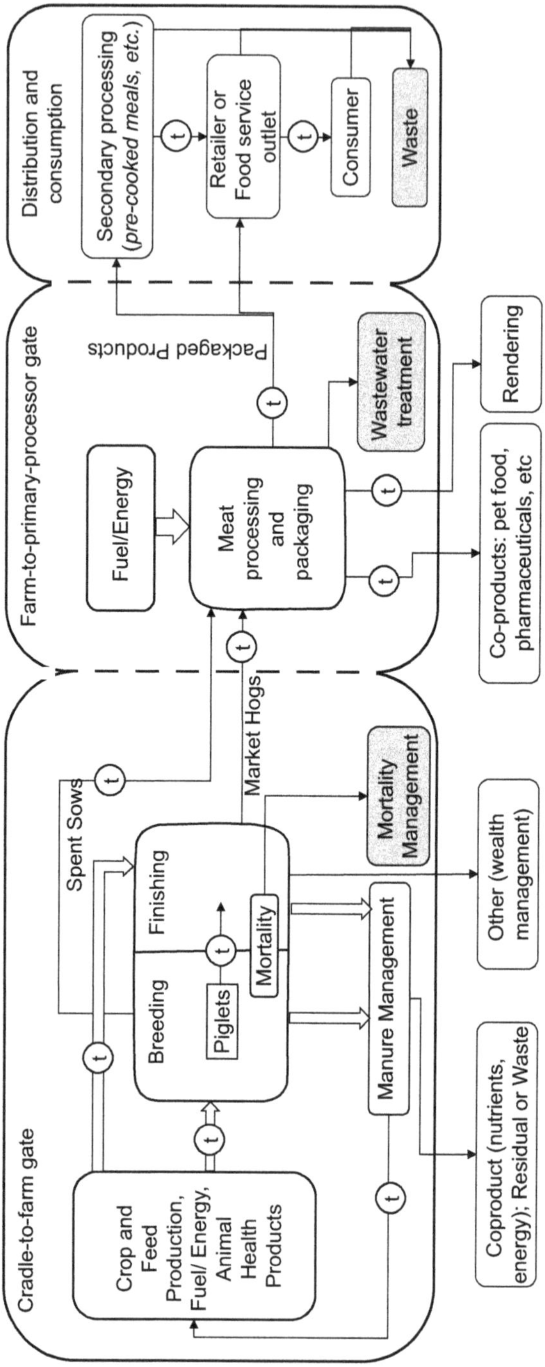

Figure 3 Schematic of system boundaries used in LCA.

Table 1 Summary of reported LCA for swine production

Study	Scope	Geography	Functional unit	GWP	AP	EP
Basset-Mens and Van der Werf (2005)	Cradle-to-farm gate	France	1 kg live weight	2.3 kg CO_2e	0.0435 kg SO_2e	0.0208 kg PO_4e
Williams et al. (2006)	Cradle-to-farm gate	UK	1000 kg carcass deadweight	6400 kg CO_2e	394 kg SO2e	100 kg PO_4e
Dalgaard et al. (2007)	Cradle to delivery to Britain port	Denmark	1 kg pork	3.6 kg CO_2e	45 g SO_2e	232 g NO_3e
Blonk et al. (2008)	Cradle to consumption	Netherlands	1 kg packaged pork	4.5 kg CO_2e	N/A	N/A
Kingston et al. (2009)	Cradle to consumption	UK	1 kg pork consumed	8.6 kg CO_2e	190 g SO_2e	57 g PO_4e
Perez (2009)	Cradle-to-farm gate	UK	1000 kg live weight	3284 kg CO_2e	43.78 kg SO_2e	192.6 kg NO_3e
Perez (2009)	Cradle-to-farm gate	Mexico	1000 kg live weight	5498 kg CO_2e	75.3 kg SO_2e	585.9 kg NO_3e
Vergé et al. (2009)	Cradle-to-farm gate	Canada	1 kg live weight	2.31 kg CO_2e	N/A	N/A
Wiedemann et al. (2010)	Cradle to abattoir gate	Australia	1 kg carcass weight	5.5 kg CO_2e	N/A	N/A
Halberg et al. (2010)	Cradle-to-farm gate	Denmark	1 kg live weight	3.32 kg CO_2e	61.4 g SO_2e	381 g NO_3e
Nguyen et al. (2010)	Cradle-to-farm gate	EU	1 kg slaughter weight	4.81 kg CO_2e	N/A	N/A
Pelletier et al. (2010)	Cradle-to-farm gate	USA	1 kg live weight	2.47 kg CO_2e	N/A	15.9 g PO_4e
Nguyen et al. (2011)	Cradle to abattoir gate	Denmark	1 kg pork abattoir	3.1 kg CO_2e	56 g SO_2e	243 g NO_3e
Devers et al. (2012)	Cradle to delivery to Belgium port	Belgium	1 kg carcass weight	2.55 kg CO_2e	0.039 kg SO_2e	0.022 kg PO_4e
Castellini et al.(2012)	Cradle-to-grave	USA	4 oz boneless pork	1.02 kg CO_2e	N/A	N/A
Dolman et al. (2012)	Cradle-to-farm gate	Netherlands	100 kg live weight	546 kg CO_2e	5.3 kg SO_2e	61.4 kg NO_3e
Reckmann et al. (2013)	Cradle to abattoir gate	Germany	1 kg slaughter weight	3.22 kg CO_2e	57.1 g SO_2e	23.3 gPO_4e
Dourmad et al. (2014)	Cradle-to-farm gate	EU	1 kg live weight	2.3 kg CO_2e	44 g SO_2e	18.5 g PO_4e
Jacobsen et al. (2014)	Cradle to abattoir gate	Belgium	1 kg deboned	4.8 kg CO_2e	N/A	N/A
Muller (2015)	Cradle-to-farm gate	South Africa	1 kg live weight	1.96 kg CO_2e	14.7 g SO_2e	3.23 g PO_4e
Winkler et al. (2015)	Cradle to abattoir gate	Austria	1 kg carcass weight	4.75 kg CO_2e	61.5 g SO_2e	381 g NO_3e
Rougour et al. (2015)	Cradle to consumption	Netherlands	1 kg carcass weight	2.97 kg CO_2e	N/A	31.1 g PO_4e
Noya et al. (2016)	Cradle-to-farm gate	Spain	1 kg live weight	6.7 kg CO_2e	N/A	N/A

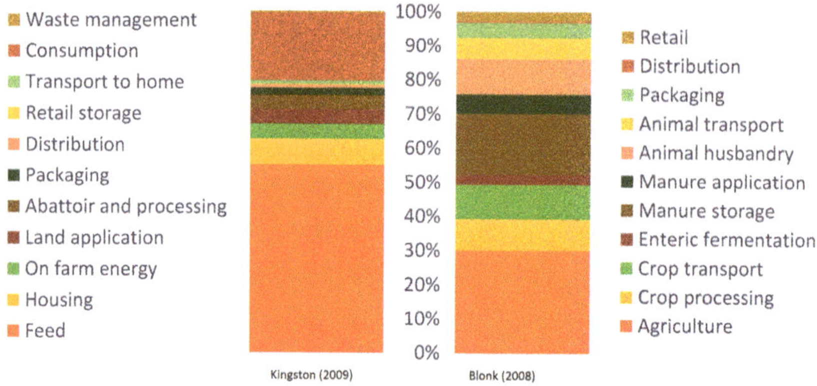

Figure 4 Cradle-to-grave contribution analysis from two studies. Each study divided the full supply chain into several stages, beginning with crop production and ending with either consumption or retail. The relative contribution of each stage is presented in the same order as the respective legend. In addition, moving up the legend generally follows the supply chain from upstream to downstream. Differences in the definitions of the various stages precludes a direct evaluation of the differences shown (Blonk et al., 2008; Kingston, Fry and Aumonier, 2009).

continual progress towards more sustainable production. In this case study, we examine the effect of changing from gestation stalls to group pens on climate change, water and energy use. The functional unit for the assessment was 1 kg of live weight market hog at the farm gate.

Assessment and comparison of alternative practices involved a two-step process. In the first step, nursery, grow-finish and sow barns were simulated using the Pig Production Environmental Footprint (PPEF) model (National Pork Board, 2017). The PPEF uses the growth and performance model developed by the National Research Council (NRC) to predict growth and feed consumption of pigs (National Research Council, 2012). The PPEF model simulates pig growth and performance (FCR), feed intake and water consumption, electricity and natural gas use in each barn, manure handling and GHG emissions over an annual cycle. Animal growth, resource consumption and emissions to the environment were simulated over a ten-year period and reported as an annual average. Simulation results were used as life cycle inventory in the second step: LCA using SimaPro V7.3 (Pre' Consultant, The Netherlands).

The NRC growth equations were calibrated to achieve 42 days in the wean-to-feeder phase and 114 days in feeder-to-finish phase. Hourly data from a Typical Meteorological Year (Wilcox and Marion, 2008) was used and the PPEF model simulated the inside barn temperature based on heating and/or cooling control systems. A swine nutrition specialist provided a recommended multi-phase diet. Dietary requirements were separately modelled for gestating sows, lactating sows and wean-to-finish pigs. The finishing diet included 10 ppm ractopamine in the final weeks. *Ad libitum* access to water was modelled and water consumption calculated in the model. Pig mortality was set to 2.9 and 3.9% for wean-to-feeder and feeder-to-finish barns, respectively (Table 2). All pigs were marketed at 125 kg.

Use of pens was assumed for gestation period only; farrowing stalls were used for both group and individual stall scenarios. The differences between productivity of sows housed in gestation stalls versus pens were number of live births, litter size, pre-weaning mortality rate and piglet birth weight. An analysis of data obtained from peer-reviewed articles

Table 2 Sow barn parameters used for uncertainty analysis for gestation stalls and group housing

Parameter	Units	Minimum impact	Maximum impact
Gestation stalls			
No. of piglets per litter surviving to weaning	Animals	9.09	7.31
No. of pigs dying per litter before weaning	Animals	1.91	3.59
Age at which piglets are removed from barn	Days	21	21
No. of days between piglet removal and insemination	Days	16	16
Piglet birth weight	kg	1.53	1.31
Gestation pens (group housing)			
No. of piglets per litter surviving to weaning	Animals	9.03	6.33
No. of pigs dying per litter before weaning	Animals	1.77	3.18
Age at which piglets are removed from barn	Days	21	21
No. of days between piglet removal and insemination	Days	14	14
Piglet birth weight	kg	1.64	1.48

Bandekar, 2015.

(Bates, Edwards and Korthals, 2003; McGlone et al., 2004; Anil et al., 2005; Harris et al., 2006; Jansen et al., 2007; Lammers et al., 2007) was performed to prepare data input requirements for both scenarios (Table 2). Pre-weaning mortality of 14 and 15% (Lammers et al., 2007) was used for stalls and group pens, respectively.

Approximately 72% of group-housed sows, compared to 68% of sows in stalls returned to oestrus within seven days post-weaning (Bates, Edwards and Korthals, 2003). Further, 94% of inseminated sows became pregnant in pen systems, while 89.4% became pregnant in stall systems. This was modelled by changing the average days between piglet removal and insemination to 14 and 16 for group and stall housing, respectively.

The difference in required infrastructure, based on differences in housing design for the two systems was included. Based on our modelling of the space requirements for sow in stalls compared to pens, the barn infrastructure requirements for pens are 65% larger. This leads to larger barns, and thus increases use of concrete, lumber and other construction materials. However, there is a reduced quantity of galvanized metal for stall construction that partially offset these impacts. In addition, the barn infrastructure is amortized over the 10-year life of the barn, leading to relatively small contribution in either system.

Three scenarios were evaluated using typical production in Iowa, Oklahoma and North Carolina to represent differences in manure management and climate. As presented in Table 3, using group gestation pens rather than individual stalls resulted in a decrease in GWP, energy use and water use compared to the baseline. GWP, energy use and water use decreased by approximately 0.9, 1.5 and 0.9%, respectively, when pen gestation were used. There were slight variations in the estimated improvements between the different locations; however, these differences are not considered to be significant, as shown through Monte Carlo simulations in Fig. 5. The relatively tight clustering of the data (compared to the wide baseline value) shows that while there is a directional improvement with

Table 3 Percentage changes in impact category metrics for gestation pen scenarios compared to the baseline use of individual stalls, reported per kg of live weight at the farm gate

Management strategy	Impact category	Iowa	Oklahoma	North Carolina
Gestation pens	IPCC GWP 100a (kg CO_2e)*	−0.973	−0.886	−0.782
	Non-renewable energy (MJ)	−1.499	−1.523	−1.537
	Water (m^3)	−0.972	−0.922	−0.923

Bandekar, 2015.
* Intergovernmental Panel on Climate Change 100-Year Global Warming Potential.

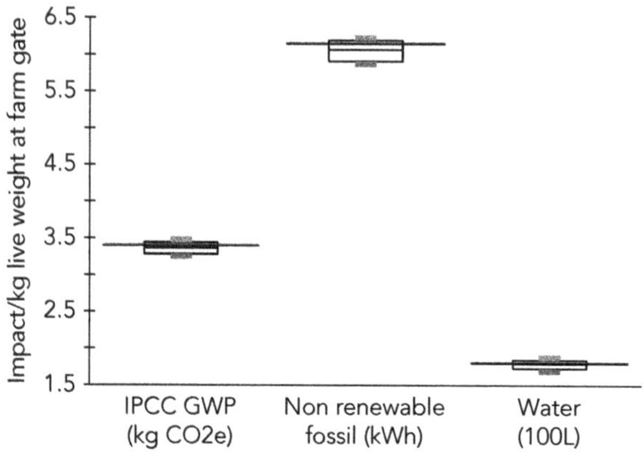

Figure 5 Monte Carlo results for comparison of gestation pens compared to gestation stalls. Boxes represent 25th and 50th percentiles of 1000 runs, grey bars represent 10th and 90th percentiles. The black horizontal bar is the baseline scenario with the use of individual stalls.

pens, that specific management may lead to either positive or negative results. The NRC growth equations for gestating sows include a smaller impact of temperature on MMER for group-housed sows, which reduces maintenance energy requirements, leaving more energy available for foetus development and weight gain. Therefore, better sustainability indices observed in this scenario were the result of better performance of sows, lower feed consumption and lower manure emissions. Use of gestation pens reduced overall feed consumption by sows by 1.92%. It also reduced CH_4 emissions by 2.9% and N_2O emissions by 2.1%.

6.2 Comparison of US and EU swine production

In a recent comparative study the production practices in the United States and EU, the importance of accounting for land transformation was highlighted (Boles et al., 2015). Information on production practices from various states and countries in the EU was gathered from publicly available sources including FAOSTAT, EUROSTAT, BPEX and NASS. These data were used to establish scenarios which were evaluated using the PPEF and aggregated to national or EU level. The influence of land transformation arises primarily from the sourcing of soymeal from South America. For the EU scenarios, EU commodity feed sources were assumed to be used for each ration component except for soy and

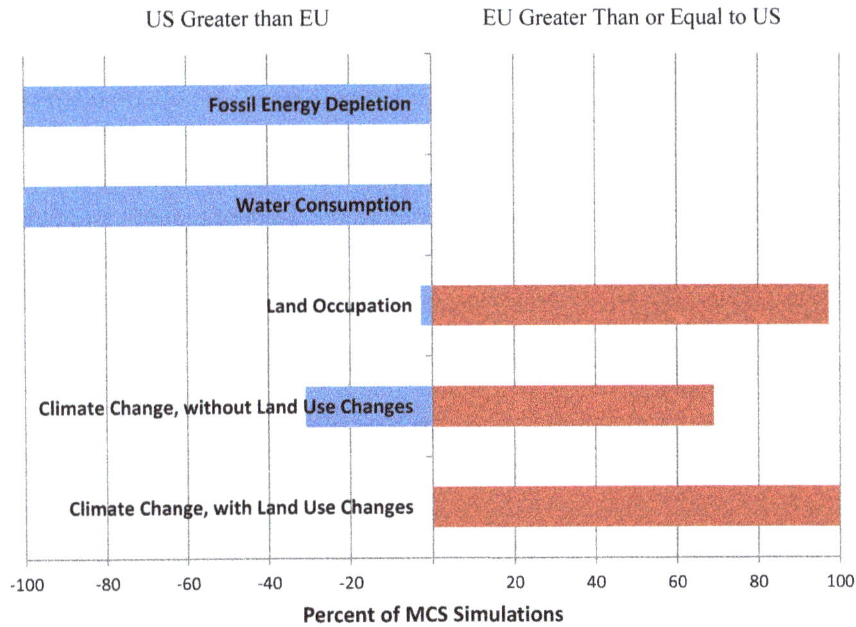

Figure 6 Uncertainty analysis results for the US and EU average production scenarios including GHG emission impacts from land use changes.

palm oil. For each European production scenario, soy was assumed to be sourced from Europe (10%), Brazil (50%) and Argentina (40%) (Blonk Consultants, 2017). Palm oil used in the EU production scenarios was assumed to be sourced from Malaysia. In the EU country scenarios, 10% of the EU corn blend was imported from the US corn industry (Blonk Consultants, 2017). North Carolina swine producers imported between 16 and 26% of soybeans used in their rations between 2011 and 2014 (U.S. Census Bureau, 2015). Monte Carlo simulations show that there are differences in the impacts of production in the two regions as shown in Fig. 6.

7 Summary and future trends

In addition to being a major source of high-quality protein, swine play other important roles, including income and wealth management in smallholder systems throughout the world (McGlone, 2013; Weiler et al., 2014). As global production continues to increase there will be an increasing need to identify opportunities to increase feeding efficiency, potentially through a combination of genetic improvements and utilization of feed that does not place additional pressure on food supplies. Economic pressures are causing a trend to larger industrialized production in all the largest pork-producing countries in order to take advantage of economies of scale. At the same time, there is increasing concern for animal welfare as a social indicator of sustainability performance and efforts will be needed to assure the consuming public that these large intensive production systems adequately address these concerns. There is no scientific basis suggesting that animal welfare is necessarily better or worse in intensively managed indoor versus extensive outdoor production systems (McGlone, 2013), but that the specific management at individual production facilities may result in different animal welfare outcomes. In the future, animal welfare issues are likely to become increasingly important (Anon., 2005).

There are also pressures on livestock production, including swine, related to feed versus food production. Because swine are monogastrics, their ability to extract nutrients from low-quality feed is less than that of ruminants, but swine are certainly capable of consuming a wide variety of alternative feeds which have little or no value in the human food supply. Further research into the issue of feed/food competition through continual improvement in nutrition and genetics, such as those reported by Mottet et al. (2017) and van Zanten et al. (2016) reported on the possibility of protein sourced from insects in swine rations will be important.

As a non-ruminant, enteric methane emissions are limited, but methane, and potentially nitrous oxide depending upon the manure management system, are significant contributors to overall GHG emissions. Manure management and feed production are the primary drivers for water quality impacts, primarily eutrophication. Continued research in manure management as well as evaluation of regulatory and economic barriers to the adoption of novel technologies (e.g. the use of anaerobic digesters) is warranted.

Finally, the concerns arising from antimicrobial use in livestock systems must be included in sustainability assessments in the future, and continued work, such as that by Vonderhoe et al. (2015) seeking alternative approaches to maintain high health status in swine herds with minimal use of antimicrobials is important and encouraging.

8 Where to look for further information

Roma et al. (2015) presented a review of LCA methodology of livestock including swine. Macleod et al. (2013) reported on global pig production, particular emphasis was given to the diversity of systems. The Food and Agriculture Organization of the United Nations is hosting the Livestock Environmental Assessment and Performance Partnership is preparing several sector-specific guidance documents for performance of LCA, with a forthcoming document containing swine-sector-specific recommendations (UN FAO, 2017). McAulfie et al. (2016) have presented an overview of LCA of swine systems with particular emphasis given to feed production, on-farm management and full system assessment.

9 References

Adeli, A., Varco, J. J. and Rowe, D. E. (2003) 'Swine effluent irrigation rate and timing effects on bermudagrass growth, nitrogen and phosphorus utilization, and residual soil nitrogen', *Journal of Environmental Quality*, 32(2), pp. 681–86. doi:10.2134/jeq2003.6810.

Anil, L., Anil, S. S., Deen, J. and et al. (2005) 'Evaluation of well-being, productivity, and longevity of pregnant sows housed in groups in pens with an electronic sow feeder or separately in gestation stalls', *American Journal of Veterinary Research*, 66(9), pp. 1630–8. doi:10.2460/ajvr.2005.66.1630.

Anon. (2005) Circa 2015–2025 Visions of the Pork Industry. *National Hog Farmer*. Available at: http://www.nationalhogfarmer.com/mag/farming_circa_visions_pork (Accessed: 31 March 2017).

APHIS and USDA. (2009) 'Escherichia Coli on U . S . Swine Sites — Antimicrobial Drug Susceptibility'. Animal and Plant Health Inspection Services. USDA. https://www.aphis.usda.gov/animal_health/nahms/swine/downloads/swine2006/Swine2006_is_ecoli.pdf.

Apple, J. K., Yancey, W. S., Hollenbeck, J. J. and et al. (2013) 'Effects of amino acid supplementation of reduced crude protein (RCP) diets on LM quality of growing-finishing swine', *Journal of Animal Science*, Volume 9 Supplement 2, p. 102.

Arogo, J., Westerman, P. W. and Heber, A. J. (2004) 'A review of ammonia emissions from confined swine feeding operations', *Transactions of the American Society of Agricultural Engineers*, 46(3), pp. 805–17.

Awad, Y. M., Kim, S. C., Abd El-Azeem, S. A. M. and et al. (2014) 'Veterinary antibiotics contamination in water, sediment, and soil near a swine manure composting facility', *Environmental Earth Sciences*, 71(3), pp. 1433–40. doi:10.1007/s12665-013-2548-z.

Bandekar, P. (2015) *Life Cycle Assessment of Alternative Swine Management Practices*. University of Arkansas.

Basset-mens, C. and van der Werf, H. M. G. (2005) 'Scenario-based environmental assessment of farming systems: The case of pig production in France', *Agriculture, Ecosystems and Environment*, 105(1–2), pp. 127–44. doi:10.1016/j.agee.2004.05.007.

Basset-mens, C., van der Werf, H. M. G., Durand, P. and et al. (2006) 'Implications of uncertainty and variability in the life cycle assessment of pig production systems', *International Journal of Life Cycle Assessment*, 11(5), pp. 298–304. doi:10.1065/lca2005.08.219.

Basset-mens, C., van der Werf, H. M. G., Robin, P. and et al. (2007) 'Methods and data for the environmental inventory of contrasting pig production systems', *Journal of Cleaner Production*, 15(15), pp. 1395–405. doi:10.1016/j.jclepro.2006.03.009.

Bates, R. O., Edwards, D. B. and Korthals, R. L. (2003) 'Sow performance when housed either in groups with electronic sow feeders or stalls', *Livestock Production Science*, 79(1), pp. 29–35. doi:10.1016/S0301-6226(02)00119-7.

Blonk, H., Kool, A., Luske, B. and et al. (2008) *Environmental Effects of Protein-rich Food Products in the Netherlands Consequences of Animal Protein Substitutes*. Available at: http://blonkconsultants.nl/en/upload/pdf/english-summaryprotein-rich-products.pdf.

Blonk Consultants (2017) *Agri-footprint Database*. PJ Gouda, the Netherlands: Blonk Consultants. Available at: http://www.agri-footprint.com/.

Boles, E. C., Sandefur, H. N., Burek, J. and et al. (2015) *Comparative Life Cycle Assessment of Pork Production in the United States and European Union Part 2 : Detailed Assessment*. Final report to National Pork Board by University of Arkansas, Fayetteville, AR.

Boyd, G., Cady, R. A., Wittig, L., Bryan, G., Anderson, D., Sutton, A. L. and Holden, P. (2012) 'A 50-Year Comparison of the Carbon Footprint and Resource Use of the US Swine Herd : 1959–2009', CAMCO. http://research.pork.org/FileLibrary/ResearchDocuments/10-174-Boyd-Camco-final-5-22-12.pdf.

Burek, J., Thoma, G. J., Popp, J. and et al. (2015) Feeding strategies to mitigate cost and environmental footprint of pig production in the US. In: *Waste To Worth 2015 Seattle WA, 30 March – 3 April 2015 – Oral Presentation*. Seattle.

Burkholder, J., Libra, B., Weyer, P. and et al. (2007) 'Impacts of waste from concentrated animal feeding operations on water quality', *Environmental Health Perspectives*, 115(2), pp. 308–12. doi:10.1289/ehp.8839.

Capper, J. L. (2011) 'The environmental impact of beef production in the United States: 1977 compared with 2007', *Journal of Animal Science*, 89(12), pp. 4249–61. doi:10.2527/jas.2010-3784.

Castellini, C., Boggia, A., Paolotti, L. and et al. (2012) Environmental impacts and life cycle analysis of organic meat production and processing. In: S. C. Ricke and et al. (Eds) *Organic Meat Production and Processing*. Oxford, UK: Wiley-Blackwell, pp. 113–36. doi:10.1002/9781118229088.ch7.

Center for Disease Dynamics Economics & Policy (2015) *The State of the World's Antibiotics 2015*. Washington, DC: Centre for Disease Dynamics, Economics & Policy, CDDEP, pp. 1–84. doi:10.1016/S1473-3099(13)70318-9.

Christianson, L. E. and Harmel, R. D. (2015) '4R water quality impacts: An assessment and synthesis of forty years of drainage nitrogen losses', *Journal of Environmental Quality*, 44(6), pp. 1852–60. doi:10.2134/jeq2015.03.0170.

Conley, D. J., Paerl, H. W., Howarth, R. W. and et al. (2009) 'Controlling eutrophication: Nitrogen and phosphorus', *Science*, 323(5917), pp. 1014–15. doi:10.1126/science.1167755.

Dalgaard, R. L. (2007) *The Environmental Impact of Pork Production From a Life Cycle Perspective*. Aarhus University.

Dalgaard, R. L. and Halberg, N. (2005) Life cycle assessment (LCA) of Danish pork. In: *Proceedings from the International Workshop on Green Pork Production*. Paris, France: Life Cycle Assessment (LCA) of Danish pork, p. 165.

Dalgaard, R. L., Halberg, N. and Hermansen, J. E. (2007) Danish pork production: An environmental assessment. *DJF Animal Science*. Available at: http://www.agrsci.dk/djfpublikation/djfpdf/djfhus82ny.pdf.

DeSutter, T. M., Pierzynski, G. M. and Ham, J. M. (2005) 'Movement of lagoon-liquor constituents below four animal-waste lagoons', *Journal of Environmental Quality*, 34(4), pp. 1234–42. doi:10.2134/jeq2004.0296.

Devers, L., Kleynhans, T. E. and Mathijs, E. (2012) 'Comparative life cycle assessment of Flemish and Western Cape pork production', *Agrekon*, 51(4), pp. 105–28. doi:10.1080/03031853.2012.741208.

de Vries, M. and de Boer, I. J. M. (2010) 'Comparing environmental impacts for livestock products: A review of life cycle assessments', *Livestock Science*. Elsevier B.V., 128(1–3), pp. 1–11. doi:10.1016/j.livsci.2009.11.007.

Dolman, M. A., Vrolijk, H. C. J. and de Boer, I. J. M. (2012) 'Exploring variation in economic, environmental and societal performance among Dutch fattening pig farms', *Livestock Science*. Elsevier, 149(1–2), pp. 143–54. doi:10.1016/j.livsci.2012.07.008.

Dong, H., Mangino, J., McAllister, T. A. and et al. (2006) Emissions from livestock and manure management. In: *2006 IPCC Guidelines for National Greenhouse Gas Inventories*. UNFCC. Available at: http://www.ipcc-nggip.iges.or.jp/public/2006gl/pdf/4_Volume4/V4_10_Ch10_Livestock.pdf.

Dourmad, J. Y., Ryschawy, J., Trousson, T. and et al. (2014) 'Evaluating environmental impacts of contrasting pig farming systems with life cycle assessment', *Animal*, 8(12), pp. 2027–37. doi:10.1017/S1751731114002134.

EPA-OW and U.S. Environmental Protection Agency (2013) *Literature Review of Contaminants in Livestock and Poultry Manure and Implications for Water Quality, EPA 820-R-13-002*. Washington, DC: United States Environmental Protection Agency.

Eriksson, I. S., Elmquist, H., Stern, S. and et al. (2005) 'Environmental Systems Analysis of Pig Production – The Impact of Feed Choice (12 pp)', *The International Journal of Life Cycle Assessment*, 10(2), pp. 143–54. doi:10.1065/lca2004.06.160.

FAO (2016) *FAOSTAT-Statistical Database, 2016*. Food and Agriculture Organization of the United Nations. Rome, IT. Available at: http://www.fao.org/faostat/en/#data/QL (Accessed: 3 January 2017).

Foley, J. A., Ramankutty, N., Brauman, K. and et al. (2011) 'Solutions for a cultivated planet', *Nature*, 478(7369), pp. 337–42. doi:10.1038/nature10452.

Fulgoni, V. L., Keast, D. R. and Drewnowski, A. (2009) 'Development and validation of the nutrient-rich foods index: A tool to measure nutritional quality of foods', *The Journal of Nutrition*, 139(8), pp. 1549–54. doi:10.3945/jn.108.101360.

García-González, M. C., Riaño, B., Teresa, M. and et al. (2016) 'Treatment of swine manure : case studies in European' s N-surplus areas', *Scientia Agricola*, 73(5), pp. 444–54. doi:10.1590/0103-9016-2015-0057.

Garcia-Launay, F., van der Werf, H. M. G., Nguyen, T. L. T. and et al. (2014) 'Evaluation of the environmental implications of the incorporation of feed-use amino acids in pig production using Life Cycle Assessment', *Livestock Science*. Elsevier, 161, pp. 158–75. doi:10.1016/j.livsci.2013.11.027.

Gerber, P. J., Hristov, a. N., Henderson, B. and et al. (2013) 'Technical options for the mitigation of direct methane and nitrous oxide emissions from livestock: A review.', *Animal : An International Journal of Animal Bioscience*, 7 Suppl 2, pp. 220–34. doi:10.1017/S1751731113000876.

Godfray, H. C. J., Beddington, J. R., Crute, I. R. and et al. (2010) 'Food security: The challenge of feeding 9 billion people', *Science*, 327(5967), pp. 812–18. doi:10.1126/science.1185383.

Gong, J., Yin, F., Hou, Y. and et al. (2014) 'Review: Chinese herbs as alternatives to antibiotics in feed for swine and poultry production: Potential and challenges in application', *Canadian Journal of Anial Science*, 94(2), pp. 223–41. doi:10.4141/cjas2013-144.

Groen, E. A., van Zanten, H. H. E., Heijungs, R. and et al. (2016) 'Sensitivity analysis of greenhouse gas emissions from a pork production chain', *Journal of Cleaner Production*. Elsevier Ltd, 129, pp. 202–11. doi:10.1016/j.jclepro.2016.04.081.

Gustafson, D., Gutman, A., Leet, W. and et al. (2016) 'Seven food system metrics of sustainable nutrition security', *Sustainability*, 8(3), p. 196. doi:10.3390/su8030196.

Halberg, N., Hermansen, J. E., Kristensen, I. S. and et al. (2010) 'Impact of organic pig production systems on CO2 emission, C sequestration and nitrate pollution', *Agronomy for Sustainable Development*, 30(4), pp. 721–31. doi:10.1051/agro/2010006.

Haley, C. A., Dargatz, D. A., Bush, E. J. and et al. (2012) 'Salmonella prevalence and antimicrobial susceptibility from the National Animal Health Monitoring System Swine 2000 and 2006 studies', *Journal of Food Protection*, 75(3), pp. 428–36. doi:10.4315/0362-028X.JFP-11-363.

Harris, M. J., Pajor, E. A., Sorrells, A. D. and et al. (2006) 'Effects of stall or small group gestation housing on the production, health and behaviour of gilts', *Livestock Science*, 102(1–2), pp. 171–9. doi:10.1016/j.livsci.2005.12.004.

Hayes, E. T., Leek, A. B. G., Curran, T. P. and et al. (2004) 'The influence of diet crude protein level on odour and ammonia emissions from finishing pig houses', *Bioresource Technology*, 91(3), pp. 309–15. doi:10.1016/S0960-8524(03)00184-6.

He, L. Y., Ying, G. G., Liu, Y. S. and et al. (2016) 'Discharge of swine wastes risks water quality and food safety: Antibiotics and antibiotic resistance genes from swine sources to the receiving environments', *Environment International*, 92–93, pp. 210–19. doi:10.1016/j.envint.2016.03.023.

Heller, M. C., Keoleian, G. A. and Willett, W. C. (2013) 'Toward a life cycle-based, diet-level framework for food environmental impact and nutritional quality assessment: A critical review', *Environmental Science and Technology*, 47(22), pp. 12632–47. doi:10.1021/es4025113.

Herrero, M., Thornton, P. K., Gerber, P. J. and et al. (2009) 'Livestock, livelihoods and the environment: Understanding the trade-offs', *Current Opinion in Environmental Sustainability*, 1(2), pp. 111–20. doi:10.1016/j.cosust.2009.10.003.

Hou, Y., Velthof, G. L. and Oenema, O. (2015) 'Mitigation of ammonia, nitrous oxide and methane emissions from manure management chains: A meta-analysis and integrated assessment', *Global Change Biology*, 21(3), pp. 1293–312. doi:10.1111/gcb.12767.

Howarth, R. W. and Marino, R. (2006) 'Nitrogen as the limiting nutrient for eutrophication in coastal marine ecosystems: Evolving views over three decades', *Limnology and Oceanography*, 51(1 part 2), pp. 364–76. doi:10.4319/lo.2006.51.1_part_2.0364.

Hribar, C. (2010) *Understanding Concentrated Animal Feeding Operations and Their Impact on Communities, National Association of Local Boards of Health*. Edited by M. Schultz. National Association of Local Boards of Health. Bowling Green, OH, Available at: http://www.cdc.gov/nceh/ehs/docs/understanding_cafos_nalboh.pdf.

International Orgainzation for Standardization (2006) *ISO 14044: Environmental Management — Life Cycle Assessment — Requirements and Guidelines, International Organization for Standardization, Geneva, Switzerland*. Geneva: International Organization for Standardization. doi:10.1007/s11367-011-0297-3.

Jacobsen, R., Vandermeulen, V., Van Huylenbroeck, G. and et al. (2014) 'Carbon footprint of pigmeat in Flanders', *International Journal of Agricultural Sustainability*, 12(1), pp. 54–70. doi:10.1080/14735903.2013.798896.

Jansen, J., Kirkwood, R. N., Zanella, A. J. and et al. (2007) 'Influence of gestation housing on sow behavior and fertility', *Journal of Swine Health and Production*, 15(3), pp. 132–6.

Jones, A. C. and Cherruault, J. Y. (2011) *Comparative LCA of Pork Production for Midland Pig Producers*. Sustain. Derbyshire, UK, http://www.leavesley.com/assets/content/pdf/Sustain Environmental Report.PDF.

Kafle, G. K. and Chen, L. (2014) 'Emissions of odor, ammonia, hydrogen sulfide, and volatile organic compounds from shallow-pit pig nursery rooms', *Journal of Biosystems Engineering*, 39(2), pp. 76–86. doi:10.5307/JBE.2014.39.2.076.

Kingston, C., Fry, J. and Aumonier, S. (2009) Life cycle assessment of pork. *BPEX*. Available at: http://bit.ly/BPEX_Report.

Koike, S., Krapac, I. G., Oliver, H. D. and et al. (2007) 'Monitoring and source tracking of tetracycline resistance genes in lagoons and groundwater adjacent to swine production facilities over a 3-year period', *Applied and Environmental Microbiology*, 73(15), pp. 4813–23. doi:10.1128/AEM.00665-07.

Kuang, X., Hao, H., Dai, M. and et al. (2015) 'Serotypes and antimicrobial susceptibility of Salmonella spp. isolated from farm animals in China', *Frontiers in Microbiology*, 6(June), p. 602. doi:10.3389/fmicb.2015.00602.

Lammers, P. J., Honeyman, M. S., Harmon, J. D. and et al. (2010) 'Energy and carbon inventory of Iowa swine production facilities', *Agricultural Systems*. Elsevier Ltd, 103(8), pp. 551–61. doi:10.1016/j.agsy.2010.06.003.

Lammers, P. J., Honeyman, M. S., Mabry, J. W. and et al. (2007) 'Performance of gestating sows in bedded hoop barns and confinement stalls', *Journal of Animal Science*, 85(5), p. 1311. doi:10.2527/jas.2006-437.

Landers, T. F., Cohen, B., Wittum, T. E. and et al. (2012) 'A review of antibiotic use in food animals: perspective, policy, and potential.', *Public Health Reports*, 127(1), pp. 4–22. doi:10.1177/003335491212700103.

Laxminarayan, R., Duse, A., Wattal, C. and et al. (2013) 'Antibiotic resistance-the need for global solutions', *The Lancet Infectious Diseases*, 13(12), pp. 1057–98. doi:10.1016/S1473-3099(13)70318-9.

Le, P. D., Aarnink, A. J. A. and Jongbloed, A. W. (2009) 'Odour and ammonia emission from pig manure as affected by dietary crude protein level', *Livestock Science*. Elsevier B.V., 121(2–3), pp. 267–74. doi:10.1016/j.livsci.2008.06.021.

Lim, T.-T., Heber, A. J., Ni, J.-Q. and et al. (2003) 'Odor and gas release from anaerobic treatment lagoons for swine manure', *Journal of Environmental Quality*, 32(2), pp. 406–16. doi:10.2134/jeq2003.4060.

Liu, F., Fiencke, C., Guo, J. and et al. (2016) 'Bioscrubber treatment of exhaust air from intensive pig production: Case study in northern Germany at mild climate condition', *Engineering in Life Sciences*, 17(4), 458–66. doi:10.1002/elsc.201600169.

Lopez-Ridaura, S., van der Werf, H. M. G., Paillat, J. M. and et al. (2009) 'Environmental evaluation of transfer and treatment of excess pig slurry by life cycle assessment', *Journal of Environmental Management*. Elsevier Ltd, 90(2), pp. 1296–304. doi:10.1016/j.jenvman.2008.07.008.

Luby, E. M., Moorman, T. B. and Soupir, M. L. (2016) 'Fate and transport of tylosin-resistant bacteria and macrolide resistance genes in artificially drained agricultural fields receiving swine manure', *Science of the Total Environment*, 550(2), pp. 1126–33. doi:10.1016/j.scitotenv.2016.01.132.

Mackie, R. I., Koike, S., Krapac, I. and et al. (2006) 'Tetracycline residues and tetracycline resistance genes in groundwater impacted by swine production facilities', *Animal biotechnology*, 17(2), pp. 157–76. doi:10.1080/10495390600956953.

Macleod, M. M., Gerber, P. J., Mottet, A. and et al. (2013) *Greenhouse gas emissions from pig and chicken supply chains – A global life cycle assessment, Food and Agriculture Organization of the United Nations (FAO)*. Rome, IT: Food and Agriculture Organization of the United Nations. Available at: http://www.fao.org/3/a-i3460e.pdf.

Maxwell, C., Tsai T-C, Apple J., Richert B., and et al. (2015) 'Effects of Decreasing Dietary Crude Protein with Amino Acid Supplementation on Performance, Carcass, and Nutrient Excretion of Pigs.' In *15th Annual Midwest Swine Nutrition Conference Proceedings*. Indianapolis, IN: Midwest Swine Nutrition Conference.

McAuliffe, G. A., Chapman, D. V and Sage, C. L. (2016) 'A thematic review of life cycle assessment (LCA) applied to pig production', *Environmental Impact Assessment Review*, 56, pp. 12–22. doi:10.1016/j.eiar.2015.08.008.

McGlone, J. (2013) 'The future of pork production in the world: Towards sustainable, welfare-positive systems', *Animals*, 3(2), pp. 401–15. doi:10.3390/ani3020401.

McGlone, J. J., Von Borell, E. H., Deen, J. and et al. (2004) 'Review: Compilation of the scientific literature comparing housing systems for gestating sows and gilts using measures of physiology, behavior, performance, and health', *The Professional Animal Scientist*, 20(2), pp. 105–17. doi: 10.15232/S1080-7446(15)31285-7.

Melorose, J., Perroy, R. and Careas, S. (2015) *World Population Prospects: The 2015 Revision*. New York: United Nations. doi:10.1017/CBO9781107415324.004.

Mosnier, E., van der Werf, H. M. G., Boissy, J. and et al. (2011) 'Evaluation of the environmental implications of the incorporation of feed-use amino acids in the manufacturing of pig and broiler feeds using Life Cycle Assessment', *Animal : An International Journal of Animal Bioscience*, 5(12), pp. 1972–83. doi:10.1017/S1751731111001078.

Mottet, A., de Haan, C., Falcucci, A. and et al. (2017) 'Livestock: On our plates or eating at our table? A new analysis of the feed/food debate', *Global Food Security*, (10 January 2017, in press corrected proof). doi:10.1016/j.gfs.2017.01.001.

Müller, J. C. (2015) *Comparative Financial and Environmental Life Cycle Assessment of three South African Pork Production Chains*. MS Thesis, Department Agricultural Economics, Stellenbosch University, Stellenbosch, South Africa. Available at: https://scholar.sun.ac.za.

National Pork Board. (2017) 'Pork Production Environmental Footprint Calculator.' National Pork Board. Des Moines, IA: http://www.pork.org/production-topics/environmental-sustainability-efforts-pork-production/carbon-footprint-pork-production-calculator/.

National Research Council (2012) *Nutrient Requirements of Swine: Eleventh Revised Edition*. Edited by Subcommittee on Swine Nutrition et al. Washington, DC: National Academies Press. doi:10.17226/13298.

Ndegwa, P. M. M., Hristov, A. N., Arogo, J. and et al. (2008) 'A review of ammonia emission mitigation techniques for concentrated animal feeding operations', *Biosystems Engineering*, 100(4), pp. 453–69. doi:10.1016/j.biosystemseng.2008.05.010.

Nguyen, T. L. T., Hermansen, J. E. and Mogensen, L. (2010) 'Fossil energy and GHG saving potentials of pig farming in the EU', *Energy Policy*. Elsevier, 38(5), pp. 2561–71. doi:10.1016/j.enpol.2009.12.051.

Nguyen, T. L. T., Hermansen, J. E. and Mogensen, L. (2011) *Environmental Assessment of Danish Pork*. Report No. 103, Department of Agroecology, Aarhaus University, Aarhaus, Denmark. Available at: www.agrsci.au.dk.

Ni, J.-Q., Heber, A. J., Lim, T. T. and et al. (2007) 'Methane and carbon dioxide emission from two pig finishing barns.', *Journal of Environmental Quality*, 37(6), pp. 2001–11. doi:10.2134/jeq2007.0386.

Nicks, B., Laitat, M., Farnir, F. and et al. (2004) 'Gaseous emissions from deep-litter pens with straw or sawdust for fattening pigs', *Animal Science*, 78, pp. 99–107.

Noya, I., Aldea, X., Gasol, C. M. and et al. (2016) 'Carbon and water footprint of pork supply chain in Catalonia: From feed to final products', *Journal of Environmental Management*, 171, pp. 133–43. doi:10.1016/j.jenvman.2016.01.039.

Noya, I., Villanueva-Rey, P., González-García, S. and et al. (2017) 'Life Cycle Assessment of pig production: A case study in Galicia', *Journal of Cleaner Production*, 142, pp. 4327–38. doi:10.1016/j.jclepro.2016.11.160.

Ogino, A., Osada, T., Takada, R. and et al. (2013) 'Life cycle assessment of Japanese pig farming using low-protein diet supplemented with amino acids', *Soil Science and Plant Nutrition*, 59(1), pp. 107–18. doi:10.1080/00380768.2012.730476.

Pelletier, F., Godbout, S., Marquis, A. and et al. (2004) 'Greenhouse gas and odor emissions from liquid swine manure storage and manure treatment facilities in Quebec', *Soil Science*, 300(4).

Pelletier, N., Lammers, P. J. ., Stender, D. and et al. (2010) 'Life cycle assessment of high- and low-profitability commodity and deep-bedded niche swine production systems in the Upper Midwestern United States', *Agricultural Systems*. Elsevier Ltd, 103(9), pp. 599–608. doi:10.1016/j.agsy.2010.07.001.

Perez, R. O. (2009) *Analysis of Sustainability in the Pig Production Chain: Life Cycle Assessment of Contrasting Scenarios*. PhD. Dissertation, Newcastle: University School of Agriculture, Newcastle, UK.

Peters, G. M., Murphy, K., Adamsen, A. P. S., Bruun, S., Svanström, M. and Hoeve, M. (2014) 'Improvement of Odor Assessment in a Life Cycle Assessment Framework'. In: *Proceedings of the 9th International Conference on Life Cycle Assessment in the Agri-Food Sector*, R. C. Schenk and D. Huizenga (Eds), 4–9 October. San Francisco: American Center for Life Cycle Assessment.

Petersen, S. O., Blanchard, M., Chadwick, D. and et al. (2013) 'Manure management for greenhouse gas mitigation', *Animal : An International Journal of Animal Bioscience*, 7 Suppl 2, pp. 266–82. doi:10.1017/S1751731113000736.

Philippe, F. X., Laitat, M., Canart, B. and et al. (2007) 'Comparison of ammonia and greenhouse gas emissions during the fattening of pigs, kept either on fully slatted floor or on deep litter', *Livestock Science*, 111(1–2), pp. 144–52. doi:10.1016/j.livsci.2006.12.012.

Philippe, F. X. and Nicks, B. (2013) Emissions of ammonia, nitrous oxide and methane from pig houses: Influencing factors and mitigation techniques. In: *European Workshop: Reconciling the Environment with Livestock Management*. Rennes, FR: IRSTEA, pp. 1–10. Available at: http://hdl.handle.net/2268/146949.

Philippe, F. X. and Nicks, B. (2015) 'Review on greenhouse gas emissions from pig houses: Production of carbon dioxide, methane and nitrous oxide by animals and manure', *Agriculture, Ecosystems and Environment*. Elsevier B.V., 199, pp. 10–25. doi:10.1016/j.agee.2014.08.015.

Phillips, F. A., Wiedemann, S. G., Naylor, T. and et al. (2016) 'Methane, nitrous oxide and ammonia emissions from pigs housed on litter and from stockpiling of spent litter', *Animal Production Science*, 56(9), pp. 1390–403. doi:10.1071/AN15650.

Prapaspongsa, T., Christensen, P., Schmidt, J. H. and et al. (2010) 'LCA of comprehensive pig manure management incorporating integrated technology systems', *Journal of Cleaner Production*. Elsevier Ltd, 18(14), pp. 1413–22. doi:10.1016/j.jclepro.2010.05.015.

Raney, T., Gerosa, S., Khwaja, Y. and et al. (2009) *The state of food and agriculture: Livestock in balance*, Food and Agriculture Organization of the United Nations. Rome, IT. http://www.fao.org/docrep/012/i0680e/i0680e.pdf.

Reckmann, K. and Krieter, J. (2015) 'Environmental impact of the pork supply chain depending on farm performances', *The Journal of Agricultural Science*, 153(3), pp. 411–21. doi:10.1017/S0021859614000501.

Reckmann, K., Traulsen, I. and Krieter, J. (2012) 'Environmental Impact Assessment – methodology with special emphasis on European pork production', *Journal of Environmental Management*. Elsevier Ltd, 107, pp. 102–9. doi:10.1016/j.jenvman.2012.04.015.

Reckmann, K., Traulsen, I. and Krieter, J. (2013) 'Life Cycle Assessment of pork production: A data inventory for the case of Germany', *Livestock Science*, 157(2–3), pp. 586–96. doi:10.1016/j.livsci.2013.09.001.

Regulation (EC) No 1831/2003 (2003) *Official Journal of the European Union*. European Union. *doi:* http://eur-lex.europa.eu/legal-content/PT/TXT/?uri=celex:32003R1831.

Roma, R., Corrado, S., De Boni, A. and et al. (2015) Life cycle assessment in the livestock and derived edible products sector. In: B. Notarnicola and et al. (Eds) *Life Cycle Assessment in the Agri-food Sector*. Cham: Springer International Publishing, pp. 251–332. doi:10.1007/978-3-319-11940-3_5.

Rougoor, C., Elferink, E., Lap, T. and et al. (2015) *LCA of Dutch Pork Assessment of Three Pork Production Systems in the Netherlands*. Available at: http://edepot.wur.nl/353083.

Sayah, R. S., Kaneene, J. B., Johnson, Y. and et al. (2005) 'Patterns of antimicrobial resistance observed in Escherichia coli isolates obtained from domestic- and wild-animal fecal samples, human septage, and surface water', *Applied and Environmental Microbiology*, 71(3), pp. 1394–404. doi:10.1128/AEM.71.3.1394.

Schauberger, G., Lim, T.-T. T., Ni, J. and et al. (2013) 'Empirical model of odor emission from deep-pit swine finishing barns to derive a standardized odor emission factor', *Atmospheric Environment*, 66(2013), pp. 84–90. doi:10.1016/j.atmosenv.2012.05.046.

Schauberger, G., Piringer, M., Jovanovic, O. and et al. (2012) 'A new empirical model to calculate separation distances between livestock buildings and residential areas applied to the Austrian guideline to avoid odour nuisance', *Atmospheric Environment*. Elsevier Ltd, 47(February), pp. 341–7. doi:10.1016/j.atmosenv.2011.10.056.

Sharpley, A. N. and Wang, X. (2014) 'Managing agricultural phosphorus for water quality: Lessons from the USA and China', *Journal of Environmental Sciences*, 26(9), pp. 1770–82. doi:10.1016/j.jes.2014.06.024.

SNC-LAVALIN Agro and Ciraig (2009) A Life Cycle Analysis of Carbon Dioxide Equivalents (CO2e) of Alberta Barley, Wheat, Peas and Canola Meal Used in Pork Production, Slaughter and Further Processing, Report prepared for Alberta Agriculture and & Rural Development.

Steinfeld, H., Gerber, P. J., Wassenaar, T. and et al. (2006) *Livestock's Long Shadow: Environmental issues and options*. Rome: Food and Agriculture Organization of the United Nations. Available at: ftp://ftp.fao.org/docrep/fao/010/a0701e/a0701e.pdf.

Stone, J. J., Aurand, K. R., Dollarhide, C. R. and et al. (2011) 'Determination of environmental impacts of antimicrobial usage for US Northern Great Plains swine-production facilities: A life-cycle assessment approach', *International Journal of Life Cycle Assessment*, 16(1), pp. 27–39. doi:10.1007/s11367-010-0241-y.

Struve, T. (2011) *Risk assessment of antimicrobial usage in Danish pig production on the human exposure to antimicrobial resistant bacteria from pork*. PhD. Dissertation, Technical University of Denmark, Copenhagen, DK.

Stylianou, K. S., Heller, M. C., Fulgoni, V. L., Ernstoff, Alexi S., Keoleian, Gregory A. and Jolliet, Olivier. (2016) 'A life cycle assessment framework combining nutritional and environmental health impacts of diet: A case study on milk', *The International Journal of Life Cycle Assessment*, 21 (5): 734–46. doi:10.1007/s11367-015-0961-0.

Thoma, G. J., Nutter, D. W., Ulrich, R. and et al. (2011) *National Life Cycle Carbon Footprint Study for Production of US Swine*. Des Moines, IA: National Pork Board. Available at: https://porkcdn.s3.amazonaws.com/sites/all/files/documents/NPB Scan Final - May 2011.pdf.

Thorne, P. S. (2007) 'Environmental health impacts of concentrated animal feeding operations: Anticipating hazards – Searching for solutions', *Environmental Health Perspectives*, 115(2), pp. 296–7. doi:10.1289/ehp.8831.

Tilman, D., Balzer, C., Hill, J. and et al. (2011) 'Global food demand and the sustainable intensification of agriculture', *Proceedings of the National Academy of Sciences*, 108(50), pp. 20260–4. doi:10.1073/pnas.1116437108.

Tomer, M. D., Moorman, T. B., Kovar, J. L. and et al. (2016) 'Eleven years of runoff and phosphorus losses from two fields with and without manure application, Iowa, USA', *Agricultural Water Management*. Elsevier B.V., 168, pp. 104–11. doi:10.1016/j.agwat.2016.01.011.

Tong, L., Li, P., Wang, Y. and et al. (2009) 'Analysis of veterinary antibiotic residues in swine wastewater and environmental water samples using optimized SPE-LC/MS/MS', *Chemosphere*. Elsevier Ltd, 74(8), pp. 1090–7. doi:10.1016/j.chemosphere.2008.10.051.

UN FAO (2017) Livestock Environmental Assessment and Performance Partnership. Available at: http://www.fao.org/partnerships/leap/publications/en/(Accessed: 15 June 2017).

U.S. Census Bureau (2015) USA Trade Online, State Export Data (Origin of Movement). *U.S. Census Bureau*. Available at: https://usatrade.census.gov/(Accessed: 17 November 2014).

U.S. Food and Drug Administration (2013) *Guidance for Industry #213*.

USDA (2016) *Escherichia coli on U.S. Swine Sites – Antimicrobial Drug Susceptibility*. Fort Collins, CO: USDA, APHIS, VS, CEAH, National Animal Health Monitoring System.

USDA FAS (2016) Production, Supply and Distribution Online. USDA Foreign Agricultural Service. Available at: https://apps.fas.usda.gov/psdonline/app/index.html#/app/advQuery (Accessed: 4 January 2017).

Van Boeckel, T. P., Brower, C., Gilbert, M. and et al. (2015) 'Global trends in antimicrobial use in food animals', *Proceedings of the National Academy of Sciences*, 112(18), pp. 5649–54. doi:10.1073/pnas.1503141112.

van der Heyden, C., Demeyer, P. and Volcke, E. I. P. (2015) 'Mitigating emissions from pig and poultry housing facilities through air scrubbers and biofilters: State-of-the-art and perspectives', *Biosystems Engineering*, Elsevier Ltd, 134, pp. 74–93. doi:10.1016/j.biosystemseng.2015.04.002.

van Dooren, C., Marinussen, M., Blonk, H. and et al. (2014) 'Exploring dietary guidelines based on ecological and nutritional values: A comparison of six dietary patterns', *Food Policy*. Elsevier Ltd, 44(2014), pp. 36–46. doi:10.1016/j.foodpol.2013.11.002.

van Kernebeek, H. R. J., Oosting, S. J., Feskens, E. J. M. and et al. (2014) 'The effect of nutritional quality on comparing environmental impacts of human diets', *Journal of Cleaner Production*, 73(2014), pp. 88–99. doi:10.1016/j.jclepro.2013.11.028.

van Zanten, H. H. E., Bikker, P., Meerburg, B. G. and et al. (2016) Environmental consequences of using alternative protein sources in pig diets. In: *10th International Conference on Life Cycle Assessment of Food*. Dublin.

Vergé, X. P. C., Dyer, J. A., Desjardins, R. L. and et al. (2009) 'Greenhouse gas emissions from the Canadian pork industry', *Livestock Science*. Elsevier B.V., 121(1), pp. 92–101. doi:10.1016/j.livsci.2008.05.022.

Vonderohe, C., Jones, A., Richert, B. and et al. (2015) Comparison of the effects of antibiotic-free and conventional management on growth performance in swine. In: *ASAS Midwest Meeting*. Des Moines: American Society of Animal Science. Available at: https://asas.confex.com/asas/mw15/webprogram/Paper11801.html (Accessed: 7 March 2017).

Webb, J., Broomfield, M., Jones, S. and et al. (2014) 'Ammonia and odour emissions from UK pig farms and nitrogen leaching from outdoor pig production. A review', *Science of the Total Environment*, 470–471, pp. 865–75. doi:10.1016/j.scitotenv.2013.09.091.

Webb, J., Pain, B., Bittman, S. and et al. (2010) 'The impacts of manure application methods on emissions of ammonia, nitrous oxide and on crop response – A review', *Agriculture, Ecosystems and Environment*. Elsevier B.V., 137(1–2), pp. 39–46. doi:10.1016/j.agee.2010.01.001.

Weiler, V., Udo, H. M., Viets, T. and et al. (2014) 'Handling multi-functionality of livestock in a life cycle assessment: the case of smallholder dairying in Kenya', *Current Opinion in Environmental Sustainability*. Elsevier B.V., 8, pp. 29–38. doi:10.1016/j.cosust.2014.07.009.

Wiedemann, S. G., McGahan, E. J., Grist, S. and et al. (2010) *Environmental Assessment of Two Pork Supply Chains Using Life Cycle Assessment. 09/176*. Available at: http://www.rirdc.gov.au.

Wilcox, S. and Marion, W. (2008) 'Users manual for TMY3 data sets', National Renewable Energy Lab Technical Report; NREL/TP-581-43156. http://www.nrel.gov/docs/fy08osti/43156.Pdf.

Williams, A. G., Audsley, E. and Sandars, D. L. (2006) 'Determining the environmental burdens and resource use in the production of agricultural and horticultural commodities' by Williams et al. 2006 at Cranfield University and the Silsoe Research Institute', (Mlc), p. 97. Available at: http://www.cranfield.ac.uk/Research/Research-activity/Current-projects/research-projects/Environmental-Burdens-of-Agricultural-and-Horticultural-Commodity-Production-LCA-IS0205.

Winkler, T., Schopf, K., Aschemann, R. and et al. (2015) 'From farm to fork – A life cycle assessment of fresh Austrian pork', *Journal of Cleaner Production*. Elsevier Ltd, 116, pp. 80–9. doi:10.1016/j.jclepro.2016.01.005.

Xin, H., Ibarburu, M., Vold, L. and et al. (2010) *A Comparative Assessment of the Environmental Footprint of the U.S. Egg Industry in 1960 and 2010*. Final Report.

Chapter 10

Nutritional strategies to reduce emissions from waste in pig production

Phung Lê Đình, Hue University of Agriculture and Forestry, Hue University, Vietnam; and André J. A. Aarnink, Wageningen University and Research, The Netherlands

1 Introduction

In various parts of the world, pig production is highly concentrated. Pigs are reared in vast numbers in some parts of Europe, some parts of Northern and Southern America and in a large part of Southeast Asia (Fig. 1). Pig production brings significant benefits to many households all over the world, especially to smallholders in Southeast Asia. There is a tendency, however, for farm sizes to expand and the number of individual farms to decrease. In the Netherlands, for instance, the number of farms with pigs has decreased from 271 000 in 1950 to 5000 in 2016, while at the same time pig meat production has increased from 236 to 1453 million kg (CBS, 2017). In Southeast Asia there has been a rapid development in pig production: pig farms have become more specialized and have grown in size, especially around large cities (Huynh et al., 2008).

The concentration, expansion and specialization of livestock production have distinct economic advantages; however, they also have a significant impact on the immediate environment. The main concerns are the mineral load to the soil from surplus manure and the gaseous load to the air from livestock housing and the storage and application of manure (Aarnink and Verstegen, 2007). Recently, concern about the emission of particulate matter and microorganisms, and their components, into the environment has been growing (Maassen, 2016). In this chapter, we restrict ourselves to a discussion of

http://dx.doi.org/10.19103/AS.2017.0030.12
Published by Burleigh Dodds Science Publishing Limited, 2018.

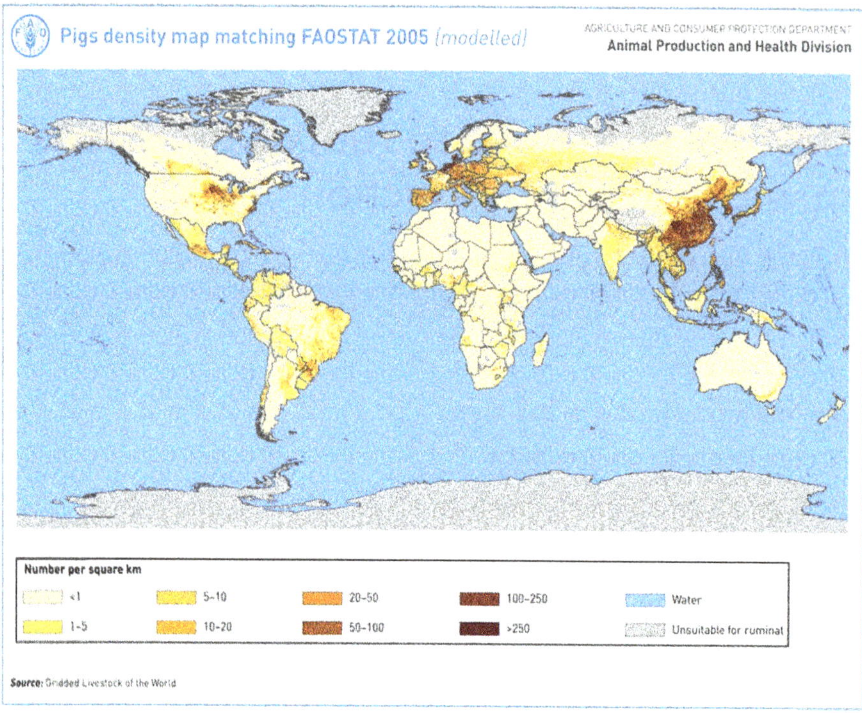

Figure 1 Pig density in the world (FAOSTAT, 2005).

direct emissions to the air from pig production that can be influenced by the pigs' diet: ammonia, odour and greenhouse gases (GHGs) (methane and nitrous oxide). Indirect emissions, especially of methane and nitrous oxide, from pig feed production are also contributing to global GHG emissions, but this topic will not be examined further in this chapter. In the next paragraphs we will explore the problems related to gaseous emissions, how these can be reduced through dietary strategies and whether these strategies might influence production results. We will then summarize how nutrition can contribute to a more sustainable means of pig production, with low emissions to the environment, and what research still needs to be done to increase our knowledge in this field. We will close with some sources for further information on the topics covered in this chapter.

2 Nutrition and ammonia emissions

Ammonia (NH_3) is an important pollutant that plays a crucial role in the acidification and eutrophication of ecosystems and in secondary fine particulate formation in the atmosphere (Krupa, 2003). In addition, ammonia is a well-known toxic gas, irritating the respiratory tract when at concentrations exceeding 15 ppm (Banhazi et al., 2008). Livestock wastes account for 39% of global NH_3 emissions (Galloway et al., 2004). In Europe, approximately

80% of NH_3 emissions originate from livestock facilities (Reidy et al., 2009). Pig production is globally responsible for approximately 15% of the NH_3 emissions associated to livestock (Olivier et al., 1998). In Europe, pig production represents nearly 25% of the livestock emissions (European Environment Agency, 2010). Ammonia released from buildings is the main source, accounting for about 50% of NH_3 from pig production (Hutchings et al., 2001; Webb and Misselbrook, 2004; Gac et al., 2007). Therefore, to protect the environment, ammonia emissions from pig production facilities need to be reduced.

In pig production, the main source of NH_3 is the rapid hydrolyzation of urea in urine by the faecal enzyme urease, which leads to the formation of ammonium (NH_4^+) in an aqueous medium (Cortus et al., 2008). Another minor source of ammonia is the degradation of undigested proteins (Zeeman, 1991; Aarnink et al., 1993). In the liquid phase, total ammoniacal nitrogen (TAN) is in a state of equilibrium between ionized NH_4^+ and unionized NH_3. This equilibrium is influenced by temperature and the pH level. Higher temperatures and pH enhance the concentration of NH_3. At pH values below 7, nearly all TAN is in an ionized form. At pH values above 7, the unionized fraction increases significantly, and at pH values of 11 or higher, TAN is mainly in the form of NH_3.

Dietary nitrogen is the main precursor of ammonia production from commercial pig farms. Ammonia emissions from pig production facilities depend on many factors, including animal housing (floor type, ambient temperature and ventilation rate); the animal itself (the animal's physiological stage, genotype/line and gender); diet (dietary crude protein (CP) and the type and level of fibre and feed additives, such as enzymes, acidifying salts, zeolites, probiotics); feeding management (liquid feeding, phase feeding) and the manure removal system (Le et al., 2005). According to Aarnink (1997), 30% of nitrogen in the diet is deposited in the body of growing-finishing pigs and 70% is excreted via urine and faeces (Fig. 2). Because of genetic improvements during recent decades, it is likely that nitrogen retention has increased and nitrogen excretion has decreased, relatively. According to

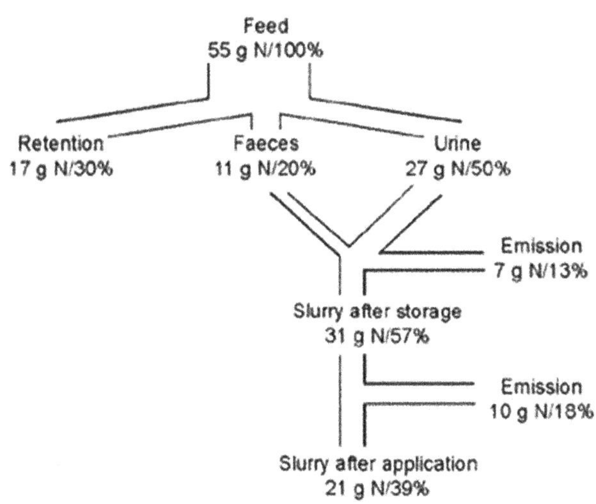

Figure 2 Nitrogen chain from feed to body and to urine and faeces in growing-finishing pigs (Aarnink, 1997).

Fig. 2, almost half of the nitrogen excreted by urine and faeces was emitted during the storage of the manure inside the pig house and during the surface application of the manure. In some countries, including the Netherlands, the surface application of manure is no longer allowed and this has significantly reduced ammonia emissions.

Feed and feeding manipulation has been shown to be effective in reducing ammonia emissions from pig production facilities (Canh, 1998). Feed and feeding strategies to reduce emissions can be grouped as follows:

1 Reducing nitrogen excretion by lowering the protein level in the diet.
2 Shifting nitrogen excretion from urea in urine to bacterial protein in faeces.
3 Lowering the pH of manure:
   a Lowering the pH of faeces
   b Lowering the pH of urine.

2.1 Lowering protein levels in the diet

Although protein is important for the maintenance, growth and development of the animal, reducing dietary CP content is one of the most significant means to mitigate ammonia emissions (Ndegwa et al., 2008). The principle of reducing nitrogen excretion and ammonia emissions through protein is to ensure that the amount of protein in a diet matches the protein requirement and to increase the efficiency of the animals' protein utilization. There is abundant literature on the impact that the reduction of dietary protein supply to pigs has on the reduction of nitrogen excretion and ammonia emissions (Canh et al., 1998c; Zijlstra et al., 2001; Zervas and Zijlstra, 2002; Le et al., 2007a, 2008a; Ndegwa et al., 2008; Le et al., 2009). Nitrogen excretion and ammonia emissions can be reduced appreciably by decreasing the CP content in diets. According to O'Connell et al. (2006), finishing pigs offered diets containing 14% CP excreted significantly less nitrogen, urinary nitrogen and faecal nitrogen than those offered diets containing 16% CP. Reduced CP diets, supplemented with crystalline amino acids (AAs), have been shown to reduce faecal nitrogen excretion by 25–30% (Cromwell and Coffey, 1993; Jongbloed and Lenis, 1993). According to Sutton et al. (1999), Shriver et al. (2003) and Aarnink and Verstegen (2007), reduced CP diets supplemented with AA decrease not only nitrogen excretion but also manure pH values, which inhibits the conversion of ammonium (NH_4^+) to ammonia, thus reducing ammonia emissions. According to Portejoie et al. (2004), lowering dietary CP from 20% to 16% and 12%, while maintaining similar levels of essential AAs, reduced total slurry ammoniacal nitrogen (N) content from 4.32 to 3.13 g N/kg and 1.92 g N/kg, respectively (the corresponding figures for total N content were 5.48, 4.30 and 3.05 g N/kg, respectively). The pH value of the slurry decreased by 1.3 units when dietary CP decreased from 20% to 12%. These changes in pH and slurry composition resulted in a 63% reduction of ammonia emissions from the slurry when dietary protein level decreased from 20% to 12%. Le et al. (2007a) found that reducing dietary CP level from 18% to 12% lowered ammonia emissions from growing-finishing pig manure by 53%. According to Canh et al. (1998c), reducing the dietary CP level from 16.5% to 12.5% can reduce ammonia emissions from pig manure by up to 50%. Kay and Lee (1997) found reductions in ammonia emissions of 58% in growing pigs and of 46% in finishing pigs when the CP content of the diets was reduced by 6% and 6.5% units, respectively. Reducing the dietary CP levels of finishing pigs from 16% to 14% reduced ammonia emissions from the

Published by Burleigh Dodds Science Publishing Limited, 2018.

chamber by 30% (Hansen et al., 2007). Generally, as a guide, for each 1% unit reduction in dietary CP level, when combined with AA supplementation, the estimated ammonia losses are reduced by 10% (Aarnink et al., 1993; Jacob, 1994; Kay and Lee, 1997; Sutton et al., 1997; Canh et al., 1998c; Le et al., 2007a).

In many studies on dietary manipulation and ammonia and odour emissions from pig manure, water consumption was restricted. When the animal has access to water *ad libitum*, the effect of lowering the CP content of the diet might be partly compensated by a lower water intake. Lowering dietary CP lowered water consumption (Portejoie et al., 2004). Ammonia emissions are directly related to the ammonium concentration in the urine and manure. When water intake is decreased, the ammonium concentration will not be as low as expected from the reduction in protein content.

It should be noted that ammonia emissions are not directly related to the protein intake, but rather to the difference between digested protein (absorbed amino acids) and the retention of AAs; so, emissions are related to the surplus digested proteins. When lowering the CP content of the diet in combination with using feed components with a higher protein digestibility, the effect on ammonia emissions will be lower. Surplus digested proteins are converted into urea by the ornithine cycle and other odour precursors in the liver. Urinary nitrogen is the main source of quickly released ammonia to the air.

2.2 Shifting nitrogen excretion from urea in urine to bacterial protein in faeces

Urinary nitrogen, which is mainly present as urea, is rapidly converted to ammonia and carbon dioxide through the enzymatic activity of urease, which is found mainly in faecal and soil bacteria (Canh et al., 1999), whereas for organic nitrogen in faeces it takes a few months, or even a year, to be converted to ammonia (Ndegwa et al., 2008). The type and level of fermentable carbohydrates (FC), also called non-starch polysaccharide (NSP), have gained much attention in dietary approaches to reducing ammonia emissions. The principle of reducing ammonia production and emissions through FC is to shift nitrogen excretion from urine to faeces and to reduce the pH of the manure. Increasing FCs in diets can result in bacterial proliferation due to an increase in the source of energy for bacteria in both the gastrointestinal tract and in the manure. Bacteria will use ammonia as a source of nitrogen for protein synthesis, thus reducing ammonia absorption into the blood and urea excretion via urine. Furthermore, urea excreted from the blood into the large intestine can be incorporated into bacterial protein. FCs in the gastrointestinal tract shift urinary nitrogen excretion to faecal nitrogen excretion in the form of bacterial protein (Younes et al., 1997), which is less susceptible to rapid hydrolysis. This phenomenon was confirmed by Morgan and Whittemore (1998), Cromwell et al. (1999) and Van der Peet-Schwering et al. (1999).

When increasing the FC content of the diet from 14% to 31%, Canh et al. (1997) found that there was a decrease in the urinary nitrogen-to-faecal nitrogen ratio from 3.8 to 1.2, while apparent nitrogen digestibility decreased from 85% to 75%. The relationship between the NSP content of the pig diet and the urinary nitrogen-to-faecal nitrogen ratio is shown in Fig. 3 (Jongbloed, 2001). This figure demonstrates that the urinary nitrogen-to-faecal nitrogen ratio decreases from approximately five to one when the NSP content of the diet is increased from 100 to 500 g/kg. The sources of the FC have an impact on this ratio and hence on ammonia emissions, because of the different components in these

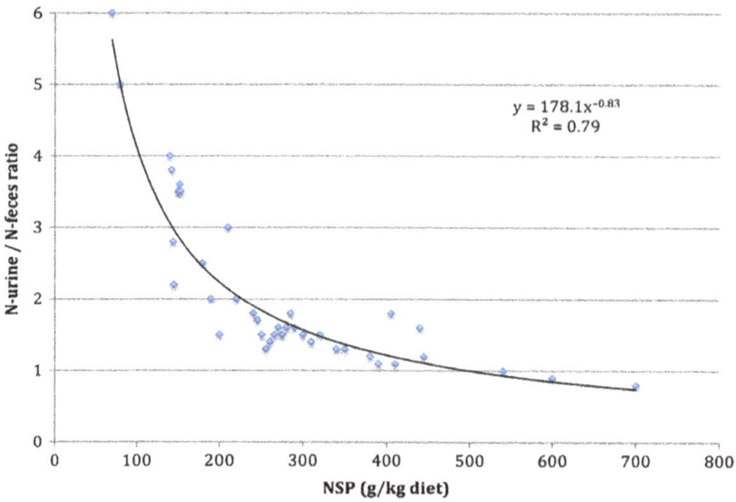

Figure 3 The relationship between the non-starch polysaccharide content of the pig diet and its urinary nitrogen-to-faecal nitrogen ratio (Jongbloed, 2001).

carbohydrates (levels of lignin, cellulose, hemicellulose, pectin and oligosaccharides) (Bakker, 1996; Canh et al., 1997, 1998e; Mroz et al., 2000; Zijlstra et al., 2001; Zervas and Zijlstra, 2002; Nguyễn et al., 2017).

2.3 Lowering the pH of faeces

Manure pH also has a direct effect on ammonia emissions since both urinary nitrogen and faecal nitrogen are further converted to ammonium (NH_4^+) at a low pH and to ammonia (NH_3) at a high pH (van der Peet-Schwering et al., 1999; Aarnink and Verstegen, 2007). The inclusion of FC in pig diets increases the concentration of volatile fatty acids (VFA) in faeces and manure storage and thereby reduces pH and thus ammonia emissions (Sutton et al., 1997; Canh et al., 1998e; Kendall et al., 1999). DeCamp et al. (2001) reported a 32% increase of total VFA concentration in the six-week-stored manure of pigs fed a diet of 10% soybean hulls when compared with no soybean hulls. In the headspace gases there was a 20% reduction in aerial ammonia. According to Le et al. (2008a), the NSP level of the diet (17%, 20%, 23%) was positively related to the VFA concentration in the slurry (6.05. 7.82, 8.28 g/kg) and negatively related to the slurry pH (7.26, 7.05, 7.02). Canh et al. (1998d, 1999) found a decrease of manure pH, with 0.12 units, and a reduction of ammonia emissions by 5.1% for every 100 g/day of extra NSP intake (range of NSP intake: 200–700 g/day; mean feed intake: 1.35 kg/day). Also, after a storage period of 16 days, the VFA content of the manure increased with the increasing NSP content of the diet. The most profound effect on the pH of the manure was found when sugar beet pulp silage (SBPS) was included in the diet of growing pigs. Levels of 0%, 5%, 10% and 15% SBPS gave NSP levels in DM (dry matter) of 27.5%, 30.5%, 34.0% and 36.5%, respectively. These levels had no effect on the ammonium content of the manure. However, the pH was significantly lowered by 1.5 units and ammonia emissions were lowered by 40% when 15% SBPS was included in the diet, compared to the control diet without SBPS.

2.4 Lowering the pH of urine

Ammonia (NH_3) in solution is in equilibrium with ammonium (NH_4^+) in a pH-dependent manner, and only NH_3 can be emitted. Thus, lowering the pH of manure results in a conversion of NH_3 into NH_4^+ and, therefore, lower NH_3 emissions. This principle has been used to develop strategies to acidify the manure in the pit to reduce ammonia emissions. However, ammonia emissions from the manure pit are not the only source of emissions. Ammonia is also emitted from urine-soiled (slatted) floors. A more effective solution is therefore to reduce the pH of urine. This will reduce ammonia emissions not only from the manure pit, but also from the soiled (slatted) floor.

One of the principles for adjusting the urine pH is to change the electrolyte balance (Na+K-Cl) of the diet (Mroz et al., 1996). A reduction of ammonia by 11% was achieved by a pH reduction in urine and manure of 0.46 and 0.17 pH units through decreasing the dietary electrolytic balance from 320 to 100 mEq./kg (Canh et al., 1998b). Some types of acid salts have been added to diets to reduce ammonia emissions based on the principle of pH reduction. Kim et al. (2004) achieved lower urine pH and ammonia emissions by replacing di-calcium-phosphate ($CaHPO_{4)}$ and calcium carbonate ($CaCO_3$) with phosphoric acid ($H_3PO_{4)}$ and calcium sulphate ($CaSO_4$), respectively. According to Canh et al. (1998b), replacing calcium carbonate with calcium sulphate, calcium chloride ($CaCl_2$) or calcium benzoate ($Ca(C_7H_5O_2)_2$) significantly reduced the pH of urine and manure and the ammonia emissions from the manure. By replacing 6 g of calcium in the form of calcium carbonate using calcium benzoate, urinary and manure pH was lowered from 6.8 to 5.3 and from 8.0 to 6.4, respectively. In that case, ammonia emissions were reduced by almost 60%. For calcium sulphate and calcium chloride, the effects were lower (an approximately 35% reduction of ammonia emissions), although the pH values of the urine were similar. This is caused by benzoate being converted into hippuric acid ($C_9H_9NO_3$) in the liver and then excreted in the urine. This hippuric acid provides an additional lowering of the manure pH compared to when calcium sulphate and calcium chloride are added. It was also confirmed by Bühler et al. (2007) that the addition of benzoic acid (1%) to the diet of pigs from 26 to 100 kg reduced urinary pH by about one pH unit and increased the concentration of urinary hippuric acid markedly. Kluge et al. (2010) reported that the relationship between benzoic acid ($C_7H_6O_2$) intake and the urinary pH response was linear. The supplementation of 1% benzoic acid lowered the urinary pH of sows from 7.20 to 5.39. In sows fed diet supplemented with 0.5% of benzoic acid, 64% of ingested benzoic acid was excreted via urine as hippuric acid and the remainder as unmetabolized benzoic acid. According to Marie-Line Daumer et al. (2007), nitrogen excretion and the urine and slurry pH of growing pigs were decreased and ammonia emissions were reduced by 20% when 1% of benzoic acid was added to the diet. According to Hansen et al. (2007), adding 3% of benzoic acid to the diet of finishing pigs reduced ammonia emissions from the chamber by 57%.

Adipic acid ($C_6H_{10}O_4$) is only partially catabolized when it is fed to animals, and a portion of it is excreted in urine, leading to a lower urinary pH and resulting in lower ammonia emissions. Van Kempen (2001) found that a diet supplemented with 1% of adipic acid reduced urinary pH from 7.7 to 5.5, and *in vitro* ammonia emissions from the urine at 1, 3, 18 and 46 h were reduced by 94%, 93%, 70% and 39%, respectively, compared to those of the standard diet. A similar experiment conducted in an animal house reduced ammonia emissions by 25%. This reduction is lower than that found for urine because of the buffering effect of the faeces in the manure, which is comprised of a mix of urine and faeces (Canh et al., 1998a).

3 Nutrition and odour emissions

The odour produced and emitted from intensive animal production systems can cause a serious nuisance to individuals living in the vicinity of livestock farms and was related by some authors to health problems (Donham, 2000; Iverson et al., 2000). To sustain livestock production, odour production and emissions must be reduced. A great number of odorous compounds are emitted from animal production facilities. O'Neill and Phillips (1992) summarized 168 odorous compounds that were identified in various studies of animal production facilities. Schiffman et al. (2001) identified 331 different compounds from pig production facilities in North Carolina. Odorous compounds can be classified into four groups: sulphurous compounds, VFA, phenols and indoles, and ammonia and volatile amines (Le et al., 2005). In animal houses, odour may come from the animals' bodies, from the floors and from the manure in the manure pit. According to Hansen (2005), the main source of odour is manure in the manure pit.

In terms of odour reduction, it would be very beneficial if a single compound or a group of compounds could be identified as an odour marker in a specific animal production system. This would allow the odour reduction solutions to focus on the odour marker. The literature does not show very consistent results for odour markers. However, in terms of odour nuisance, sulphur-containing compounds are proposed as the most important odour marker. Williams (1984) and Hobbs et al. (1997) produced a list of four major groups of odorants: VFA, indoles, phenols and sulphides. According to Curtis (1993), the odour groups are ammonia and volatile amines, sulphurous compounds, VFA, indoles and phenols, and alcohols and carbonyls. According to Le et al. (2007b), sulphur-containing compounds are the main odour markers. Since odour markers are not yet well established, evaluating odour by odour concentration, odour intensity and odour hedonic tone by olfactometry remains the best choice.

Odour compounds are mainly produced by the anaerobic digestion of organic matter present in the manure and the enzymatic hydrolysis of some urinary compounds (Aarnink and Verstegen, 2007). In particular, the incomplete anaerobic decomposition of the organic matter in the slurry results in foul odours (Webb et al., 2014). The availability, type and level of odour precursors in the digestive tract of the animals and in the manure determine the production of odorous compounds. The fundamental strategy to reduce odour at the source of production by dietary approaches is to minimize potential odour precursors from protein and its metabolic products. This strategy can be achieved using two approaches:

- Reducing the amount of protein/AAs in the diet by formulating diets that closely match the animals' requirement so that fewer excess AAs are absorbed in the small intestine and less protein enters the large intestine. Fewer odour precursors are then excreted via urine and faeces.
- Balancing diets in such a way that non-utilized nutrients are optimally used for bacterial biomass synthesis so that fewer odour precursors are produced and converted into odorous compounds.

3.1 Reducing the amount of dietary crude protein

A good base for reducing nitrogen excretion and odour production is by reducing the amount of protein in the diet. Excessive protein from the diet is excreted in three forms relevant for odour production: (1) urea, glucuronides and sulphate in urine; (2) non-digested proteins in faeces and (3) bacterial proteins in faeces. These excreta, especially the first two forms, are major precursors of odour formation. Thus, reducing dietary protein is expected to reduce odour compounds from pig manure. Hobbs et al. (1996) showed that five out of ten odorous compounds in the manure of growing pigs and nine out of ten odorous compounds in the manure of finishing pigs declined when pigs were fed reduced CP diets supplemented with AA, compared with pigs that were fed commercial diets. They also reported reductions of total VFA, branched-chain VFA, p-cresol, indole and 3-methyl indole in manure from pigs that were fed low-protein diets (14% and 13% CP for grower and finisher diets, respectively) compared with pigs that were fed high-protein diets (21% and 19% CP for grower and finisher diets, respectively). Sutton et al. (1998) reported a 62% reduction of volatile organic sulphur compounds (dimethyl sulphide, dimethyl disulphide, dimethyl trisulphide, dimethyl sulphoxide and carbon disulphide) in 53-kg gilts when their diet with 13% CP was compared with a diet with 8% CP and supplemented AA. According to Stevens et al. (1993), increasing the protein content of diets increased the excretion of sulphurous compounds capable of producing sulphide under anaerobic conditions.

The relationship between each odour compound or groups of odour compounds and odour nuisance is not yet well established. The reduction of a certain odour compound or groups of odour compounds may not always result in odour nuisance reduction. Therefore, it is necessary to evaluate the effect of dietary solutions on odour production and emissions by using olfactometry. Le et al. (2007a) compared three levels of dietary CP (12%, 15% and 18%) and measured odour concentration and emissions of the air collected above the manure pits by olfactometry according to the guidelines of the European Committee for Standardization (European Committee for Standardization, 2003). Odour concentration and odour emissions from pig manure were highest with the 18% CP treatment at 31 888 ou_E/m^3 and 4.46 $ou_E/s/m^2$, respectively. The 12% CP treatment had the lowest odour concentration and lowest odour emissions at 7259 ou_E/m^3 and 1.03 $ou_E/s/m^2$, respectively. Further analyses showed that the 12% CP treatment had 80% lower odour concentration and odour emissions than the 18% CP treatment. The 15% CP treatment had intermediate results, but did not differ from the 12% CP treatment or from the 18% CP treatment. When considering the hedonic tone of the odour emitting from the manure of pigs fed the three CP levels, the unpleasantness of the odour was similar when diets with 12% CP or 15% CP were fed, while the odour was clearly more unpleasant when the diet with 18% CP was fed to the animals.

They also found that reduced dietary CP levels decreased total nitrogen, methyl sulphide, carbon disulphide, ethanethiol, phenol, 4-ethyl phenol, indole and 3-methyl indole concentrations in the manure. VFAs and cresol concentrations in the manure of pigs fed different dietary CP levels were similar. They concluded that reducing dietary CP levels and at the same time providing essential AA is an option to reduce odour concentration and emissions from pig manure. This study confirmed the findings of Hayers et al. (2003), who found that odour emissions were reduced by 31% and 33% by decreasing dietary CP content from 19% to 16% and 13%, respectively. However, Obrock et al. (1997) found no difference in odour concentration between finishing pigs fed 13% and 9% CP with AA-supplemented diets. Possibly, the protein levels in the study by Obrock et al. (1997) were so low that they neither did cause any difference in odour concentration from the pig manure nor had sufficiently high levels of FCs. The latter are considered to have a significant effect on odour concentration and emissions.

The types of protein have significant effects on odour production. According to van Heugten and van Kempen (2002), diets containing fishmeal and a high sulphur content, as a result of adding up to 12% feather meal to the diet, showed a high odour concentration. So, sulphurous compounds can be considered as important odour markers. This was confirmed in a study by Le et al. (2007b). The experiment included three treatments: (1) a 15% CP basal diet with three times the requirement of sulphur-containing AA (SAA) (14.2 g/kg diet); (2) a basal diet with two times the requirement of Tryptophan (Trp) and Phenylalanine (Phe)+Tyrosine (Tyr) (2.9 and 20.4 g/kg diet, respectively) (TAA); and (3) a basal diet with AA supplementation to levels sufficient for the maximum protein gain (NOAA – diet without feed-use amino acids). Air samples were collected from the manure pits for odour concentration and odour emissions, odour intensity and hedonic tone measurements. Results showed that odour concentration and odour emissions from pig manure were highest from pigs given the SAA treatment, at 111, 302 ou_E/m^3 and 15.48 $ou_E/s/m^2$, respectively. The NOAA treatment had the lowest odour concentration and odour emissions, at 13 224 ou_E/m^3 and 1.88 $ou_E/s/m^2$, respectively. Supplementing crystalline Trp, Tyr and Phe in excess of recommended requirements did not affect odour concentration and odour emissions. In practice, diets are formulated with a wide range of raw materials. Some ingredients are rich in sulphur, for example, blood meal, soybean meal and molasses. To reduce odour from pig manure, these ingredients should be used at an appropriate level. In addition, sulphur-containing AA should be formulated so that they meet the animals' dietary requirement, but no more.

3.2 Balancing diets for bacterial biomass synthesis

The ratio of CP to FCs may affect microbial activities in the large intestine of animals and in manure stores (Bergman, 1990; Gibson and Roberfroid, 1995; Reid and Hillman, 1999). There is an optimum ratio between CP and FC at which odour production is minimized. At this ratio, not much protein is used as an energy source for bacteria. It is used to synthesize microbial biomass, so less protein is broken down into odour precursors and less odour is produced. This hypothesis was confirmed by Le et al. (2008a). They conducted a study to determine the effects of dietary levels of CP and levels of FCs and their interaction on odour concentration and emissions. There were two dietary CP levels (low, 12%; high, 18%) and three digestible FC levels: (low, 95.5; medium, 145.5 and high, 195.5 g/kg feed). Air samples were collected directly above the manure in the manure pit. The dietary CP level and FC level did not affect odour concentration and emissions, but their interaction did. At a high dietary CP level, an increased level of FC decreased odour emissions, while at a low CP level, an increased amount of FC increased odour emissions from the pig manure. Odour production from pig manure is low when pigs are fed diets with a low CP content and a low-FC diet or a high-CP combined with a high-FC diet. The finding implies that CP and FC do affect odour production from pig manure in combination with each other. To increase odour reduction, a certain dietary factor must be considered in association with other dietary factors. In practice, pig diets contain much more protein than the pigs' requirement, and the effect of excess CP on odour production can be lowered by increasing the level of dietary FC.

Another potential approach to managing odour from pig manure is by utilizing different feed formulation procedures. In many parts of the world, pig diets are formulated on a total, or on an apparent, ileal digestible AA basis. A standardized ileal digestibility formulation can be used to minimize excess AA in the diet. This formulation takes into account endogenous losses caused by specific feed ingredients and uses these values to determine AA values for particular feedstuffs (Rademacher, 2000). When developing a diet based on standardized ileal AA, less AA is needed in the diet to meet the pigs' nutrient requirements. In addition, formulating diets that have optimal ratios of ileal-absorbable AA will reduce excess AA in the diet. Other dietary approaches to reduce excess protein are phase feeding and individual feeding. These approaches also aim to maximize nutrient utilization or to formulate diets that closely match the animals' dietary requirements and thus reduce excess nutrients that result in odour production.

3.3 Parallel reduction of odour and ammonia emission from pig manure

From previous paragraphs it is clear that dietary solutions can be used to reduce odour emissions or ammonia emissions. A question of concern in many studies is whether dietary solutions that simultaneously reduce both odour and ammonia emissions exist. Inconsistent findings have been reported in the literature in this regard. From various experiments on the effects of the CP level, the fermentable protein level, types of AA and FCs, Le et al. (2007a,b,c, 2008a,b, 2009) found a low and not significant linear correlation ($P > 0.05$) between ammonia and odour emissions (from the slurry), ranging from 0.1 to −0.3. This means that the dietary strategies which reduce ammonia emissions may not be very effective for reducing odour emissions. This was confirmed by Hansen et al. (2007). They found that reducing dietary CP levels from 16% to 14%, adding 3% of benzoic acid and including 15% of inulin reduced ammonia emissions (measured from the outgoing air of the chamber) by 30%, 57% and 34%, respectively. However, it was not possible to influence odour emissions (odour concentration measured by olfactometry) in any of the aforementioned dietary alternatives.

The relationship between ammonia and odour emissions was also studied by Leek et al. (2007) in an experiment on finishing pigs which were fed four diets containing 13%, 16%, 19% and 21% of CP, respectively. They found that reducing the dietary CP concentration linearly decreased ammonia emissions. There was a quadratic relationship between the dietary CP concentration and odour emissions (measured from the outgoing air of the sealed pens). Odour emissions decreased between 21% and 16% CP and increased between 16% and 13% CP. They concluded that reducing dietary CP levels could be used effectively to reduce ammonia emissions and odour emissions, although no significant advantage was to be gained in odour emissions from reducing the CP level below 16%. Hayes et al. (2004) assessed the influence of CP content in finishing pig diets on odour and ammonia emissions from the animal house. Four diets containing 13%, 16%, 19% and 22% CP, respectively, were used. The odour emissions per livestock unit (500 kg) for the 13%, 16%, 19% and 22% CP were 77.6, 80.0, 115.8 and 102.9 Ou_E/s/LU, respectively ($P < 0.01$). The ammonia emissions per animal for the 13%, 16%, 19% and 22% CP diets were 3.11, 3.89, 5.89 and 8.27 g/day per animal, respectively ($P < 0.001$). They concluded that the manipulation of dietary CP levels offers a start-of-pipe solution to reduce odour and ammonia emissions from finishing pig houses; however, more work needs to be done in order to confirm this conclusion.

Similarly, manure pH is a driving factor behind the correlation between ammonia and odour emissions. High manure pH increases ammonia emissions. At the same time, at a high pH, hydrogen sulphide is in an ionized form, which means that the emission of hydrogen sulphide is low. The role of manure pH should therefore be further investigated.

An experiment should be conducted to determine the effects on ammonia and odour emissions of acidifying or alkalizing the manure from pigs. In such an experiment, diets should be equal in CP and FC content, which affect both ammonia and odour emissions. Different levels of manure pH can be created by using a different dietary electrolyte balance (dEB) or by supplementing diets with calcium salts, for example, calcium carbonate, calcium benzoate or calcium chloride. Salts with sulphur should not be used in such experiments because they can produce odour. An acid or a base can also be artificially added to different types of manure collected from pigs fed diets with different combinations of CP and FC content.

4 Nutrition and greenhouse gas emissions

According to the Food and Agriculture Organization (FAO), in 2005 livestock contributed 14.5%, or 7.1 gigatonnes of CO_2-eq per year to human-induced GHG emissions (Gerber et al., 2013). The same report concluded that pigs contribute 9% to livestock GHG emissions and methane is the main contributor to these emissions (44%), followed by nitrous oxide (29%) and carbon dioxide (27%). Methane and nitrous oxide have a high global warming potential because they absorb infrared radiation much more strongly than carbon dioxide (Lashof and Ahuja, 1990). It is estimated that the impact of one molecule of methane and nitrous oxide on global warming is 28 and 265 times that of carbon dioxide, respectively (Stocker, 2014).

In pig production, two main sources of methane can be distinguished: (1) the enteric fermentation of carbohydrates in the intestines of the pigs and (2) the fermentation (anaerobic digestion) of stored manure. Monteny et al. (2001) estimated, based on Dutch data at that time, an average methane emission in pig production of 4.8 kg/year/pig (all categories), of which 1.5 kg results from enteric fermentation and 3.3 kg results from the manure. In a study concerning the frequent and separate removal of urine and faeces with a manure belt system, an average methane emission in growing-finishing pigs of 0.94 kg/year was found (Aarnink et al., 2007). This amount is most probably only from enteric fermentation. Enteric fermentation in pigs is mainly determined by the amount of FCs in the diet. In a Dutch study, the average daily intake of NSPs was 0.17 kg/day in weaned piglets, 0.57 kg/day in growing-finishing pigs and 0.98 kg/day in pregnant sows (Aarnink et al., 2017). When a linear relationship between NSP intake and methane production in pigs is assumed and when taking the 0.94 kg/year in growing-finishing as a reference, we estimate that enteric methane emissions are 0.28 kg/year for weaned piglets and 1.6 kg/year for pregnant sows.

Table 1 provides the methane and nitrous oxide emissions for different pig houses in the Netherlands. The total average methane emissions were measured at 15.7 kg/year in four traditional houses for growing-finishing pigs with a storage period of the manure of approximately six months and at 2.1 kg/year in four houses for growing-finishing pigs with a storage period of the manure of approximately one month (Mosquera et al., 2011a). Total average methane emissions of 5.1 kg/year were found in two partially slatted houses and of 1.3 kg/year in two fully slatted houses for weaned piglets, both with a storage period of the manure of approximately six weeks (Winkel et al., 2011). For pregnant sows in partially slatted houses, average methane emissions were 21.6 kg/year for individually housed sows and 25.2 kg/year for group housed sows, both with a storage period of the manure of approximately six months (Mosquera et al., 2011b).

Published by Burleigh Dodds Science Publishing Limited, 2018.

Table 1 Methane and nitrous oxide emissions in pig production (per pig place[a]) in a study in the Netherlands

Category	No[b]	Storage period (months)	Methane (kg/y)	Nitrous oxide (g/y)	Reference
Growing-finishing, traditional housing	4	6	15.7 (10.9)	8.0 (5.3)	Mosquera et al. (2011a)
Growing-finishing, low emission, dry feed	2	1	1.9 (0.2)	7.5 (2.4)	Mosquera et al. (2011a)
Growing-finishing, low emission, wet feed	2	1	2.3 (1.0)	8.8 (8.7)	Mosquera et al. (2011a)
Weaned piglets, partially slatted	2	1.5	5.1 (3.4)	0.9 (0.9)	Winkel et al. (2011)
Weaned piglets, fully slatted	2	1.5	1.3 (0.2)	4.2 (5.4)	Winkel et al. (2011)
Pregnant sows, individually housed	2	6	21.6 (11.4)	17.4 (15.7)	Mosquera et al. (2011b)
Pregnant sows, group housed	2	6	25.2 (0.5)	124.1 (25.4)	Mosquera et al. (2011b)

Standard deviation between different animal houses is given in brackets.
[a] Calculated with an inoccupation period of 3% for growing-finishing pigs and pregnant sows and of 9% for weaned piglets.
[b] Number of pig houses.

Emissions of nitrous oxide are generally low in slurry-based systems. In deep-bedded systems, nitrous oxide emissions can be considerable and can even be higher than the ammonia emissions (Groenestein and Van Faassen, 1996). Nitrous oxide is only formed when ammonia can be oxidized. In slurry, oxygen concentrations are too low for this process. In litter systems, however, generally, enough oxygen is available. Nitrous oxide is not formed under optimal oxidation processes (Monteny et al., 2001); it is only formed when temporarily low oxygen concentrations cause denitrification of the nitrates in the manure. Nitrous oxide emissions were very similar within different housing and feeding types of growing-finishing pigs (Table 1). Average emission levels of 8.1 g/year/pig place were found in eight houses (Mosquera et al., 2011a). In two partially and two fully slatted houses for weaned piglets, nitrous oxide emissions were 0.9 and 4.2 g/year, respectively (Winkel et al., 2011). In pregnant sows, nitrous oxide emissions were, on average, 17.4 and 124.1 g/year for individually and group-housed sows, respectively (Mosquera et al., 2011b). It is not clear why differences in nitrous oxide emissions occur between the different housing systems of weaned piglets and pregnant sows, but Table 1 shows that variations between similar houses are also high. From the data in Table 1 it can be concluded that nitrous oxide emissions are generally very low in slurry systems (almost a factor of 1000 lower than the methane emissions).

From the above discussion it can be concluded that housing is the main approach through which methane and nitrous oxide emissions could be reduced. Methane emissions can be significantly reduced if the manure is stored for a short period only, or removed completely, so that little bacterial biomass can develop inside the manure pit to anaerobically convert carbohydrates into methane. Nitrous oxide emissions can be prevented by slurry systems (low oxygen concentrations) or by bedding systems in which oxidation processes are optimized (high oxygen concentrations).

Nutritional means to lower methane emissions should focus on lowering the FCs in the diet. While FCs play a crucial role in determining levels of ammonia emissions and odour as well, these concentrations should be optimized and should be harmonized with the protein concentrations.

Nutritional means to lower nitrous oxide emission should focus on lowering the excretion of urea. To realize this, similar measures as those for reducing ammonia emissions can be taken. This means carefully tailoring the daily protein intake with the requirements of the pigs. Limiting the intake of AAs can also be helpful in this respect.

5 Effectiveness of dietary solutions for reduction of gaseous emissions

The effects of dietary manipulation on animal performance is a major concern since reducing the protein level in diets may reduce animal productivity. As reviewed by Bindelle et al. (2008), dietary fibre can reduce the digestibility of most nutrients and the effect is highly dependent on the physico-chemical characteristics of the fibre (Wenk, 2001). This interaction can change the viscosity of digesta, resulting in changes in passage rate, thus affecting the digestibility of nutrients. In particular, protein digestibility may be affected in this way, as reviewed previously, but results of various studies are equivocal (Dikeman and Fahey, 2006). However, it is well documented that reducing the dietary CP level at the same time as supplementing the diet with AA, especially essential AA, will maintain animal performance. Oldenburg and Heinrichs (1996) found no negative effects on the productivity and leanness of pigs between 50 and 110 kg when protein levels in diets were reduced from 17% to 13.5%. According to Canh et al. (1998c), lowering dietary CP (16.5%, 14.5% and 12.5%) and supplementing the diet with AA can help maintain a normal growth rate. In an experiment in which dietary protein level was reduced from 19% to 15% in starter diets, from 16% to 12% in grower diets and from 14% to 11% in finisher diets, with or without AA supplements, Kerr et al. (1995) found that a reduction in pig performance and carcass muscle can be prevented by supplementing the diet with proper amounts of AA. According to Lopez et al. (1994) and Hahn et al. (1995), pigs fed reduced CP diets (a reduction of 3.5% to 4% units) supplemented with AA had similar carcass characteristics to pigs fed diets with a normal CP level. Le et al. (2007a, 2009) found similar average daily gains of pigs fed diets with CP levels of 12%, 15% and 18% supplemented with essential AA.

One study at four commercial pig farms examined the effects on ammonia emissions from pig houses in which a low-protein diet was fed, with and without the partial replacement of calcium carbonate by calcium chloride. The calcium excretion via urine was found to increase when the diet with calcium chloride was fed (Smits et al., 2012). Furthermore, the ash, calcium and phosphorus concentrations of the bones appeared to be slightly decreased in pigs fed the diet with calcium chloride. These results show that the acidification of the diet should be done carefully, preventing any disturbance of the acid–base balances of the pigs.

6 Conclusion

From the above discussion, it can be concluded that ammonia emissions can be reduced by lowering the CP content in the pigs' diet. Ammonia emissions are reduced when the surplus digested protein (the difference between digested protein and protein deposition in the body) is lowered. In some studies that adopted an *ad libitum* water intake, it was noticed that the pigs' water intake decreased when the protein levels of the diet were lowered. Because ammonia emissions are related to the ammonium concentration, and not to ammonium excretion, this will lower the effect of the CP reduction. On the other hand, lowering the CP content will also lower the pH of the manure. This increases the protein reduction effect on ammonia emissions.

Lowering the protein content of the diet also reduces the production and emissions of different odorous compounds. However, in this respect, the type of proteins in the diet is important, especially as proteins with a high sulphuric AA content generate high levels of odour. These sulphuric AAs should therefore be carefully tailored to the requirements of the pigs to reduce odour emissions.

FCs also play an important role in ammonia and odour emissions. Increasing the levels of FCs in the diet causes a shift in nitrogen excretion from urine to faeces. Furthermore, it increases the concentrations of VFAs in faeces and manure, causing a lower manure pH. Both effects reduce ammonia emissions. For odour emissions, it is important to optimize the ratio between CP and FC in the large intestine. At the optimal ratio, a maximum amount of CP is used for microbial biomass, so less can be broken down into different odorous compounds.

Methane and nitrous oxide emissions are mainly affected by the storage and management of manure. Methane emissions can be reduced to some extent by reducing the amount of FC in the diet. However, this may contradict the nutritional strategies that are used to reduce ammonia and odour emissions.

7 Future trends

The authors suggest the following as areas for further study in future research:

- The role of manure pH in the emission of different odorous compounds and total odour.
- The optimal ratio between CP and FCs in diets for the different growth stages of pigs to achieve low ammonia and odour emissions, with acceptable levels of enteric methane production.
- A nutritional means to lower the pH of manure without negatively affecting the acid–base balance of the pigs.
- The relationship between concentrations of the most important odorous compounds and the odour concentration determined by olfactometry.
- The effect of various nutritional means (additive or interactive) on odour, ammonia and GHG emissions.
- The effects of nutritional strategies on manure characteristics, and subsequently on odour, ammonia and GHG emissions, at the animal-house level.

8 Where to look for further information

For more information about climate change and the contribution of livestock to GHG emissions, see:

Stocker, T. (2014). *Climate Change 2013: The Physical Science Basis: Working Group I Contribution to the Fifth Assessment Report of the Intergovernmental Panel on Climate Change.* Cambridge University Press.

Gerber, P., Steinfeld, H., Henderson, B., Mottet, A., Opio, C., Dijkman, J., Falcucci, A. and Tempio, G. (2013). *Tackling Climate Change through Livestock – A Global Assessment of Emissions and Mitigation Opportunities*, 115pp. FAO, Rome, Italy.

Garnett, T. (2009). Livestock-related GHG emissions: impacts and options for policy makers. *Environmental Science & Policy*, 12(4), 491–503.

For more information about ammonia emissions from livestock, see:

Sommer, S. G., Zhang, G. Q., Bannink, A., Chadwick, D., Misselbrook, T., Harrison, R., Hutchings, N. J., Menzi, H., Monteny, G. J., Oenema, O. and Webb, J. (2006). Algorithms determining ammonia emission from buildings housing cattle and pigs and from manure stores. *Advances in Agronomy*, 89, 261–335.

Webb, J., Menzi, H., Pain, B., Misselbrook, T., Dämmgen, U., Hendriks, H., and Döhler, H. (2005). Managing ammonia emissions from livestock production in Europe. *Environmental Pollution*, 135(3), 399–406.

For more information about odour emissions from livestock, see:

Hobbs, P., Webb, J., Mottram, T., Grant, B., and Misselbrook, T. (2004). Emissions of volatile organic compounds originating from UK livestock agriculture. *Journal of the Science of Food and Agriculture*, 84(11), 1414–20.

Schiffman, S. S., Bennett, J. L. and Raymer, J. H. (2001). Quantification of odors and odorants from swine operations in North Carolina. *Agricultural and Forest Meteorology*, 108(3), 213–40.

9 References

Aarnink, A. J. A. (1997). Ammonia emission from houses for growing pigs as affected by pen design, indoor climate and behaviour. In *Animal Sciences*, Wageningen Agricultural University, Wageningen, The Netherlands, p. 173.

Aarnink, A. J. A. and Verstegen, M. W. A. (2007). Nutrition, key factor to reduce environmental load from pig production. *Livestock Science* 109(1): 194–203.

Aarnink, A. J. A., Hoeksma, P. and Ouwerkerk, E. N. J. (1993). Factors affecting ammonium concentration in slurry from fattening pigs. In *Nitrogen Flow in Pig Production and Environmental Consequences*, pp. 413–20 (M. W. A. Verstegen, L. A. den Hartog and G. J. M. van Kempen (Eds)). EAAP Pub No. 69, Pudoc, Wageningen, The Netherlands.

Aarnink, A. J. A., Huis in 't Veld, J., Hol, A. and Vermeij, I. (2007). *Kempfarm vleesvarkensstal: milieu-emissies en investeringskosten*, 32pp. Animal Sciences Group, Lelystad.

Aarnink, A. J. A., Van de Pas, P. A., Van der Peet-Schwering, C. M. C., Hol, A., Binnendijk, G. P., Le Dinh, P., Hafner, S. D. and Ogink, N. W. M. (2017). *Rekentool voor het bepalen van de effecten van voer- en management-maatregelen op de ammoniakemissie bij varkens: ontwikkeling en validatie*. Wageningen Livestock Research, Wageningen, The Netherlands.

Bakker, G. C. M. (1996). Interaction between carbohydrates and fat in pig diets; impact on energy evaluation of feeds, PhD thesis, Wageningen Agricultural University, Wageningen, The Netherlands, p. 191.

Banhazi, T. M., Seedorf, J., Rutley, D. L. and Pitchford, W. S. (2008). Identification of risk factors for sub-optimal housing conditions in Australian piggeries: Part 2. Airborne pollutants. *Journal of Agricultural Safety and Health* 14: 21–39.

Bergman, E. N. (1990). Energy contributions of volatile fatty acids from the gastrointestinal tract in various species. *Physiological Review* 70(2): 567–90.

Bindelle, J., Leterme, P. and Buldgen, A. (2008). Nutritional and environmental consequences of dietary fibre in pig nutrition: A review. *Biotechnology, Agronomy, Society and Environment* 12: 69–80.

Bühler, K., Wenk, C., Broz, J. and Gebert, S. (2007). Influence of benzoic acid and dietary protein level on performance, nitrogen metabolism and urinary pH in growing-finishing pigs. *Journal of Archives of Animal Nutrition* 60(5): 382–9.

Canh, T. T. (1998). Ammonia emission from excreta of growing-finishing pigs as affected by dietary composition, PhD Dissertation, Wageningen Agricultural University, Wageningen, The Netherlands, p. 163.

Canh, T. T., Verstegen, M. W. A., Aarnink, A. J. A. and Schrama, J. W. (1997). Influence of dietary factors on nitrogen portioning and composition of urine and faeces of fattening pigs. *Journal of Animal Science* 75: 700–6.

Canh, T. T., Aarnink, A. J. A., Verstegen, M. W. A. and Schrama, J. W. (1998a). Influences of dietary factors on the pH and ammonia emissions of slurry from growing-finishing pigs. *Journal of Animal Science* 76(4): 1123–30.

Canh, T. T., Aarnink, A. J. A., Mroz, Z., Jongbloed, A. W., Schrama, J. W. and Verstegen, M. W. A. (1998b). Influences of electrolyte balance and acidifying calcium salts in the diet of growing-finishing pigs on urinary pH, slurry pH and ammonia volatilisation from slurry. *Livestock Production Science* 56: 1–13.

Canh, T. T., Aarnink, A. J. A., Schutte, J. B., Sutton, A., Langhout, D. J. and Verstegen, M. W. A. (1998c). Dietary protein affects nitrogen excretion and ammonia emission from slurry of growing–finishing pigs. *Livestock Production Science* 56(3): 181–91.

Canh, T. T., Schrama, J. W., Aarnink, A. J. A., Verstegen, M. W. A., van Klooster, C. E. and Heetkamp, M. J. W. (1998d). Effect of dietary fermentable fibre from pressed sugar-beet pulp silage on ammonia emission from slurry of growing-finishing pigs. *Animal Science* 67(3): 583–90.

Canh, T. T., Sutton, A. L., Aarnink, A. J. A., Verstegen, M. W. A., Schrama, J. W. and Bakker, G. C. M. (1998e). Dietary carbohydrates alter faecal composition and pH and ammonia emission from slurry of growing pigs. *Journal of Animal Science* 76(7): 1887–95.

Canh, T. T., Verstegen, M. W. A., Mui, N. B., Aarnink, A. J. A., Schrama, J. W., van't Klooster, C. E. and Duong, N. K. (1999). Effect of nonstarch polysaccharide-rich by-product diets on nitrogen excretion and nitrogen losses from slurry of growing-finishing pigs *Asian-Australian Journal of Animal Science* 12(4): 573–8.

CBS (2017). CBS StatLine. Centraal Bureau voor de Statistiek, Den Haag. http://statline.cbs.nl.

Cortus, E. L., Lemay, S. P., Barber, E. M., Hill, G. A. and Godbout, S. (2008). A dynamic model of ammonia emission from urine puddles. *Biosystems Engineering* 99: 390–402.

Cromwell, G. L. and Coffey, R. D. (1993). Future strategies to diminish nitrogen and phosphorus in swine manure. In *Proceedings of NPPC Environmental Symposium 'Meeting the Environmental Challenge'*, Minneapolis, MN. NPPC, Ames, IA, pp. 20–32.

Cromwell, G. L., Turner, L. W., Gates, R. S., Taraba, J. L., Lindemann, M. D., Traylor, S. L., Dozier III, W. A. and Monegue, H. J. (1999). Manipulation of swine diets to reduce gaseous emissions from manure that contributes to odour. *Journal of Animal Science* 77(Suppl. 1): 69 (Abstr.).

Curtis, S. E. (1993). *Environmental Management in Animal Agriculture.* Iowa State University Press, Ames, IA.

Daumer, M. -L, Guiziou F. and Dourmad, J. Y. (2007). Influence de la teneur en protéines de l'aliment et de l'addition d'acide benzoïque et de phytase microbienne sur les caractéristiques des effluents chez le porc à l'engraissement. *Journées Recherche Porcine* 39: 13–22.

DeCamp, S. A., Hill, B. E., Hankins, S. L., Bundy, D. C. and Powers, W. J. (2001). Effects of soybean hulls in commercial diet on pig performance, manure composition, and selected air quality parameters in swine facilities. *Journal of Animal Science* 79(Suppl. 1): 252 (Abstr.).

Dikeman, C. L. and Fahey, G. C. (2006). Viscosity as related to dietary fiber: A Review. *Critical Reviews in Food Science and Nutrition* 46: 649–63.

Donham, K. J. (2000). The concentration of swine production. Effects on swine health, productivity, human health and the environment. *Veterinary Clinician of North America: Food Animal Practitioner* 16(3): 559–97.

European Committee for Standardization. (2003). CEN standard 13725, Air Quality – Determination of Odour Concentration by Dynamic Olfactometry. European Committee for Standardization, Brussels, Belgium.

European Environment Agency. (2010). European community emission inventory report 1990–2008 under the UNECE convention on long-range transboundary air pollution. European Environment Agency, Copenhagen, Denmark.

FAOSTAT, 2005. Pigs density map. Agriculture and consumer protection department, Animal Production and Health Division, FAO, Rome.

Gac, A., Beline, F., Bioteau, T. and Maguet, K. (2007). A French inventory of gaseous emissions (CH4, N2O, NH3) from livestock manure management using a mass-flow approach. *Livestock Science* 112: 252–60.

Galloway, J. N., Dentener, F. J., Capone, D. G., Boyer, E. W., Howarth, R. W., Seitzinger, S. P., Asner, G. P., Cleveland, C. C., Green, P. A., Holland, E. A., Karl, D. M., Michaels, A. F., Porter, J. H., Townsend, A. R. and Vorosmarty, C. J. (2004). Nitrogen cycles: Past, present, and future. *Biogeochemistry* 70: 153–226.

Gerber, P., Steinfeld, H., Henderson, B., Mottet, A., Opio, C., Dijkman, J., Falcucci, A. and Tempio, G. (2013). *Tackling Climate Change through Livestock – A Global Assessment of Emissions and Mitigation Opportunities*, FAO, Rome, Italy. p. 115

Gibson, G. R. and Roberfroid, M. B. (1995). Dietary modulation of the human colonic microbiota: Introducing the concept of prebiotics. *Journal of Nutrition* 125(6): 1401–12.

Groenestein, C. M. and Van Faassen, H. G. (1996). Volatilization of ammonia, nitrous oxide and nitric oxide in deep-litter systems for fattening pigs. *Journal of Agricultural Engineering Research* 65: 269–74.

Hahn, J. D., Biehl, R. R. and Baker, D. H. (1995). Ideal digestible lysine level for early- and late-finishing swine. *Journal of Animal Science* 73(3): 773–84.

Hansen, C. F. (2005). *Grisenes indflydelse pa lugtemissionen*, 10pp. Landsudvalget for svin, danske slagterier.

Hansen, C. F., Sørensen, G. and Lyngbye, M. (2007). Reduced diet crude protein level, benzoic acid and inulin reduced ammonia, but failed to influence odour emission from finishing pigs. *Livestock Science* 109(1): 228–31.

Hayers, E. T., Curran, T. P. and Dodd, V. A. (2003). Odour and ammonia emissions from pig and poultry units. In *American Society of Agricultural Engineers, Annual International Meeting*, Las Vegas, NV.

Hayes, E. T., Leek, A. B. G., Curran, T. P., Dodd, V. A., Carton, O. T., Beattie, V. E. and O'Doherty, J. V. (2004). The influence of diet crude protein level on odour and ammonia emissions from finishing pig houses. *Bioresource Technology* 91(3): 309–15.

Hobbs, J. P., Brian, F. P., Roger, M. K. and Lee, P. A. (1996). Reduction of odorous compounds in fresh pig slurry by dietary control of crude protein. *Journal of the Science of Food and Agriculture* 74: 508–14.

Hobbs, P. J., Misselbrook, T. H. and Pain, B. F. (1997). Characterisation of odorous compounds and emissions from slurries produced from weaner pigs fed dry feed and liquid diets. *Journal of the Science of Food and Agriculture* 78(4): 437–45.

Hutchings, N. J., Sommer, S. G., Andersen, J. M. and Asman, W. A. H. (2001). A detailed ammonia emission inventory for Denmark. *Atmospheric Environment* 35: 1959–68.

Huynh, T. T. T., Aarnink, A. J. A., Drucker, A. and Verstegen, M. W. A. (2008). Pig Production in Cambodia, Laos, Philippines and Viet Nam: A review. *Asian Journal of Agriculture and Development* 3(1&2): 69–90.

Iverson, M., Kirychuk, S., Drost, I. L. and Jacobson, L. (2000). Human health effects of dust exposure in animal confinement buildings. *Journal of Agricultural Safety and Health* 6: 283–6.

Jacob, J. P., Blair, R., Benett, D. C., Scott, T. and Newbery, R. (1994). The effects of dietary protein and amino acid levels during the grower phase on nitrogen excretion of broiler chickens. In *Proceeding of the 29th Pacific Northwest Animal Nutrition Conference*, Vancouver, BC, Canada, p. 137.

Jongbloed, A. W. (2001). Hebben voermaatregelen ter verlaging van de ammoniakemissie een negatief effect op de vertering en benutting van nutriënten? (Do feeding measures for lowering ammonia emission have a negative effect on the digestibility and utilisation of nutrients?). In *Veevoeding en Ammoniakemissie uit Varkensstallen. Samenvattingen van de themamiddag*, Vol. 2174, pp. 11–23 (A. W. Jongbloed and M. C. Blok (Eds)). Productschap Diervoeder, The Hague, The Netherlands.

Jongbloed, A. W. and Lenis, N. P. (1993). Excretion of nitrogen and some minerals by livestock. In *Nitrogen Flow in Pig Production and Environmental Consequences*, pp. 22–36 (M. W. A. Verstegen, L. A. den Hartog and G. J. M. van Kempen (Eds)). EAAP Pub No. 69, Pudoc, Wageningen, The Netherlands.

Kay, R. M. and Lee, P. A. (1997). Ammonia emission from pig buildings and characteristics of slurry produced by pigs offered low crude protein diets. In *International Symposium on Ammonia and Odour Control from Animal Production Facilities*, Vol. I, pp. 253–9 (Eds J. A. M. Voermans and G. J. Monteny). Vinkeloord, The Netherlands. NVTL, Rosmalen, The Netherlands.

Kendall, D. C., Richert, B. T., Sutton, A. L., Frank, J. W., DeCamp, S. A., Bowers, K. A., Kelly, D. T. and Cobb, M. (1999). Effects of fibre addition (10% soybean hulls) to a reduced crude protein diet supplemented with synthetic amino acids versus a standard commercial diet on the performance, pit composition, odour and ammonia levels in swine buildings. *Journal of Animal Science* 77(Suppl. 1): 176 (Abstr.).

Kerr, B. J., McKeith, F. K. and Easter, R. A. (1995). Effect on performance and carcass characteristics of nursery to finisher pigs fed reduced crude protein, amino acid-supplemented diets. *Journal of Animal Science* 73: 433–40.

Kim, I. B., Ferket, P. R., Powers, W. J., Stein, H. H. and Van Kempen, T. A. T. G. (2004). Effects of different dietary acidifier sources of calcium and phosphorus on ammonia, methane and odorant emission from growing-finishing pigs. *Asian-Australasian Journal Animal Science* 17: 1131–8.

Kluge, H., Broz, J. and Eder, K. (2010). Effects of dietary benzoic acid on urinary pH and nutrient digestibility in lactating sows. *Livestock Science* 134: 119–21.

Krupa, S. V. (2003). Effects of atmospheric ammonia (NH3) on terrestrial vegetation: A review. *Environmental Pollution* 124: 179–221.

Lashof, D. A. and Ahuja, D. R. (1990). Relative contributions of greenhouse gas emissions to global warming. *Nature* 344(6266): 529–31.

Le, P. D., Aarnink, A. J. A., Ogink, N. W. M., Becker, P. M. and Verstegen, M. W. A. (2005). Odour from animal production facilities: Its relation to diet. *Nutrition Research Reviews* 18(1): 3–30.

Le, P. D., Aarnink, A. J. A., Jongbloed, A. W., van de Peet-Schwering, C. M. C., Verstegen, M. W. A. and Ogink, N. W. M. (2007a). Effects of dietary crude protein level on odour from pig manure. *Animal* 1: 734–44.

Le, P. D., Aarnink, A. J. A., Jongbloed, A. W., van der Peet Schwering, C. M. C., Ogink, N. W. M. and Verstegen, M. W. A. (2007b). Effects of crystalline amino acid supplementation to the diet on odour from pig manure. *Journal of Animal Science* 85(3): 791–801.

Le, P. D., Aarnink, A. J. A. and Verstegen, M. W. A. (2007c). Dietary composition influences odour and ammonia emissions from pig manure In *Ammonia Emissions in Agriculture*, pp. 78–81 (G.-J. Monteny and E. Hartung (Eds)). Wageningen Academic Publishers, Wageningen.

Le, D. P., Aarnink, A. J. A., Jongbloed, A. W., van de Peet-Schwering, C. M. C., Verstegen, M. W. A. and Ogink, N. W. M. (2008a). Interactive effects of dietary crude protein and fermentable carbohydrate levels on odour from pig manure. *Livestock Production Science* 114: 48–61.

Le, P. D., Aarnink, A. J. A., Jongbloed, A. W., van der Peet Schwering, C. M. C., Ogink, N. W. M. and Verstegen, M. W. A. (2008b). Content of dietary fermentable protein and odour from pig manure. *Animal Feed Science and Technology* 146(1–2): 98–112.

Le, P. D., Aarnink, A. J. A. and Jongbloed, A. W. (2009). Odour and ammonia emission from pig manure as affected by dietary crude protein level. *Livestock Science* 121(2–3): 267–74.

Leek, A. B. G., Hayes, E. T., Curran, T. P., Callan, J. J., Beattie, V. E., Dodd, V. A. and O'Doherty, J. V. (2007). The influence of manure composition on emissions of odour and ammonia from finishing pigs fed different concentrations of dietary crude protein. *Bioresource Technology* 98(18): 3431–9.

Lopez, J., Goodband, R. D., Allee, G. L., Jesse, G. W., Nelssen, J. L., Tokach, M. D., Spiers, D. and Becker, B. A. (1994). The effects of diets formulated on an ideal protein basis on growth performance, carcass characteristics, and thermal balance of finishing gilts housed in a hot, diurnal environment. *Journal of Animal Science* 72(2): 367–79.

Maassen, C. B. M. (2016). *Veehouderij en gezondheid omwonenden [Livestock and public health of residential people]*, RIVM, Bilthoven, The Netherlands, p. 133.

Monteny, G. J., Groenestein, C. M. and Hilhorst, M. A. (2001). Interactions and coupling between emissions of methane and nitrous oxide from animal husbandry. *Nutrient Cycling in Agroecosystems* 60: 123–32.

Morgan, C. A. and Whittemore, C. T. (1998). Dietary fibre and nitrogen excretion and retention by pigs. *Animal Feed Science and Technology* 19: 185–9.

Mosquera, J., Hol, J. M. G., Winkel, A., Lovink, E., Ogink, N. W. M. and Aarnink, A. J. A. (2011a). Fijnstofemissie uit stallen: vleesvarkens, 37pp. Rapport 292, herziene versie, Lelystad: Livestock Research.

Mosquera, J., Hol, J. M. G., Winkel, A., Nijeboer, G. M., Ogink, N. W. M. and Aarnink, A. J. A. (2011b). Fijnstofemissie uit stallen: dragende zeugen, Rapport 294, herziene versie. Lelystad: Livestock Research. p. 26.

Mroz, Z., Jongbloed, A. W., Canh, T. T. and Lenis, N. P. (1996). Lowering ammonia volatilization from pig excreta by manipulating dietary acid-base difference. In *Proceedings of the 8th Animal Science Congress of AAAP Tokyo*. Chiba (Tokyo), Japan.

Mroz, Z., Jongbloed, A. W., Partanen, K. H., Vreman, K., Kemme, P. A. and Kogut, J. (2000). The effect of calcium benzoate in diets with or without organic acids on dietary buffering capacity, apparent digestibility, retention of nutrients, and manure characteristics in swine. *Journal of Animal Science* 78: 2622–32.

Ndegwa, P. M., Hristov, A. N., Arogo, J. and Sheffield, R. E. (2008). A review of ammonia emission mitigation techniques for concentrated animal feeding operations. *Biosystem Engineering* 100: 453–69.

Nguyễn, H. Q., Lê Đình, P., Chim, C. and Veerle, F. (2017). Effect of different fiber sources in practical diets for growing pigs on animal performance, nitrogen balance and ammonia emission from slurry. *Animal Science and Feed Technology*, Submitted.

Obrock, H. C., Miller, P. S. and Lewis, A. J. (1997). The effect of reducing dietary crude protein concentration on odour in swine facilities, Nebraska Swine Report, University of Nebraska-Lincoln, Lincoln, NE, pp. 14–16.

O'Connell, J. M., Callan, J. J. and O'Doherty, J. V. (2006). The effect of dietary crude protein level, cereal type and exogenous enzyme supplementation on nutrient digestibility, nitrogen excretion, faecal volatile fatty acid concentration and ammonia emissions from pigs. *Animal Feed Science and Technology* 127(1): 73–88.

Oldenburg, J. and Heinrichs, P. (1996). Quantitative Aspecte einer proteinreduzierten Schweinemast. *Lohamann Information* 1: 13–16.
Olivier, J. G. J., Bouwman, A. F., Van der Hoek, K. W. and Berdowski, J. J. M. (1998). Global air emission inventories for anthropogenic sources of NOx, NH_3 and N_2O in 1990 *Environmental Pollution* 102: 135–48.
O'Neill, D. H. and Phillips, V. R. A. (1992). Review of the control of odour nuisance from livestock buildings: Part 3: Properties of the odorous substances which have been identified in livestock wastes or in the air around them. *Journal of Agricultural Engineering Research* 53(1): 23–50.
Portejoie, S., Dourmad, J., Martinez, J. and Lebreton, Y. (2004). Effect of lowering dietary crude protein on nitrogen excretion, manure composition and ammonia emission from fattening pigs. *Livestock Production Science* 91(1–2): 45–55.
Rademacher, M. (2000). How can diets be modified to minimize the impact of pig production on the environment? In *Amino News*, Huls AG technical Bulletin, Degussa, Frankfurt, Germany. Vol. 1, p. 10.
Reid, C. A. and Hillman, K. (1999). The effects of retrogradation and amylose/amylopectin ratio of starches on carbohydrate fermentation and microbial populations in the porcine colon. *Animal Science* 68(3): 503–10.
Reidy, B., Webb, J., Misselbrook, T. H., Menzi, H., Luesink, H. H., Hutchings, N. J., Eurich-Menden, B., Doher, H. and Dammgen, U. (2009). Comparison of models used for national agricultural ammonia emission inventories in Europe: Litter-based manure systems. *Atmospheric Environment* 43: 1632–40.
Schiffman, S. S., Bennett, J. L. and Raymer, J. H. (2001). Quantification of odors and odorants from swine operations in North Carolina. *Agricultural and Forest Meteorology* 108(3): 213–40.
Shriver, J. A., Carter, S. D., Sutton, A. L., Richert, B. T., Senne, B. W. and Pettey, L. A. (2003). Effects of adding fiber sources to reduced-crude protein, amino acid-supplemented diets on nitrogen excretion, growth performance, and carcass traits of finishing pigs. *Journal of Animal Science* 81(2): 492–502.
Smits, M. C. J., Hol, J. M. G., Van der Peet-Schwering, C. M. C. and Aarnink, A. J. A. (2012). Invloed van eiwitgehalte en toevoeging calciumchloride op de ammoniakemissie uit vleesvarkensstallen, Wageningen UR Livestock Research, Wageningen, The Netherlands, p. 38.
Stevens, R. J., Laughlin, R. J. and Frost, J. P. (1993). Effects of diet and storage time on the concentration of sulphides on dairy cow slurry. *Bioresource Technology* 45(1): 13–16.
Stocker, T. (2014). *Climate Change 2013: The Physical Science Basis: Working Group I Contribution to the Fifth Assessment Report of the Intergovernmental Panel on Climate Change*. Cambridge University Press, Cambridge, UK and New York, NY.
Sutton, A. L., Kephart, K. B., Patterson, J. A., Mumma, R., Kelly, D. T., Bous, E., Don, B. S., Jones, D. D. and Heber, A. J. (1997). Dietary manipulation to reduce ammonia and odorous compounds in excreta and anaerobic manure storage. In *International Symposium on Ammonia and Odour Control from Animal Production Facilities*, Vol. I, pp. 245–52 (J. A. M. Voermans and G. J. Monteny (Eds)). Vinkeloord, The Netherlands. NVTL, Rosmalen, The Netherlands.
Sutton, A. L., Patterson, J. A., Adeola, O. L., Richert, B. A., Kelly, D. T., Heber, A. J., Kephart, K. B., Mumma, R. and Bogus, E. (1998). Reducing sulfur-containing odours through diet manipulation. In *Animal Production Systems and the Environment: An International Conference on Odour, Water Quality, Nutrient Management and Socioeconomic Issues*, Iowa State University, Des Moines, IA, pp. 125–30.
Sutton, A. L., Kephart, K. B., Verstegen, M. W. A., Canh, T. T. and Hobbs, P. J. (1999). Potential for reduction of odorous compounds in swine manure through diet modification. *Journal of Animal Science* 77(2): 430–9.
van der Peet-Schwering, C. M. C., Aarnink, A. J. A., Rom, H. B. and Dourmad, J. Y. (1999). Ammonia emissions from pig houses in The Netherlands, Denmark and France. *Livestock Production Science* 58: 265–9.

van Heugten, E. and Van Kempen, T. A. (2002). Growth performance, carcass characteristics, nutrient digestibility and fecal odorous compounds in growing-finishing pigs fed diets containing hydrolyzed feather meal. *Journal of Animal Science* 80(1): 171–8.

van Kempen, T. A. T. G. (2001). Dietary adipic acid reduces ammonia emission from swine excreta. *Journal of Animal Science* 79: 2412–17.

Webb, J. and Misselbrook, T. H. (2004). A mass-flow model of ammonia emissions from UK livestock production. *Atmospheric Environment* 38: 2163–76.

Webb, J., Broomfield, M., Jones, S. and Donovan, B. (2014). Ammonia and odour emissions from UK pig farms and nitrogen leaching from outdoor pig production. A Review. *Science of the Total Environment* 470: 865–75.

Wenk, C. (2001). The role of dietary fibre in the digestive physiology of the pig. *Animal Feed Science and Technology* 90: 21–33.

Williams, A. G. (1984). Indicators of piggery slurry odour offensiveness. *Agricultural Waste* 10(1): 15–36.

Winkel, A., Mosquera, J., Hol, J. M. G., Van Hattum, T. G., Lovink, E., Ogink, N. W. M. and Aarnink, A. J. A. (2011). Fijnstofemissie uit stallen: biggen, Rapport 293, herziene versie. Livestock Research, Lelystad. p. 27.

Younes, H., Remesy, C., Behr, S. and Demigne, C. (1997). Fermentable carbohydrate exerts a urea-lowering effect in normal and nephrectomized rat. *American Journal of Physiology* 272: 515–21.

Zeeman, G. (1991). Mesophilic and psychrophilic digestion of liquid manure, PhD thesis, Wageningen Agricultural University, Wageningen, The Netherlands, p. 116.

Zervas, S. and Zijlstra, R. T. (2002). Effects of dietary protein and oat hull fibre on nitrogen excretion patterns and plasma urea in grower pigs. *Journal of Animal Science* 80: 3238–46.

Zijlstra, R. T., Oryschak, M. A., Zervas, S. and Ekpe, D. E. (2001). Diet manipulation to reduce nutrient content in swine manure. In *Focus on the Future Conference*. Red Deer, Alberta, Canada. http://triumph.usask.ca/psc/pdf/psc/69.pdf.

Chapter 11

Organic pig production systems, welfare and sustainability

Sandra Edwards, University of Newcastle, UK; and Christine Leeb, University of Natural Resources and Life Sciences, Austria

1 Introduction

Organic production systems may also be referred to as 'biological' or 'ecological' systems in some countries. On a worldwide basis, the International Federation of Organic Agriculture Movements (IFOAM) sets baseline standards for the various national organisations, and monitors national standards to assist with international trade. This organisation has defined Principles of Organic Agriculture (IFOAM, 2005) as:

- The Principle of Health
- The Principle of Ecology
- The Principle of Fairness
- The Principle of Care

Lampkin (1998) presented an elaboration of these principles as:

- To work with natural systems rather than seeking to dominate them
- To encourage and enhance biological cycles within the farming system, involving microorganisms, soil flora and fauna, plants and animals
- To maintain and increase the long-term fertility of soils

http://dx.doi.org/10.19103/AS.2017.0030.13

- To use as far as possible renewable resources in locally organised agricultural systems
- To work as much as possible within a closed system with regard to organic matter and nutrient elements
- To give all livestock conditions of life that allow them to perform all aspects of their innate behaviour
- To avoid all forms of pollution that may arise from agricultural techniques
- To maintain the genetic diversity of the agricultural system and its surroundings, including the protection of plant and wildlife habitats
- To allow agricultural producers an adequate return and satisfaction from their work including a safe working environment
- To consider the wider social and ecological impact of the farming system
- To produce food of high nutritional quality in sufficient quantity

From these objectives, it is apparent that organic production is a whole farm commitment with many interacting variables. In discussing organic pig production as an enterprise, its place within the whole farm system is therefore an important issue.

To be sustainable, organic pig production must meet the expectations of consumers and the wider society in relation to these production objectives. Only then can it expect to command the market premium which is necessary for the organic meat produced, in order to cover the higher costs inherent in such production requirements. Unless this can be achieved, the economic viability of the organic pig enterprise will be challenged and its long-term survival will be at risk.

2 Standards for organic pig production

In order for food to be labelled as organic, the farm where it is produced must be registered with an approved certification body. These organisations formulate the production rules which must be adhered to and provide a regular independent inspection to ensure that these rules are being correctly implemented on the farm. All European certifying bodies must, at a minimum, adhere to the legal standards set down by the European Council in Regulation (EC) No. 834/2007 (European Council, 2007). This sets out the principles, aims and overarching rules of organic production and defines how organic products should be labelled. Commission Regulation (EC) No. 889/2008 (European Commission, 2008) gives further details on the implementation of this regulation. In some countries, certification is done by a governmental body, but in others, it is implemented through private schemes (Früh et al., 2014). Within the UK, for example, there are currently nine different Certification Bodies, which have been approved by government, of which the best known and oldest is operated by the Soil Association. These private schemes may adopt standards which are stricter than those of the European regulation if they choose to do so. In the USA, organic pig production is regulated under the Organic Food Production Act of 1990 (USDA, 1990) and USDA National Organic Program 2010 [USDA (NOP), 2010] which, in comparison with the wider European Union (EU) requirements, places emphasis more on feeding and medication practices in organic certification, although it also prohibits confinement and requires access to outdoors. A summary of the regulatory framework in other countries can be found in Blair (2007).

It is not possible to have an organic pig enterprise in isolation, since pigs must be kept within an organic whole-farm system. However, it is beyond the scope of this chapter to give full organic standards for all farm enterprises. In this section, the key requirements for organic pig production in the EU which differ from conventional production practice will be outlined.

2.1 Land and housing

Permanent indoor housing of organic livestock is not permitted. Animals can be housed indoors for a maximum of 20% of their lifetime, but at other times must be kept either in fields or in housing where they have permanent access to an outdoor run. The land which outdoor pigs occupy must itself be of organic status and, in order to avoid problems of excessive nitrogen (N) deposition, the EC Regulations provide recommendations on the livestock carrying capacity of organic farms. The total amount of manure applied on the farm may not exceed 170 kg N/ha/annum. Table 1 summarises the resulting stocking density recommendations relating to pigs. These are for calculation of whole farm livestock carrying capacity, and do not specifically refer to individual paddocks.

In general, breeding sows and boars are often kept in outdoor paddocks with simple shelters in the same way as conventional outdoor production. Weaned and growing pigs can be kept in similar paddock systems, in outdoor hut-and-run systems, or in more permanent housing with an outdoor exercise and dunging area. All housing must have a bedded lying area and slats, while permitted, must not exceed 50% of the total floor area (25% in some schemes). The space requirements for organic pigs are greater than those conventionally used, as shown in Table 2. The use of tethers, stalls and farrowing crates is forbidden.

2.2 Feed

All feed supplied to organic pigs should itself be produced to organic standards. One of the greatest current challenges for organic pig producers is to find enough organically produced protein feed of appropriate quality. Most farms use diets based on organic cereals (wheat, barley, oats) and protein from peas, beans or soya in countries where these can be grown (Blair, 2007). There is a prohibition on any use of genetically modified (GM) crops or derivatives from GM organisms. The use of feed additives is also very limited; in the EU, Regulation 889/2008 has annexes listing permitted feed ingredients and additives. Mineral and vitamin supplements should only be used to correct known deficiencies, whilst all growth promoters (antibiotics, probiotics) are prohibited. All pigs must be given access to roughage or fodder, and farmers use such things as grazed grass/clover swards, conserved silage from such swards, from whole crop cereals or from maize, hay or straw,

Table 1 Stocking rates for pigs at pasture (EC Regulation 889/2008)

Class of pig	Maximum number of pigs/ha (equivalent to 170 kg N/ha/year)
Breeding sows	6.5
Piglets	74
Fattening pigs	14
Other pigs	14

Table 2 Minimum space requirements for organic pigs (EC Regulation 889/2008)

Class of pig	Indoor area (m^2/head)	Outdoor exercise area (m^2/head)
Dry sows	2.5	1.9
Boars	6.0	8.0
Lactating sow and litter – up to 40 days	7.5	2.5
Piglets – up to 30 kg	0.6	0.4
Fattening pigs – up to 50 kg	0.8	0.6
– up to 85 kg	1.1	0.8
– up to 110 kg	1.3	1.0

and root crops such as fodder beet. There is considerable scope for utilising by-products from other organic farm enterprises, such as surplus potatoes, carrots or green vegetable trimmings.

2.3 Breeding

The use of traditional, slower growing breeds is recommended to promote animal welfare under extensive conditions and to maintain genetic diversity within the species. Traditional breeds, such as the Saddleback and Tamworth breeds in the UK or the Mora Romagnola and Cinta Senese in Italy, are often used but suffer from poor carcass quality, with excess fatness as a consequence of their early maturing characteristics. Many organic farms therefore use the modern improved breeds developed for conventional systems (Früh et al., 2014). Replacement breeding stock should ideally come from within the herd or from other organic sources. This is easy to manage if purebred animals are being kept, but to exploit the well-documented benefits of heterosis requires organic breeders of F1 gilts or use of a criss-cross breeding system (Kelly et al., 2001). More sophisticated breeding structures for organic production would offer production and economic benefits (Leenhouwers et al., 2011), but dedicated purebred organic lines are currently not cost-effective because the sow population size is not sufficiently large.

2.4 Management and husbandry

Organic production systems seek to provide more natural conditions for the animals and forbid many of the practices which have become common in intensive husbandry systems. Piglets cannot be weaned at less than 6 weeks of age, and 8 weeks is recommended by some private schemes such as the UK Soil Association. Practices such as teeth clipping and tail docking should not be carried out, although castration is permitted when necessary by some schemes, according to the EU regulation with analgesia and/or anaesthesia. Nose ringing of sows to reduce pasture damage is prohibited except in Denmark.

2.5 Health management

The objective of organic production is to maintain good health through the adoption of effective management practices. For example, an emphasis is placed on good pasture management and regular rotation of paddocks to control parasite build-up. Organic production forbids the routine use of antibiotics, although it does not preclude their use for individual animals where there is a veterinary need. In such circumstances, longer withdrawal periods are specified (twice that required by law) before the animal can be slaughtered for meat. Animals which have received repeated antibiotic treatment cannot be sold for organic meat. According to the EU Directive, with the exception of vaccinations, treatments for parasites and compulsory eradication schemes, where an animal receives more than three courses of treatment with chemically synthesised allopathic veterinary medicinal products or antibiotics within 12 months, or more than one course of treatment if their productive lifecycle is less than 1 year, the livestock concerned, or products derived from them, may not be sold as organic.

3 Current organic production

The number of farms committed to organic production has been growing rapidly within Europe, with around 270 000 organic producers and a 20% increase in organically certified land area between 2010 and 2015. France, Germany, Italy and Spain each have more than 1 m ha of the 11 m total within the EU (Eurostat, http://ec.europa.eu/eurostat/statistics-explained/index.php/Organic_farming_statistics).

The number of organic pigs in Europe numbered about 1 million in 2016, with around 70 000 breeding sows reported by Eurostat. The biggest production is found in Denmark, France and Germany, with Austria, the Netherlands and Italy also having significant numbers (Fig. 1). In the USA, around 12 000 organic pigs were reported for 2011 (USDA, https://www.ers.usda.gov/data-products/organic-production/). Data on population sizes in other countries of the world are hard to find, but are generally of smaller scale.

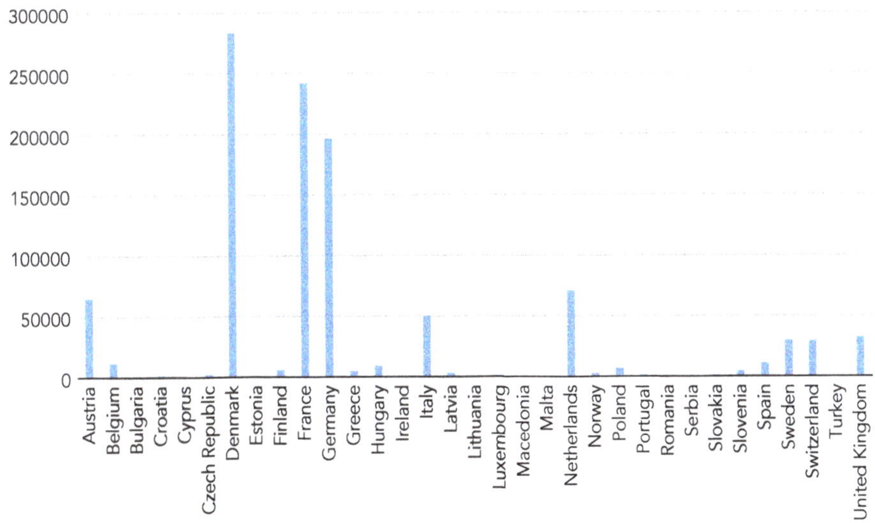

Figure 1 Organic pig numbers in different European countries in 2016 (Eurostat).

3.1 The diversity of organic systems

Although all farms must conform to organic standard requirements, there is variation in the way that these are interpreted by different national or private certification bodies which, together with the different climatic and geological constraints in different regions, has resulted in differences in typical keeping systems. Within Europe, these differences have been described by Früh et al. (2014). For example, in the UK, all production stages generally take place in outdoor paddocks for both sows (Fig. 2) and growing pigs (Fig. 3). In contrast, in Germany, Austria and Switzerland, most organic pigs are kept in permanent buildings with access to an outdoor area on concrete (Figs. 4 and 5). Other countries use a combination of these systems. This may depend on stage of production, with breeding sows mainly kept outdoors and fattening pigs in permanent buildings in Denmark and France, or on season of the year, with animals often outdoors in the summer but housed in the winter in Sweden. In the USA, pigs are often kept in hoop barns with deep straw bedding (Honeyman, 2005).

3.2 Production results of organic pig herds

There is relatively little reliable information on production levels in organic herds. Many herds are small and do not keep detailed records. Information from 74 organic pig farms in eight different European countries was gathered in the recent EU ProPIG project (Leeb et al., 2015). This showed great variability between farms, depending on breed, production system and quality of management, making average performance figures somewhat meaningless (Table 3).

While some herds achieved very good sow output levels, within the constraints of the high weaning age, others showed evidence of problems related to rebreeding of sows and mortality of piglets. Sow feed use was generally higher than that in conventional production, as would be expected with a long lactation period. Finishing herd performance also showed relatively poor feed conversion ratios, as would be expected from animals in

Figure 2 An outdoor system for organic sows. Photograph: C. Leeb.

Figure 3 An outdoor system for organic growing pigs. Photograph: C. Leeb.

housing with outdoor access and no environmental control, and fed lower nutrient density diets incorporating forages.

3.3 Product quality from organic pig systems

There are a number of reasons why consumers choose to purchase organic pig meat despite its higher price (Dransfield et al., 2005). Some of these relate to a belief that the product is more ethical because of better animal welfare and environmental impact than conventional production systems. These claims are discussed in the following sections. However, other reasons for purchase relate to a belief that the product is safer and healthier for human consumption. This derives partly from the ban on growth-promoting agents

Figure 4 A system for organic lactating sows with permanent housing and a concrete outside run. Photograph: C. Leeb.

Figure 5 A system for organic finishing pigs with permanent housing and a concrete outside run. Photograph: C. Leeb.

and very tight restrictions on antibiotic use in organic standards, leading consumers to believe that the risk from chemical residues and antibiotic resistant bacteria will be greatly reduced. Reviews of comparisons of microbiological safety of organic and conventionally produced meat (van Loo et al., 2012; Luecke, 2016) demonstrate conflicting results in different studies. Five studies reviewed by van Loo et al. (2012) and a further two studies reported by Luecke (2016) indicate a lower prevalence of antibiotic-resistant bacteria in organically reared pigs. However, since organic pigs must have outdoor access, they may have a higher risk of exposure to zoonotic agents from the environment or transmitted by wildlife. All five studies cited in the reviews reported a higher percentage of organic pigs carrying antibodies against *Toxoplasma gondii* than conventionally produced pigs. In contrast, *Salmonella* prevalence has been reported to be higher (two studies), lower (one study) or no different (two studies) in organic compared to conventional production, highlighting the varied farm-specific risk factors which contribute to this outcome.

Table 3 Performance in organic pig herds in Europe recorded in the EU ProPIG project (Rudolph, 2015)

	Minimum	Maximum
Piglets born per litter	6.0	16.5
Piglets weaned per sow per year	5.0	25.0
Piglet weight at weaning (kg)	5.5	28.0
Liveweight at slaughter (kg)	86	200
Feed per sow (kg/year)	675	2236
Fattening pig feed conversion ratio (kg feed/kg gain)	2.4	7.5

There is also a belief by some consumers that organic pig meat has an improved composition in relation to human nutrition and health. In a systematic review and meta-analysis of published studies comparing composition of organic and conventional meat, Srednicka-Tober et al. (2016) found very few studies relating to pig meat (4–6 studies for different parameters). These related primarily to fat quality and indicated that the concentrations of saturated fatty acids, deemed detrimental to health, were lower in organic pork, whereas concentrations of polyunsaturated fatty acids, perceived as beneficial to health, were found to be higher. These results may be explained by the ingestion of forage by organic pigs and the higher vegetable oil content in diets with cold-pressed rather than chemically extracted soya. However, there were differences in outcome between different studies, probably reflecting farm-specific differences in diet and husbandry of the animals, and any conclusions must be treated with caution.

Some consumers also believe that the organoleptic quality and eating experience of pig meat are improved by organic rearing. There is little objective evidence that this is the case, although many aspects of production which might be modified in organic systems, including use of traditional breeds, growth rate, carcass adiposity, choice of feed ingredients and pre-slaughter stress, have the potential to influence meat quality (Edwards, 2005). It has also been demonstrated that sensory evaluation of attributes of product quality can be influenced by cognitive factors (beliefs and attitudes), meaning that the perception of eating quality may be either positively or negatively influenced, depending on the ethical attitudes of consumers to the system in which the meat was produced (Oude Ophuis, 1994).

3.4 Economic results of organic pig herds

Information on the economics of organic pig production is also relatively sparse, since most current herds are small and do not carry out detailed performance or enterprise recording. Data gathered across different European countries during the EU Q-Porkchains project (Bonneau et al., 2011) indicated that herds producing pigs under organic standards had higher production costs than conventional systems (by 61%) or other non-organic differentiated production systems (by 39%), but also achieved much higher selling prices (by 69% and 34%, respectively) and thus a greater margin per kg of pig produced (by 188% and 4%, respectively). A more sophisticated analysis was made in the project of the economic sustainability of a subsample of 23 organic farms using the IDEA methodology, which goes further than the traditional short-term economic performance by also taking into account the farm's degree of independence, its capacity to pass on a viable business and the efficiency of its production process. This again showed that there was great variability between the farms, and indicated that economic sustainability indicators covered a range similar to that found in conventional pig production (Ilari-Antoine et al., 2014).

4 Animal welfare in organic pig production

There have been relatively few objective comparisons of pig welfare in organic and conventional production systems. This subject was reviewed in detail in the EU CorePig project and considered for each production stage in a series of papers published in 2014 (Edwards et al., 2014b). Debate about the relative welfare implications of different systems often stems from the multifactorial nature of animal welfare. If considered according to the welfare framework of the Five Freedoms (FAWC, 1979), the following general considerations about organic production systems might apply:

- Freedom from hunger and thirst – Possible negative effects if pigs are reliant on natural foraging or given poorer quality diets because of ingredient restrictions, but benefits for satiety from forage feeding.
- Freedom from thermal and physical discomfort – possible negative effects from climatic extremes in outdoor paddocks or less sophisticated housing, but benefits from the requirement for greater space and provision of bedding.
- Freedom from injury and disease – possible negative effects from reduced biosecurity, exposure to environmental/wildlife challenges and the possibility of reduced therapeutic treatment, but benefits from reduced population density and the possible use of more robust breeds.
- Freedom from fear and stress – possible negative effects from reduced human assistance for subordinate animals in extensive systems and stress at the time of slaughter from unfamiliarity with confinement and handling, but benefits from greater space allowance, enriched behavioural development and prohibition of some painful management procedures.
- Freedom to express normal behaviour – clear positive effects from a more diverse environment, greater space and forage feeding.

The way in which these trade-offs are perceived may be viewed differently by different people, depending on their perspective of what is important for animal welfare. Fraser (2003) highlighted how such perceptions vary. Those who equate animal welfare with 'natural living' are likely to believe organic production to be superior, whilst those who equate animal welfare with 'biological functioning' may perceive the greater environmental control and production management offered by conventional systems to give better welfare. Those who equate animal welfare with 'affective state' would be looking for objective evidence of a difference between the systems in signs of negative, but also positive, emotions experienced by the animals.

Recently, objective welfare data on organic pig farms have been gathered in two major EU projects, to provide the first large-scale objective assessment of organic pig welfare. The health and welfare of organic pigs from 101 farms across six EU countries, using selected animal-based parameters from the Welfare Quality® protocol, were assessed in the CorePig project (Dippel et al., 2014). Data were collected on sows, suckling and weaned piglets and growing-finishing pigs in 3–20 farms per country. In a follow-up study, a survey of 74 organic farms in eight different European countries in the EU ProPIG project gathered data in relation to different organic management systems (Leeb et al., 2015). Data were collected on both resource provision (such as flooring, quality and quantity of litter, type and number of drinkers, feeding system and rations), and welfare outcomes including clinical indicators (such as body condition score, lesions, lameness, diarrhoea, respiratory problems, ectoparasites), data on veterinary treatments (such as for reproductive problems, diarrhoea, respiratory problems) and production performance (conception, litter size, feed efficiency, mortality).

4.1 The breeding sow

Welfare considerations for the pregnant and lactating sow in organic production systems were reviewed by Edwards et al. (2014a). There was relatively little information on the comparative prevalence of sow health and welfare problems in organic and conventional units, as few large-scale published surveys were found. In general, papers cited in the review suggested that relatively few health and welfare problems were seen. In the CorePig study (Dippel et al., 2014), the most prevalent problems identified in sows were thinness (median farm prevalence 18.8%, range 0–81.0), injuries on the anterior part of the body (15.5%, 0–66.7), injuries on hind part of the body (7.9%, 0–50), obesity (4.9%, 0–50.0) and vulva lesions (3.5%, 0–42.9). In the subsequent ProPIG survey (Rudolph et al., 2017), median prevalence of most adverse welfare outcomes was zero in both outdoor and indoor-with-run systems. Although median lameness and Mastitis-Metritis-Agalactia (MMA) prevalence were zero in outdoor systems, slightly higher values were recorded in indoor-with-run systems (7% and 16% respectively), suggesting attention to these issues is required.

Possible risk factors for health and welfare conditions in organic sows, extrapolated from knowledge gained in conventional systems, suggested issues of particular importance to include outdoor access, roughage feeding, later weaning, less sophisticated diets and lack of good health management strategies. When considering the presence of hazards with the potential to give rise to sow health and welfare problems, organic sows therefore have more behavioural freedom, but may be exposed to greater climatic challenges, parasite infestation and risk of body condition loss. The most widely discussed problem for organic sows is that of parasites, both ectoparasites and endoparasites. These have been reported in many studies to be more prevalent than in conventional systems due to the housing conditions and restrictions on prophylactic chemical measures (Baumgartner et al., 2003; Eijck and Borgsteede, 2005), although endoparasite prevalence is lower in adult sows than in younger pigs (Katakam et al., 2016). The ProPIG survey found very few sows in all systems with signs of ectoparasites, which can partly be explained by high levels of antiparasitic treatments reported, especially in indoor-with-run systems (Rudolph, 2015).

4.2 The suckling piglet

Welfare considerations for the suckling piglet in organic production systems were reviewed by Prunier et al. (2014a). This particularly highlighted the issue of piglet mortality, with many studies reporting losses in excess of 20% and even 30%. In the CorePig survey the median prevalence of piglet mortality at birth and during lactation was 29.3% (with an interquartile range of 17.2–40.5%). In the ProPIG survey, total piglet losses were also high in all types of organic system (medians of 19–22%; Rudolph et al., 2017).

Among the most important hazards that can be identified from knowledge in conventional pig production are a high litter size, elevated sow parity, excessive or insufficient ambient temperature in less sophisticated housing and with degraded pastures for outdoor paddocks, longer suckling period, sometimes with grouping of sows during lactation, and insufficient protection against predators. Issues with particular importance for organic production include the selection of genotypes poorly adapted to organic farming, particularly prolific sow lines with large litter size, the microclimate surrounding newborn piglets, the hygiene management and the supervision of animals around birth.

Because of its economic and welfare importance, the topic of piglet mortality has been the subject of a significant amount of research in recent years. Prunier et al. (2014b) analysed piglet mortality in relation to data on farm characteristics and production practice in a subset of 42 CorePig herds. Production records showed a detrimental effect of high litter size at birth on piglet mortality and that high standard deviation in litter size may exacerbate this problem. In consequence, particular research effort on this topic is now underway in Denmark, where use of the highly prolific conventional Danish sow breed is normal in organic production (Rangstrup-Christensen et al., 2017). Breed and housing comparisons are in progress and will hopefully highlight possibilities for future improvement.

Other aspects of suckling piglet welfare were also considered in the 2014 review, which concluded that piglets on organic farms benefit from large space, longer lactation and environmental enrichment that allow them to perform their natural behaviour. In the CorePig study, the median prevalence for adverse welfare outcomes in terms of groups affected per farm was 0% for all parameters. However, one issue that was highlighted was the widespread use of surgical castration of male pigs in organic systems, often without provision of pain relief. Although studies are investigating the possibility for use of entire males in organic production (Thomsen et al., 2015), the high levels of boar taint which result on some farms currently preclude this. This is likely to be a particular problem with early-maturing traditional breeds, in extensive systems with slower growing pigs which are older at slaughter, higher slaughter weight in organic pigs and where diets contain less-idealised protein sources which give rise to additional taint compounds.

4.3 The weaned piglet

Welfare considerations for the weaned piglet in organic production systems were reviewed by Leeb et al. (2014). The predominant problems identified related to health issues including post-weaning diarrhoea, post-weaning multisystemic wasting syndrome and endoparasite infestation. It was concluded that diseases around weaning are multifactorial in nature, with several factors contributing simultaneously as stressors at the time of weaning. Although piglets benefit from later weaning age than in conventional systems, distress and frustration arising from separation from the mother, a new environment, mixing and fear of humans still occur. The limitations on use of animal-derived raw materials with high protein quality, and of synthetic amino acids, can exacerbate digestive problems at this time.

In the CorePig study, the median prevalence of adverse welfare outcomes in terms of groups affected per farm was 0%, but with a range of 0–100% for all parameters (Dippel et al., 2014). Results from the ProPIG survey showed that, whilst most welfare outcomes once again had zero median prevalence, diarrhoea in weaned piglets was zero in outdoor systems but 25% in indoor-with-run systems (Rudolph et al., 2017). Respiratory problems and runt pigs also had zero median prevalence in outdoor systems but showed a higher prevalence in indoor-with-run systems, highlighting the importance of hygiene management in these systems.

4.4 The finishing pig

Welfare considerations for the finishing pig in organic production systems were reviewed by Lindgren et al. (2014). The results of comparisons between organic and conventional production indicated much lower antibiotic use, in line with organic standards. For example, conventional Danish herds consumed three times as much antibiotics (anthelmintics not included) as the organic herds, whilst there was no difference in mortality rate. Slaughter data reported in the review indicated that organic pigs had fewer respiratory problems, skin lesions (including abscesses and hernias) and tail wounds compared to conventional pigs. On the other hand, carcass defects associated with joint lesions and white liver spots were more common among organic pigs. Data from large-scale studies reported subsequent to the review have not always supported these conclusions. A survey of lesions found during meat inspection in Danish pigs over a 1-year period, and covering more than 200000 organic/free-range pigs and more than 1 million conventionally raised pigs, reported significant differences in many health and welfare outcome indicators (Alban et al., 2015). Although the majority of the lesion types were recorded infrequently (<4%), a total of 13 lesion types (including old fractures, tail lesions, osteomyelitis) were more frequent among organic/free-range pigs than among conventional pigs, four lesion types were equally frequent (chronic pneumonia, chronic pleuritis, fresh fracture and head abscess) and four lesion types (abscess in leg/toe, hernia and scar/hock lesion) were recorded less frequently among organic/free-range pigs compared with conventionally raised pigs. Subsequently, Kongsted and Sørensen (2017) performed a similar analysis, but differentiated organic from conventionally produced free-range pigs. Both conventional free-range and organic free-range production systems were associated with increased prevalence of several types of lesion when compared with conventional indoor systems. These included white liver spots, tail lesions, arthritis, skin lesions, bone fractures and septicaemia. In contrast, leg swellings, hernias and hoof abscess were less prevalent in the free-range systems than in conventional indoor systems. Comparison of organic and conventional free-range production systems did not reveal any statistically significant differences and there was a marked herd effect. These results suggested possibilities for herd-level management interventions of the problems studied. Results from on-farm assessments in the ProPIG survey showed that, whilst most adverse welfare outcomes once again had zero median prevalence, some differences between different types of organic system were apparent. Respiratory problems in fattening pigs has zero median prevalence in outdoor systems, but 67% in indoor-with-run systems (Rudolph et al., 2017). Tail lesions also had a slightly higher prevalence indoors; however, mortality of fattening pigs was higher in outdoor systems (3%) than in indoor-with-run systems (1%).

The risk of parasitic infections in organic fattening pigs has been repeatedly highlighted from both liver and lung lesion prevalence at the abattoir and from faecal egg counts on farm. In a 1-year field study on five Danish organic pig farms (Katakam et al., 2016), the average prevalence across farm and season of *Ascaris suum* was 48% in weaners and 28% in finishers. However, the prevalence varied with season, increasing towards the end of the year when 83–96% of finishing pigs had fresh white liver spots. Although deep litter bedding appeared to be a less important source of *A. suum* eggs than previously believed, long-term pasture rotation to eliminate pasture contamination was found not to be possible, as eggs can remain viable in the environment for up to at least 9 years and re-infect new animals. To control endoparasites, outdoor areas should therefore be rotated with as long an interval as possible, by including the pigs in the farm crop rotation.

Other non-health aspects of finishing pig welfare can benefit from the housing conditions specified by organic standards. Greater space allowance in combination with outdoor housing, with functional wallows and access to grass and roots, or outdoor runs and roughage, can enhance many aspects of pig welfare and reduce injurious pen-mate-directed oral behaviours and aggression.

5 Environmental impact of organic pig production

Minimising negative effects on the environment is a high-profile claim of organic farming, but is often called into question in the case of pig production. There is much debate about the most appropriate way to assess this criterion, and different approaches can lead to different conclusions.

5.1 Resource use on organic farms with pigs

One of the less well-studied subject areas to date is the optimal pattern of land use and rotations for a sustainable pig enterprise within an organic farm. In the UK, a typical organic rotation with pigs might be a root crop, spring wheat, winter cereal, spring peas and spring cereal undersown with grass/clover ley for subsequent pig grazing (Martins et al., 2002). This gives a combination of legume crops which fix atmospheric N and build soil fertility, and crops which exploit the nutrients deposited by the pigs during their 3- to 4-month occupation of each area of the herbage ley.

Based on a Danish model of nutrient balances (Kristensen and Kristensen, 1997), a system has been devised for a farm with pigs as the only livestock, based on 0.7 livestock units per ha (where 1 LSU = one sow and 20 progeny for pork production). Such a farm would have 80% of its area in cereal production and 20% as grass clover leys, but would not be totally self-sufficient, needing to import 25% of its pig feed and also some manure (45 kg N/ha). The nutrient balance for such a farm is shown in Table 4.

The problem posed in such a system is the low level of N fixation with a relatively low percentage of the farm in grass/clover ley. This gives lower soil fertility and lower yields of subsequent cereal crops. Manure production to storage, for use on such crops, is low because only the slaughter pigs are kept in housing systems where manure can be

Table 4 Nutrient balance for an organic farm with pigs as the only livestock enterprise (after Kristensen and Kristensen, 1997)

Farm characteristics:			
Import of feed (%)		25	
Import of animal manure (kg N/ha)		45	
Area with grass/clover (%)		20	
Herd characteristics:			
Sows/100 ha		71	
Total feeding level (SFU/ha)		4277	
Meat produced (kg/ha)		1237	
Manure produced (kg N/ha)		91	
Land use and production:	% ha	DM/ha	N/ha
• grass/clover	20	3379	142
• whole crop cereals	11	4290	83
• spring cereals	48	2738	54
• winter cereals	21	3888	70
Average/ha		3279	78
N supply from cropping:	N applied	N residual	N available
Average/ha	47	30	77

captured. Because of the limited ability of pigs to utilise dietary fibre, the dry matter (DM) yield of the grass/clover ley, even at this low % of total crop area, cannot be fully utilised. Thus, the whole system is dependent on some import of both feed and manure. For this reason, farms with mixed pig and ruminant livestock enterprises are more biologically efficient and sustainable. Pigs might play a role in soil tillage within the cropping rotation (Andresen, 1998), and utilise crop residues to enhance nutrient cycling (Edwards, 2002; Blair, 2007).

5.2 Factors affecting the environmental impact of organic pig systems

Since the biggest contribution to the environmental impact from pig production comes from inefficiency of feed use, less sophisticated diets and poorer climatic control resulting in poorer feed efficiency will contribute to a higher environmental burden. These circumstances can often pertain to organic production. A particular problem relates to diets with poorly digestible protein ingredients or protein with a poorly balanced amino acid profile for pig meat production. These result in increased N excretion, with the potential for leaching into groundwater and watercourses where it can cause eutrophication, and increased volatilisation as gaseous ammonia which contributes to acid rain production. Since synthetic amino acids cannot be used in organic diets to idealise the protein composition, crude protein content and N excretion are frequently increased. Another nutrient that can give environmental problems is phosphorus (P), which is a potent source of eutrophication of waterways. The ability of pigs to utilise P in plant sources in the feed is limited because it is often bound within phytate molecules. In conventional production, phytase enzyme addition is widely used to enhance P utilisation, with the digestibility increased from 30% to 60–70% (Dourmad and Jondreville, 2007), but this enzyme is not permitted in organic diets. As a result, P excretion into the environment will be higher in organic systems.

Where pigs are kept on pasture, risks to the environment from excreted nutrients are particularly high, as pollution hotspots develop in favoured excretory areas (Watson et al., 2003) and the destruction of vegetation resulting from pig rooting activities removes the possibility of nutrient capture by herbage. To try and address this issue, some Danish organic farms use nose rings for the sows to prevent rooting. While aiding the environment, this practice has significant animal welfare implications and, as such, is not permitted by many certification bodies. A more recent approach to the problem is to try and integrate outdoor pigs with agroforestry, where the more robust trees can capture a greater proportion of excreted nutrients (Horsted et al., 2012).

In less-extensive organic systems, the housing and manure systems will influence the form in which excreted N enters the environment. In housing with solid or slatted floors and slurry collection, emission of ammonia is the main problem, whereas in deep litter systems, the N emissions also include nitrous oxide, a potent greenhouse gas (Rigolot et al., 2010). The ammonia emission in deep litter systems is generally higher than that in slurry systems, increasing from 14% to 25% of excreted N. If hygiene becomes poor in deep-litter bedded systems, ammonia emissions are particularly elevated. When pigs are kept in indoor-with-run systems, the deposition of excreta on the outdoor area can also result in high ammonia emissions, especially in warm weather. Nitrogen losses from an indoor-with-run organic system in Sweden were estimated to be 26–27% of N excreted, giving approximately three to four times higher ammonia emission than standard values from conventional pigs (Olsson et al., 2014). A larger fouled area, particularly in the outdoor run, partly explains this result. Good design of the outdoor area, with provision of peat as a rooting material, can decrease ammonia emission (Olsson et al., 2016). However, the problem with higher ammonia emission in organic pig production is largely due to the fact that organic pigs have more space in which to move around and a greater area of solid flooring, creating a conflict between welfare and environmental objectives.

5.3 Life Cycle Assessment of organic pig systems

Life Cycle Assessment (LCA) provides a standardised framework to quantify the environmental impact within the life cycle of a product (Basset-Mens et al., 2007). It relates the environmental impact, most commonly expressed in terms of greenhouse gas emissions (GHGEs), acidification potential (AP) and eutrophication potential (EP), to a functional unit of the product, such as kg of pig liveweight at slaughter or kg of pork. Nitrogen excretion from animals, and the associated ammonia emission, is one reason for acidification. Eutrophication is influenced by both N and P excretion, whilst GHGEs contributing to global warming include carbon dioxide (CO_2), methane (CH_4) and nitrous oxide (N_2O). Different system boundaries may be chosen for the analysis, such as cradle-to-farm gate, cradle-to-abattoir product or cradle-to-supermarket shelf. These differences in functional unit and system boundary can make comparison of different studies difficult.

Many of the early LCA studies were based on model scenarios or data from only a small number of farms. More recently, larger-scale comparisons of organic and conventional pig meat production systems based on actual farm survey data have been published. In general, emissions are most influenced by feed production burdens, with direct emissions from excreta in the fattening stage often making the next most important contribution. The first study to use real data from a large-scale survey of different farm types was carried out in the EU Q-Prochains project (Dourmad et al., 2014). They evaluated 15 contrasting pig production systems from five European countries, with data collected on 5–10 farms per system. Organic systems showed a GHGE of 2.4 kg CO_2-eq, an AP of 57 g SO_2-eq and an EP of 16 g PO_4-eq per kg pig liveweight. Animal housing, feed production, manure storage and manure spreading all resulted in higher absolute impact values in organic systems than in conventional ones. Compared with conventional systems, impacts per kg LW for climate change (GHGE), acidification, eutrophication, energy use and land occupation were +4%, −16%, +29%, +11% and +121% for organic systems, respectively. However, their analysis highlighted that the ranking of system impacts depends on which functional unit is used in a study. The degree of intensification was inversely proportional to environmental impacts expressed per kg of pig live weight produced, but proportional to them when expressed per ha of land used. The mean land occupation for organic systems (9.1 m^2/kg LW) was close to the values published previously for such systems by Basset-Mens and Van der Werf (2005; 9.9 m^2/kg LW) and Halberg et al. (2010; 6.9–9.2 m^2/kg LW).

When compared to previous studies based on scenario modelling, most studies, with the exception of Williams et al. (2006), have indicated higher environmental burdens per unit of pig meat produced with organic production (Table 5). However, Halberg et al. (2010) noted that, due to the integration of grass-clover in their organic crop rotations, organic systems had an estimated net soil carbon sequestration which, when included in the LCA, gave lower GHGE compared with conventional pig production. Differences between studies in the environmental impact indicators are mainly explained by differing

Table 5 Results from Life Cycle Assessment of organic pig production in comparison with conventional production (cf CON)

	Greenhouse gas emissions		Acidification potential		Eutrophication potential	
	kg CO_2-eq	cf CON (%)	g SO_2-eq	cf CON (%)	g PO_4-eq	cf CON (%)
Williams et al. (2006) UK	5.6	−11	129	−67	57	−43
Basset-Mens and Van Der Werf (2005) France	4.0	+73	37	+17	22	+4
Halberg et al. (2010) Denmark	2.8–3.3	+7–22	50–61	+35–45	25–38	+21–65
Dourmad et al. (2014) Five EU countries	2.4	+4	57	−16	16	+29
Kool et al. (2009) Four EU countries	4.3–5.0	+14–26				

1 kg liveweight was used as the functional unit, except in the study of Williams et al. which used 1 kg carcass weight as the functional unit.

measured or assumed levels of animal performance and, in the case of the study by Basset-Mens and Van der Werf (2005), due to higher N_2O emissions associated with the use of straw bedding.

The system used for organic pig production can influence the level of impacts which are calculated in LCA. Halberg et al. (2010) used a scenario analysis to compare the typical Danish system of sows kept in huts on grassland and finishing pigs raised in indoor-with-run housing, with the alternatives of also rearing the fattening pigs on grassland all year round, or in a bedded tent with restricted access to a grazing area. Higher burdens were calculated for all indicators in the free-range finishing system. In the ProPIG project (Rudolph, 2015), comparison was also made of different types of organic production system using data collected on individual farms. This showed no difference between systems in GHGE, but higher AP in indoor-with-run systems than in systems with mixed indoor and paddock housing. EP was lower in these mixed systems than in all-paddock systems.

It has to be emphasised that in all studies, variability in impact indicators has been high, indicating the important role of farm-specific features and practices. Generalisations at the system level should therefore be treated with extreme caution.

6 Conclusion

The demand for organic pig meat, along with other organic produce, has been stimulated by consumer concerns about animal welfare, the environment and food safety. This demand cannot currently be met and attractive price premiums are available within Europe. Organic pig production, because of the competition for cereals with a potential organic human food market, is still very small in scale compared to conventional production, but substantial expansion is occurring at the present time in many countries. Restrictions imposed by organic production standards on feeding, housing and management practices impose significant constraints on biological performance, but an extensive programme of research in different countries has been initiated to optimise future production systems. Great diversity in production methods and sustainability outcomes exists both between and within countries at the present time. The consequences for both real and perceived product quality are inconclusive. Whilst some aspects of animal welfare are enhanced by the more extensive production systems, others are threatened by lower control of biosecurity and climatic environment. Animal welfare objectives may also sometimes be in conflict with minimising negative environmental impact. Objective assessment of environmental impact can yield different conclusions depending on the criteria and methods used, with environmental burdens per unit of product generally increased, but burdens per unit of land area decreased. As a result of the trade-offs between indicators and large variation between farms, overall sustainability evaluations show great overlap between organic production systems, other types of non-organic differentiated production systems and conventional systems (Bonneau et al., 2011, 2014).

7 Future trends in research

Historically, organic pig production was subject to little research, but a large research effort began in the 1990s, with major national projects initiated in Scandinavian countries, the UK and Austria. Initially, these focused more on understanding the systems in use and identifying aspects of production practice in need of improvement. In 1996, a major research portfolio related to organic agriculture was initiated in Denmark, with a specific project to investigate organic pig production (at that time accounting for only 0.02% of Danish pig production). This work focused on the effect of roughage feeding on performance, welfare and meat quality of slaughter pigs, the management of pasture for maximising N utilisation, minimising pollution and parasite control, housing systems for growing pigs and animal health management.

Since 2007, coordinated transnational programmes to research aspects of health management, feeding, welfare and environmental impact have been carried out under the EU CORE Organic ERA-Net initiative and continue in a very recent project POWER (looking specifically at challenges for young pigs and combined indoor–outdoor systems). Many of the research collaborations initiated in this programme are continuing under the new SusAn ERA-Net, with a particular focus on on-farm measurement of indicators of welfare outcomes and environmental impact to better understand how further system improvements can be made. There is a particular need to investigate new ways to reduce piglet mortality, increase efficiency of feed use and optimise the integration of the pig enterprise within the overall farming system to enhance nutrient cycling and conservation.

8 Where to look for further information

For international news on organic agriculture and links to national bodies, the IFOAM website is a good starting place – https://www.ifoam.bio/en/sector-platforms/iaha-animal-husbandry-alliance.

Many useful publications from research into organic pig production are archived at the searchable 'Organic Eprints' website (http://orgprints.org/). Organic Eprints is an international open access archive for papers and projects related to research in organic food and farming, containing full-text papers in electronic form, abstracts and other metadata, and information on organisations, projects and facilities in the context of organic farming research.

More information on recent EU research projects exploring aspects of organic pig production can be found at the project websites:

CorePig – https://www.coreorganic.org/core1/research/projects/corepig/index.html
ProPIG – http://www.coreorganic2.org/coreorganic2.asp
ICOPP – http://www.organicresearchcentre.com/icopp/?page=home
LowInputBreeds – http://www.lowinputbreeds.org/home.html
Information on diets and feeding practice for organic pigs, together with other useful information, can be found in the textbook by Blair (2007).
Information on health management for organic pigs can be found in the ProPIG handbook for farmers which, together with other literature for organic pig farmers, is available from FIBL – https://shop.fibl.org/chde/k4.html?cat=31

9 References

Alban, L., Petersen, J. V. and Busch, M. E. 2015. A comparison between lesions found during meat inspection of finishing pigs raised under organic/free-range conditions and conventional, indoor conditions. *Porcine Health Management*, 1: 4.

Andresen, N. K. 1998. Soil tillage effects of pigs on grassland: consequences for the following winter wheat crop. In: *Proceedings of the 12th International IFOAM Scientific Conference*. Edited by D. Ferguelman and W. Lockeretz. Mar del Plata, Argentina, pp. 250–3.

Basset-Mens, C. and Van Der Werf, H. M. G. 2005. Scenario-based environmental assessment of farming systems: The case of pig production in France. *Agriculture, Ecosystems and Environment*, 105: 127–44.

Basset-Mens, C., Van Der Werf, H. M. G., Robin, P., Morvan, T., Hassouna, M., Paillat, J. M. and Vertès, F. 2007. Methods and data for the environmental inventory of contrasting pig production systems. *Journal of Cleaner Production*, 15: 1395–405.

Baumgartner, J., Leeb, T., Gruber, T. and Tiefenbacher, R. 2003. Husbandry and animal health on organic pig farms in Austria. *Animal Welfare*, 12: 631–5.

Blair, R. B. 2007. *Nutrition and Feeding of Organic Pigs*. CABI, Wallingford, UK.

Bonneau, M., Antoine-Ilari, E., Phatsara, C., Brinkmann, D., Hviid, M., Christiansen, M. G., Fàbrega, E., Rodríguez, P., Rydhmer, L., Enting, I., de Greef, K., Edge, H., Dourmad, J.-Y. and Edwards, S. 2011. Diversity of pig production systems at farm level in Europe. *Journal on Chain and Network Science*, 11: 115–35.

Bonneau, M., Gonzàlez, J., Rydhmer, L., Ilari-Antoine, E., Dourmad, J. Y., de Greef, K., Houwers, H. W. J., Cinar, M. U., Fàbrega, E., Zimmer, C., van der Oever, B. and Edwards, S. A. 2014. Evaluation of the sustainability of contrasted pig farming systems: Integrated evaluation. *Animal*, 8: 2058–68.

Dippel, S., Leeb, C., Bochicchio, D., Bonde, M., Dietze, K., Gunnarsson, S., Lindgren, K., Sundrum, A., Wiberg, S., Winckler, C. and Prunier, A. 2014. Health and welfare of organic pigs in Europe assessed with animal-based parameters. *Organic Agriculture*, 4: 149–61.

Dourmad, J.-Y. and Jondreville, C. 2007. Impact of nutrition on nitrogen, phosphorus, Cu and Zn in pig manure, and on emissions of ammonia and odours. *Livestock Science*, 112: 192–8.

Dourmad, J. Y., Ryschawy, J., Trousson, T., Bonneau, M., Gonzalez, J., Houwers, H. W. J., Hviid, M., Zimmer, C., Nguyen, T. L. T. and Morgensen, L. 2014. Evaluating environmental impacts of contrasting pig farming systems with life cycle assessment. *Animal*, 8: 2027–37.

Dransfield, E., Ngapo, T. M., Nielsen, N. A., Bredahl, L., Sjödén, P. O., Magnusson, M., Campo, M. M. and Nute, G. R. 2005. Consumer choice and suggested price for pork as influenced by its appearance, taste and information concerning country of origin and organic pig production. *Meat Science*, 69: 61–70.

Edwards, S. A. 2002. *Feeding Organic Pigs. A Handbook of Raw Materials and Recommendations for Feeding Practice*. University of Newcastle, Newcastle upon Tyne, UK, 59p.

Edwards, S. A. 2005. Product quality attributes associated with outdoor pig production. *Livestock Production Science*, 94: 5–14.

Edwards, S., Mejer, H., Roepstorff, A. and Prunier, A. 2014a. Animal health, welfare and production problems in organic pregnant and lactating sows. *Organic Agriculture*, 4: 93–105.

Edwards, S. A., Prunier, A., Bonde, M. and Stockdale, E. A. 2014b. Organic pig production in Europe – animal health, welfare and production challenges (Guest Editorial). *Organic Agriculture*, 4: 79–81.

Eijck, I. A. J. M. and Borgsteede, F. H. M. 2005. A survey of gastrointestinal pig parasites on free-range, organic and conventional pig farms in The Netherlands. *Veterinary Research Communications*, 29: 407–14.

European Commission. 2008. Commission Regulation (EC) No. 889/2008 of 5 September 2008 laying down detailed rules for the implementation of Council Regulation (EC) No. 834/2007 on organic production and labelling of organic products with regard to organic production, labelling and control. Official Journal of the European Union, L250(18 September 2008): 42–3.

European Council. 2007. Council Regulation (EC) No. 834/2007 of 28 June 2007 on organic production and labelling of organic products and repealing Regulation (EEC) No. 2092/91. *Official Journal of the European Union* L189(20 July 2007): 1–23.

FAWC. 1979. Press statement. Farm Animal Welfare Council, Tolworth, UK. http://webarchive.nationalarchives.gov.uk/20121010012428/http://www.fawc.org.uk/pdf/fivefreedoms1979.pdf

Fraser, D. 2003. Assessing animal welfare at the farm and group level: The interplay of science and values. *Animal Welfare*, 12: 433–43.

Früh, B., Bochicchio, D., Edwards, S., Hegelund, L., Leeb, C., Sundrum, A., Werne, S., Wilberg, S. and Prunier, A. 2014. Description of organic pig production in Europe. *Organic Agriculture*, 4: 83–92.

Halberg, N., Hermansen, J. E., Kristensen, I. S., Eriksen, J., Tvedegaard, N. and Petersen, B. M. 2010. Impact of organic pig production systems on CO_2 emission, C sequestration and nitrate pollution. *Agronomy for Sustainable Development*, 30: 721–31.

Honeyman, M. S. 2005. Extensive bedded indoor and outdoor pig production systems in USA: Current trends and effects on animal care and product quality. *Livestock Production Science*, 94: 15–24.

Horsted, K., Kongsted, A. G., Jorgensen, U. and Sorensen, J. 2012. Combined production of free-range pigs and energy crops-animal behaviour and crop damages. *Livestock Science*, 150: 200–8.

IFOAM. 2005. Principles of organic agriculture. International Federation of Organic Agriculture Movements, Bonn, Germany. https://www.ifoam.bio/sites/default/files/poa_english_web.pdf

Ilari-Antoine, E., Bonneau, M., Klauke, T. N., Gonzalez, J., Dourmad, J. Y., De Greef, K., Houwers, H. W. J., Fabrega, E., Zimmer, C., Hviid, M., Van der Oevera, B. and Edwards, S. A. 2014. Evaluation of the sustainability of contrasted pig farming systems: Economy. *Animal*, 8: 2047–57.

Katakam, K. K., Thamsborg, S. M., Dalsgaard, A., Kyvsgaard, N. C. and Mejer, H. 2016. Environmental contamination and transmission of *Ascaris suum* in Danish organic pig farms. *Parasites and Vectors*, 9: 80.

Kelly, H. R. C., Browning, H. M., Martins, A. P., Pearce, G. P., Stopes, C. and Edwards, S. A. 2001. Breeding and feeding pigs for organic production. In: *Proceedings of the NAHWOA workshop*, Wageningen, 24–27 March. Edited by M. Hovi, and T. Baars. University of Reading, Reading, UK, pp. 86–93.

Kongsted, H. and Sørensen, J. T. 2017. Lesions found at routine meat inspection on finishing pigs are associated with production system. *Veterinary Journal*, 223: 21–6.

Kool, A., Blonk, H., Ponsioen, T., Sukkel, W., Vermeer, H., de Vries, J. and Hoste, R. 2009. Carbon footprint of conventional and organic pork. Assessments of typical production systems in the Netherlands, Denmark, England and Germany. Research report, bioKennis, Wageningen, pp. 1–93.

Kristensen, I. S. and Kristensen, T. 1997. Animal production and nutrient balances on organic farming systems. Prototypes. *Proceedings of the 3rd ENOF Workshop: 'Resource use in organic farming'*, Ancona, Italy, 4–5 June 1997, pp. 189–202.

Lampkin, N. 1998. *Organic Farming*. Farming Press, Ipswich, UK.

Leeb, C. and ProPIG Consortium. 2015. Final report for the CORE Organic II funded project 'ProPIG': Farm specific strategies to reduce environmental impact by improving health, welfare and nutrition of organic pigs. Available from organic eprints at: http://orgprints.org/29937/1/COII_ProPIG_Final%20Report_Leeb%20et%20al.pdf

Leeb, C., Hegelund, L., Edwards, S., Mejer, H., Roepstorff, A., Rousing, T., Sundrum, A. and Bonde, M. 2014. Animal health, welfare and production problems in organic weaner pigs. *Organic Agriculture*, 4: 123–33.

Leenhouwers, J. I., Ten Napel, J., Hanenberg, E. H. A. T. and Merks, J. W. M. 2011. Breeding replacement gilts for organic pig herds. *Animal*, 5: 615–21.

Lindgren, K., Bochicchio, D., Hegelund, L., Leeb, C., Mejer, H., Roepstorff, A. and Sundrum, A. 2014. Animal health and welfare in production systems for organic fattening pigs. *Organic Agriculture*, 4: 135–47.

Luecke, F.-K. 2016. Microbiological safety of organic and conventional foods. In: *Foodbalt 2017–11th Baltic Conference on Food Science and Technology: Food Science and Technology in a Changing World*. Edited by E. Straumite,. Latvia University of Agriculture, Jelgava, Latvia, pp. 12–16.

Martins, A., Kelly, H., Day, J., Stopes, C., Browning, H. and Edwards, S. 2002. *Optimising Organic Pig Production – a Guide to Good Practice*. Agricultural Development and Advisory Service, Kings Lynn, UK.

Olsson, A.-C., Jeppsson, K.-H., Botermans, J., von Wachenfelt, H., Andersson, M., Bergsten, C. and Svendsen, J. 2014. Pen hygiene, N, P and K budgets and calculated nitrogen emissions for organic growing–finishing pigs in two different housing systems with and without pasture access. *Livestock Science*, 165: 138–46.

Olsson, A.-C., Botermans, J., Andersson, M., Jeppsson, K.-H. and Bergsten, C. 2016. Use of different rooting materials to improve hygiene and to lower ammonia emission within the outdoor concrete area in organic growing–finishing pig production. *Livestock Science*, 191: 64–71.

Oude Ophuis, P. A. M. 1994. Sensory evaluation of free range and regular pork meat under different conditions of experience and awareness. *Food Quality and Preference*, 5: 173–8.

Prunier, A., Lubac, S., Mejer, H., Roepstorff, A. and Edwards, S. 2014a. Health, welfare and production problems in organic suckling piglets. *Organic Agriculture*, 4: 107–21.

Prunier, A., Dippel, S., Bochicchio, D., Edwards, S., Leeb, C., Lindgren, K., Sundrum, A., Dietze, K. and Bonde, M. 2014b. Characteristics of organic pig farms in selected European countries and their possible influence on litter size and piglet mortality. *Organic Agriculture*, 4: 163–73.

Rangstrup-Christensen, I., Krogh, M. A., Pedersen, L. J. and Sørensen, J. T. 2017. Sow-level risk factors for stillbirth of piglets in organic sow herds. *Animal*, 11: 1078–83.

Rigolot, C., Espagnol, S., Robin, P., Hassouna, M., Bélin, F., Paillat, J. M. and Dourmad, J.-Y. 2010. Modelling of manure production by pigs and NH_3, N_2O and CH_4 emissions. Part II: effect of animal housing, manure storage and treatment practices. *Animal*, 4: 1413–24.

Rudolph, G. 2015. Effect of husbandry system on animal health, welfare and environmental impact of organic pigs in selected European countries. PhD thesis, University of Natural Resources and Life Sciences (BOKU), Vienna, Austria. http://epub.boku.ac.at/obvbokhs/content/titleinfo/1930150

Rudolph, G., Dippel, S., Bochicchio, D., Edwards, S., Früh, B., Holinger, M., Holmes, D., Illmann, G., Knop, D., Prunier, A., Rousing, T., Winckler, C. and Leeb, C. 2017. Does animal health and welfare of organic pigs differ between husbandry systems? In: *Proceedings W.A.F.L. 2017, 7th*

International Conference on the Welfare Assessment of Animal Welfare and Group Level, 5–8 September 2017, Ede, the Netherlands.

Srednicka-Tober, D., Baranski, M., Seal, C., Sanderson, R., Benbrook, C., Steinshamn, H., Gromadzka-Ostrowska, J., Rembialkowska, E., Skwarlo-Sonta, K., Eyre, M., Cozzi, G., Larsen, M. K., Jordon, T., Niggli, U., Sakowski, T., Calder, P. C., Burdge, G. C., Sotiraki, S., Stefanakis, A., Yolcu, H., Stergiadis, S., Chatzidimitriou, E., Butler, G., Stewart, G. and Leifert, C. 2016. Composition differences between organic and conventional meat: A systematic literature review and meta-analysis. *British Journal of Nutrition*, 115: 994–1011.

Thomsen, R., Edwards, S. A., Jensen, B. B., Rousing, T. and Sorensen, J. T. 2015. Weight and season affects androstenone and skatole occurrence in entire male pigs in organic pig production. *Animal*, 9: 1577–86.

USDA. 1990. Organic Foods Production Act of 1990. Title 21 of Food, Agriculture, Conservation, and Trade Act of 1990, codified at 7 U.S.C. 6501–6522.

USDA National Organic Program (NOP). 2010. Agricultural Marketing Service 7, Code of Federal Regulations (CFR), Part 205.

Van Loo, E. J., Alali, W. and Ricke, S. C. 2012. Food safety and organic meats. *Annual Review of Food Science and Technology*, 3: 203–25. Edited by M. P. Doyle, and T. R. Klaenhammer,.

Watson, C. A., Atkins, T., Bento, S., Edwards, A. C. and Edwards, S. A. 2003. Appropriateness of nutrient budgets for environmental risk assessment: A case study of outdoor pig production. *European Journal of Agronomy*, 20: 117–26.

Williams, A. G., Audsley, E. and Sandars, D. L. 2006. Determining the environmental burdens and resource use in the production of agricultural and horticultural commodities. Defra Research Project IS0205, Cranfield University and Defra, Bedford, UK.

Index

Lightning Source UK Ltd.
Milton Keynes UK
UKHW02n2205070618
323883UK00001B/8/P

9 781786 760883